# Angular 2 实例(影印版)
# Angular 2 by Example

Chandermani Arora, Kevin Hennessy 著

南京　东南大学出版社

**图书在版编目(CIP)数据**

Angular 2 实例:英文/(印)钱德玛尼·阿罗拉(Chandermani Arora),(美)凯文·汉尼斯(Kevin Hennessy)著.—影印本.—南京:东南大学出版社,2017.10

书名原文:Angular 2 by Example

ISBN 978-7-5641-7359-3

Ⅰ.①A… Ⅱ.①钱… ②凯… Ⅲ.①超文本标记语言-程序设计-英文 Ⅳ.①TP312.8

中国版本图书馆 CIP 数据核字(2017)第 193635 号

图字:10-2017-117 号



Angular 2 实例(影印版)

出版发行:东南大学出版社
地　　址:南京四牌楼 2 号　　邮编:210096
出 版 人:江建中
网　　址:http://www.seupress.com
电子邮件:press@seupress.com
印　　刷:常州市武进第三印刷有限公司
开　　本:787 毫米×980 毫米　16 开本
印　　张:32
字　　数:627 千字
版　　次:2017 年 10 月第 1 版
印　　次:2017 年 10 月第 1 次印刷
书　　号:ISBN 978-7-5641-7359-3
定　　价:96.00 元

本社图书若有印装质量问题,请直接与营销部联系。电话(传真):025-83791830

# Credits

**Authors**
Chandermani Arora
Kevin Hennessy

**Copy Editor**
Safis Editing

**Reviewer**
Josh Kurz

**Project Coordinator**
Devanshi Doshi

**Commissioning Editor**
Sarah Crofton

**Proofreader**
Safis Editing

**Acquisition Editor**
Kirk D'Costa

**Indexer**
Mariammal Chettiyar

**Content Development Editor**
Samantha Gonsalves

**Graphics**
Kirk D'Penha

**Technical Editor**
Madhunikita Sunil Chindarkar

**Production Coordinator**
Shantanu N. Zagade

# About the Authors

**Chandermani Arora** is a software craftsman, with a passion for technology and expertise on the web stack.

With more than a decade of experience under his belt, he has architected, designed, and developed solutions of all shapes and sizes on the Microsoft platform.

He has been building apps on Angular 1 from its early days. Such is his love for the framework that every engagement that he is a part of has an Angular footprint.

Being an early adopter of the Angular 2 framework, he tries to support the platform in every possible way – be it writing blog posts on various Angular topics or helping his fellow developers on StackOverflow, where he is often seen answering questions on the Angular2 channel.

An ex-MSFT, he now works for Technovert where he leads a bunch of awesome developers who build cloud-scale web applications using Angular and other new age frameworks.

He is also the author for the first edition of this book, *AngularJS by Example*.

> *Writing this book has just been a surreal experience, and I would like to thank my Technovert family who have supported me in all possible ways, be it helping me with the sample apps, reviewing the content, or offloading some of my professional commitment to make sure I get enough time for book writing. And finally I want to express my gratitude towards my family. I know your blessings are always there with me.*

**Kevin Hennessy** is a Senior Software Engineer with Applied Information Sciences. He has 18 years of experience as a developer, team lead, and solutions architect, working on web-based projects, primarily using the Microsoft technology stack. Over the last several years, he has presented and written about single-page applications and JavaScript frameworks, including Knockout, Meteor, and Angular 2. Most recently, he spoke about Angular 2 at the All Things Open Conference. His corporate blog is `http://blog.appliedis.com/?s=Kevin+Hennessy`.

> *I would like to acknowledge my wife, Mary Gene Hennessy. Her unstinting love and support (and editorial suggestions) through the period of late nights and weekends I spent writing this book, have made me ever more aware and appreciative of how truly amazing it is to be married to her.*

# About the Reviewer

**Josh Kurz** is a Technical Architect at Turner Broadcasting System. He has written a book on AngularJS, called *Mastering AngularJS Directives*, and he has contributed to many open source projects.

*I would like to thank my baby girl Evelyn for being the sweetest girl in the world.*

# Table of Contents

# Preface

Angular 2 is here, and we are super exited! This book allows us to reach out to you and lend a helping hand in your quest to learn Angular 2.

While the growth of Angular 1 was organic, the same cannot be said about Angular 2. It rides on the popularity of its predecessor and has already generated phenomenal interest among the developer community. Everyone expects a super awesome future proof framework! And we believe Angular 2 has taken steps in the right direction, which will make it the ubiquitous platform for web and mobile development.

If you are an Angular 1 developer, then there is loads of exciting stuff to learn, and for developers getting started there is a whole new world to explore.

Getting started with Angular 2 can be overwhelming even for a seasoned Angular 1 developer. Too many terms will be thrown at you, such as TypeScript, Transpiler, Shim, Observable, Immutable, Modules, Exports, Decorators, Components, Web Component, Shadow DOM, and more. But relax! We are trying to embrace the modern web and everything new here is to make our life easier. A number of these concepts are not specific to Angular itself but highlight the direction in which web platform development is moving. We will try our best to present these concepts in a clear and concise manner, helping everyone understand how these pieces fit into this big ecosystem.

Learning by examples has its advantages; you immediately see the concept explained in action. This book follows the same pattern as its predecessor. Using the Do It Yourself (DIY) approach, we build multiple simple and complex applications using Angular 2.

## Readers coming from the previous version

Angular 2 is a completely new framework and the only thing that it shares with its predecessor is its name! Very few Angular 1 core concepts have made it to Angular 2. Given this fact, this book too is a complete rewrite with all new content. We may be building the same application, but this time we build it using Angular 2.

# Why a new version of Angular?

To be frank, that is a question that many Angular developers have asked since Angular 2 was first announced at the ng-europe conference in October 2014. Angular 1 is a hugely popular JavaScript framework. Over 1 million developers worldwide have used it. Many of them have contributed add-ons/extensions that enhance and strengthen the framework. So why is there a need for a different, new version?

There are several answers to that question. But fundamentally, they all revolve around the fact that Angular 1 is six years old – which is a lifetime in terms of web technology. For example, Angular 1 predates much of what has developed around mobile technology. In addition, a new version of JavaScript (ES2015) was approved in 2015 that revolutionizes JavaScript programming. And finally, Angular1 was not designed for use with emerging web standards such as Web Components.

With newer frameworks such as Facebook's React that have been engineered to maximize performance and emphasize mobile-first development, the need for change became more compelling. Angular 2 responds to this challenge by adopting the latest web technologies and incorporating them into a framework for the modern browser.

# Angular 2 design

Highlighting some of what is not in Angular 1 leads logically to what the design of Angular 2 is all about. Angular 2 has a mobile-first design. It is therefore engineered for a *small footprint*, meaning that the data that flows from the server to the browser is minimized as much as possible. The framework itself has been broken into a collection of *modules* so that only the code needed to run the application is loaded. Moreover, a simplified and more coherent syntax makes it easier to learn and also provides better support for tooling and automation.

Each of the emerging technologies being used in Angular 2 provides key ingredients for realizing these goals. Web Components enable Angular 2 applications to be built out of reusable building blocks that encapsulate their internal logic. ES2015 provides classes and a solid system for loading Angular modules. TypeScript brings types that enable a simpler and more robust syntax for building large-scale applications.

# Why use TypeScript?

The examples in this book all use TypeScript. As mentioned, Angular2 allows us to write code in both ES5 (standard JavaScript) and ES2015, along with TypeScript. There are several reasons why we chose TypeScript. To start with, the Angular 2 team itself is using TypeScript to build the framework. Angular 2 code written in TypeScript is far terser than the alternatives. The use of TypeScript also enables IDEs to provide better IntelliSense and code completion support than what is available for JavaScript.

One final point – we think it is easier to learn about Angular 2 using TypeScript. Since this book is about teaching you this new technology, it seemed to be the best selection for the widest audience. As a superset of JavaScript, it offers JavaScript developers an easy migration path to working with types in their Angular applications. And for those developers who are moving to Angular 2 from more traditional object-oriented languages, it offers the familiarity of types and classes.

# What this book covers

`Chapter 1`, *Getting Started*, introduces you to the Angular framework. We create a super simple app in Angular that highlights some core features of the framework.

`Chapter 2`, *Building Our First App - 7 Minute Workout*, teaches us how to build our first real Angular app. In the process, we learn more about one of the primary building blocks of Angular – components. We are also introduced to Angular's templating constructs, its data binding capabilities and Angular services.

`Chapter 3`, *More Angular 2 – SPA, Routing, and Data Flows in Depth*, covers the routing constructs in the framework where we build multiple pages for a 7 Minute Workout. The chapter also explores a number of patterns around inter-component communication.

`Chapter 4`, *Building Personal Trainer*, introduces a new exercise where we morph the 7 Minute workout into a generic Personal Trainer app. This app has the ability to create new workout plans other than the original 7 Minute Workout. This chapter covers Angular's form capabilities, and how we can use them to build custom workouts.

`Chapter 5`, *Supporting Server Data Persistence*, deals with saving and retrieving workout data from the server. We augment Personal Trainer with persistence capabilities as we explore Angular's HTTP client library and how it uses RxJS Observables.

`Chapter 6`, *Angular 2 Directives in Depth*, goes deep into the inner workings of Angular 2 directives and components. We build a number of directives to support Personal Trainer.

`Chapter 7`, *Testing Personal Trainer*, introduces you to the testing world in Angular. You build a suite of unit and end-to-end tests that verify the working of Personal Trainer.

`Chapter 8`, *Some Practical Scenarios*, provides some practical tips and guidance around scenarios that we might encounter while developing apps on this framework. We cover scenarios such as authentication and authorization, localization, performance, and the most important case, migrating apps from Angular 1 to Angular 2.

# What you need for this book

We will be building our apps in the TypeScript language, therefore it would be preferable if you have an IDE that makes development with TypeScript easy. IDEs such as Atom, Sublime, WebStorm, and Visual Studio (or VS Code) are great tools for this purpose.

All the code enlisted in this book is written and tested for Angular 2.0.0.

# Who this book is for

This book is for readers with no prior experience in Angular. We start from Angular 2 basics and gradually build your understanding of the framework by working through the multiple exercises in the book.

To get the most out of this book, you should have experience in developing on web platforms using HTML, CSS, JavaScript, and a little bit of TypeScript. Angular 1 experience may be advantageous but not necessary for this book.

If you lack TypeScript experience, we highly recommend you visit the TypeScript website: `http://www.typescriptlang.org` and look at the tutorial, handbook, and samples. For a JavaScript developer, it does not take much time to get up and running with TypeScript.

# Conventions

In this book, you will find a number of text styles that distinguish between different kinds of information. Here are some examples of these styles and an explanation of their meaning.

Code words in text, database table names, folder names, filenames, file extensions, pathnames, dummy URLs, user input, and Twitter handles are shown as follows: "Mount the downloaded `WebStorm-10*.dmg` disk image file as another disk in your system."

A block of code is set as follows:

```
@Directive({
 selector: '[a2beBusyIndicator]',
})
export class BusyIndicator {
 constructor(private _control: NgControl) { }
}
```

When we wish to draw your attention to a particular part of a code block, the relevant lines or items are set in bold:

```
<div class="panel-body">
 {{description}}
 </div>
 <div class="panel-body">
 {{steps}}
 </div>
```

Any command-line input or output is written as follows:

```
npm install -g angular2
```

**New terms** and **important words** are shown in bold. Words that you see on the screen, for example, in menus or dialog boxes, appear in the text like this: "The shortcuts in this book are based on the `Mac OS X 10.5+` scheme."

Warnings or important notes appear in a box like this.

Tips and tricks appear like this.

# Reader feedback

Feedback from our readers is always welcome. Let us know what you think about this book-what you liked or disliked. Reader feedback is important for us as it helps us develop titles that you will really get the most out of. To send us general feedback, simply e-mail `feedback@packtpub.com`, and mention the book's title in the subject of your message. If there is a topic that you have expertise in and you are interested in either writing or contributing to a book, see our author guide at `www.packtpub.com/authors`.

# Customer support

Now that you are the proud owner of a Packt book, we have a number of things to help you to get the most from your purchase.

## Downloading the example code

You can download the example code files for this book from your account at `http://www.packtpub.com`. If you purchased this book elsewhere, you can visit `http://www.packtpub.com/support` and register to have the files e-mailed directly to you.

You can download the code files by following these steps:

1. Log in or register to our website using your e-mail address and password.
2. Hover the mouse pointer on the **SUPPORT** tab at the top.
3. Click on **Code Downloads & Errata**.
4. Enter the name of the book in the **Search** box.
5. Select the book for which you're looking to download the code files.
6. Choose from the drop-down menu where you purchased this book from.
7. Click on **Code Download**.

Once the file is downloaded, please make sure that you unzip or extract the folder using the latest version of:

- WinRAR / 7-Zip for Windows
- Zipeg / iZip / UnRarX for Mac
- 7-Zip / PeaZip for Linux

The code bundle for the book is also hosted on GitHub at `https://github.com/PacktPublishing/Angular2-By-Example-Second-Edition`. We also have other code bundles from our rich catalog of books and videos available at `https://github.com/PacktPublishing/`. Check them out!

Additionally, the code bundle written by the authors is available on GitHub in the following repository link: `https://github.com/chandermani/angular2byexample`.

# Downloading the color images of this book

We also provide you with a PDF file that has color images of the screenshots/diagrams used in this book. The color images will help you better understand the changes in the output. You can download this file from `https://www.packtpub.com/sites/default/files/downloads/Angular2ByExampleSecondEdition_ColorImages.pdf`.

# Errata

Although we have taken every care to ensure the accuracy of our content, mistakes do happen. If you find a mistake in one of our books-maybe a mistake in the text or the code-we would be grateful if you could report this to us. By doing so, you can save other readers from frustration and help us improve subsequent versions of this book. If you find any errata, please report them by visiting `http://www.packtpub.com/submit-errata`, selecting your book, clicking on the **Errata Submission Form** link, and entering the details of your errata. Once your errata are verified, your submission will be accepted and the errata will be uploaded to our website or added to any list of existing errata under the Errata section of that title.

To view the previously submitted errata, go to `https://www.packtpub.com/books/content/support` and enter the name of the book in the search field. The required information will appear under the **Errata** section.

# Piracy

Piracy of copyrighted material on the Internet is an ongoing problem across all media. At Packt, we take the protection of our copyright and licenses very seriously. If you come across any illegal copies of our works in any form on the Internet, please provide us with the location address or website name immediately so that we can pursue a remedy.

Please contact us at `copyright@packtpub.com` with a link to the suspected pirated material.

We appreciate your help in protecting our authors and our ability to bring you valuable content.

# Questions

If you have a problem with any aspect of this book, you can contact us at `questions@packtpub.com`, and we will do our best to address the problem.

# 1
# Getting Started

Developing applications in JavaScript is always a challenge. Due to its malleable nature and lack of type checking, building a decent-sized application in JavaScript is difficult. Moreover, we use JavaScript for all types of processes, such as user interface (UI) manipulation, client-server interaction, and business processing/validations. As a result, we end up with spaghetti code that is difficult to maintain and test.

Libraries, such as jQuery, do a great job of taking care of various browser quirks and providing constructs that can lead to an overall reduction in the lines of code. However, these libraries lack any structural guidance that can help us when the codebase grows.

In recent years, JavaScript frameworks have emerged to manage this complexity. Many of these frameworks, including earlier versions of Angular, use a design pattern called **Model-View-Controller** to separate the elements of the application into more manageable pieces. The success of these frameworks and their popularity in the developer community have established the value of using this pattern.

Web development, however, is constantly evolving and a lot has changed since Angular was first introduced in 2009. Technologies such as Web Components, the new version of JavaScript (ES2015), and TypeScript have all emerged. Taken together, they offer the opportunity to build a new, forward-looking framework. And with this new framework comes a new design pattern—the component pattern.

This chapter is dedicated to understanding the component pattern and how to put it into practice as we build a simple app using Angular.

The topics that we will cover in this chapter are as follows:

- **Angular basics**: We will briefly talk about the component pattern that is used to build Angular applications

- **Building our first Angular app**: We will build a small game, *Guess the Number!*, in Angular
- **An introduction to some Angular constructs**: We will review some of the constructs that are used in Angular, such as interpolation, expressions, and the data binding syntax
- **Change detection**: We will discuss how change detection is managed in an Angular app
- **App initialization**: We will talk about the app initialization process in Angular; this is also known as **bootstrapping**
- **Tools and resources**: Lastly, we will provide some resources and tools that will come in handy during Angular development and debugging

# Angular basics

Let's get started by looking at how Angular implements the component pattern.

# The component pattern

Angular applications use the component pattern. You may not have heard of this pattern, but it is all around us. It is used not only in software development but also in manufacturing, construction, and other fields. Put simply, it involves combining smaller, discrete building blocks into larger finished products. For example, a battery is a component of an automobile.

In software development, components are logical units that can be combined into larger applications. Components tend to have internal logic and properties that are shielded or hidden from the larger application. The larger application then consumes these building-blocks through specific gateways, called interfaces, which expose only what is needed to make use of the component. In this way, the component's internal logic can be modified without affecting the larger application, as long as the interfaces are not changed.

Getting back to our battery example, the car consumes the battery through a series of connectors. If the battery dies, however, it can be replaced by an entirely new battery, as long as that battery has the same connectors. This means that the builder of the car does not have to worry about the internals of the battery, which simplifies the process of building the car. Even more importantly, the car owner does not have to replace his or her car every time the battery dies.

To extend the analogy, manufacturers of batteries can market them for a range of different vehicles, for example, ATVs, boats, or snowmobiles. So the component pattern enables them to realize even greater economies of scale.

# Using the component pattern in web applications

As web applications continue to become more sophisticated, the need to be able to construct them out of smaller and discrete components becomes more compelling. Components allow applications to be built in a way that prevents them from becoming messes of spaghetti code. Instead, component-based design allows us to reason about specific parts of the application in isolation from the other parts, and then we can stitch the application together into a finished, whole through agreed-upon points of connection.

Also, maintenance costs are less because each component's internal logic can be managed separately without affecting the other parts of the application. And putting applications together using self-describing components makes the application easier to understand at a higher level of abstraction.

# Why weren't components used before in Angular?

If this idea makes so much sense, why was the component pattern not adopted in earlier versions of Angular? The answer is that the technologies that existed when Angular was first released did not fully support the implementation of this pattern in web applications.

Earlier versions of Angular, however, made substantial steps in the direction of enabling more intelligent web application design and organization. For example, they implemented the MVC pattern, which separates an application into a model, view, and controller (you will see the use of the MVC pattern continuing within the components that we will build in Angular).

With the MVC pattern, the model is the data, the view is a web page (or a mobile app screen or even a Flash page), and the controller populates the view with data from the model. In this way, separation of concerns is achieved. Following this pattern along with an intelligent use of directives will get you pretty close to components.

So, the earlier versions of Angular allowed applications to be designed and built more logically. However, this approach was limited by the fact that the technologies used were not truly isolated. Instead, they all ended up being rendered without any true separation from other elements on the screen.

# What's new that enables Angular to use this pattern?

By contrast, the newest version of Angular embraces recently emerging technologies, which make it possible to implement the component pattern more fully. These technologies include Web Components, ES2015 (the new version of JavaScript), and TypeScript. Let's discuss what each of these technologies brings to the mix that makes this possible.

## Web Components

Web Components is an umbrella term that actually covers four emerging standards for web browsers:

- Custom elements
- Shadow DOM
- Templates
- HTML imports

More information on Web Components can be found at `http://webcomponents.org`.

Let's now discuss each of these in detail:

- Custom elements enable new types of element to be created other than the standard HTML tag names such as `<div>` and `<p>`. The ability to add custom tags provides a location on the screen that can be reserved for binding a component. In short, this is the first step towards separating a component from the rest of the page and making it possible to become truly self-contained.

- Shadow DOM provides a hidden area on the page for scripts, CSS, and HTML. Markup and styles that are within this hidden area will not affect the rest of the page, and equally important they will not be affected by the markup and styles on other parts of the page. Our component can use this hidden area to render its display. So, this is the second step in making our component self-contained.

- Templates are repeatable chunks of HTML that have tags that can be replaced with dynamic content at runtime using JavaScript. Many JavaScript frameworks already support some form of templating. Web Components standardize this templating and provide direct support for it in the browser. Templates can be used to make the HTML and CSS inside the Shadow DOM used by our component dynamic. So, this is the third step in making our component.
- The final standard that makes up Web Components is HTML imports. They provide a way to load resources such as HTML, CSS, and JavaScript in a single bundle. Angular does not use HTML imports. Instead, it relies on JavaScript module loading, which we will discuss a little later in this chapter.

## Angular and Web Components

Web Components are not fully supported in current web browsers. For that reason, Angular components are not strictly Web Components. It is probably more accurate to say that Angular components implement the design principles behind Web Components. They also make it possible to build components that can run on today's browsers.

At the time of writing this book, Angular supports evergreen browsers, such as Chrome, Firefox, Edge, as well as IE 7 and later. It also has mobile support for Android 4.1 and later. For a list of browsers supported by Angular, visit `https://github.com/angular/angular`.

Therefore, throughout the rest of this book, we will focus on building Angular components and not Web Components. Despite this distinction, Angular components align closely with Web Components and can even interoperate with them. As browsers begin to support Web Components more fully, the differences between Angular components and Web Components will begin to disappear. So, if you want to begin adopting the Web Component standards of the future, Angular provides you with the opportunity to do so today.

## Language support in Angular

You can develop components with ES5, but Angular enhances the ability to develop components by adding support for key features that are found in the latest languages, such as ES2015 and TypeScript.

## ES2015

ES2015 is the new version of JavaScript; it was approved in June 2015. It adds many improvements to the language, which we will see throughout this book, but the two that interest us the most at this point are the following:

- Classes
- Module loading

**Classes** did not previously exist in JavaScript. The key advantage of using them, now that they do exist, is that they provide a convenient container for the code in our component.

To be clear, JavaScript classes do not introduce something that is completely new. The **Mozilla Developer Network** (**MDN**) describes them as *"syntactical sugar over JavaScript's existing prototype-based inheritance. . . [that] provide a much simpler and clearer syntax to create objects and deal with inheritance."* For more information visit `https://developer.mozilla.org/en-US/docs/Web/JavaScript/Reference/Classes`.

We'll ecplore these throughout the examples in this book. If you have not worked with object-oriented languages, you may not be familiar with classes, so we will cover them as we work through the examples in this chapter.

ES2015 also introduces a new approach to **module loading**. A module provides a way for JavaScript files to be encapsulated. When they are encapsulated, they do not pollute the global namespace and can interact with other modules in a controlled manner. We will cover modules in more details in later chapters.

Once we have our modules defined, we need a way to load them into our application for execution. Module loading allows us to select just what we need for our application from the modules that make up Angular and other components that we create or use.

Currently, a range of approaches and libraries exists to support module loading in JavaScript. ES2015 adds a new, consistent syntax for loading modules as part of the language. The syntax is straightforward and involves prefixing modules with the `export` keyword (or using the default export) and then using `import` to consume them elsewhere in our application.

ES 2015 module loading enables us to combine components into useful bundles or features that can be imported or exported within our applications. In fact, modules are at the core of Angular itself. We will see that modules are used extensively both in Angular itself and in the applications that we are building throughout this book.

It is important to understand that, while Angular uses syntax that has similarities to ES2015 module-loading syntax, Angular modules (which we will discuss a little later in this chapter) are not the same as JavaScript modules. For further details on these differences see the Angular documentation at `https://angular.io/docs/ts/latest/guide/architecture.html`. From this point on we will be focusing on Angular modules.

Because ES2015 is not fully supported by today's browsers, we will need to convert ES2015 into ES5 in order to use features such as classes and module loading in our applications. We do this through a process called **transpilation**.

Transpilation is like compilation, except that instead of converting our code into a machine language as compilation does, transpilation converts one type of source code to another type of source code. In this case, it converts ES2015 to ES5. There are several tools called **transpilers** that enable us to do that. Common transpilers include Traceur and Babel. TypeScript (which we will discuss next) is also a transpiler, and it is the one that we will use for the examples in this book.

Once ES2015 is transpiled to ES5, we can then use a module loader such as **SystemJS** to load our modules. SystemJS follows the ES2015 syntax for module loading and gives us the ability to do module loading in today's browsers.

## TypeScript

TypeScript was created by Microsoft as a superset of JavaScript, which means that it contains the features of ES2015 (such as classes and module loading) and adds the following:

- Types
- Decorators

**Types** allow us to mark variables, properties, and parameters in our classes to indicate that they are numbers, strings, Booleans, or various structures such as arrays and objects. This enables us to perform type checking at design time to make sure that the proper types are being used in our application.

**Decorators** are simple annotations that we can add to our classes using the @ symbol along with a function. They provide instructions (called metadata) for the use of our classes. In the case of Angular, decorators allow us to identify our classes as Angular components. Decorators can specify modules to be used with a component and how to implement various bindings and directives, including attaching an HTML view to the component. We will cover much more about the use of decorators as we go through this book.

Decorators form a part of the ES2017 proposal and are not part of ES2015. They were added to TypeScript as part of a collaboration between Microsoft and Google. As mentioned earlier, TypeScript compiles into ES5, so we are able to use both types and decorators in browsers that do not fully support ES2015 or the proposed standard for decorators.

As mentioned previously, it is not necessary to use either ES2015 or TypeScript with Angular. However, we think that you will see the advantages of using them as we work through the examples in this book.

## Putting it all together

By following the Web Component standards and adding support for ES2015 and TypeScript, Angular gives us the ability to create web applications that implement the component design pattern. These components help realize the vision behind the standards of building large-scale applications through collections of self-describing and self-contained building blocks.

We hope you will see in the examples in this book that Angular enables components to be constructed in a straightforward and declarative way that makes it easier for developers to implement them. As we proceed through the examples in this book, we will highlight where each of these technologies is being used.

# Angular modules

Components are the basic building block of an Angular application. But how do we then organize these building blocks into complete applications? Angular modules provide the answer to this question. They enable us to combine our components into reusable groups of functionality that can be exported and imported throughout our application. For example, in a more sophisticated application we would want to have modules for things such as authentication, common utilities, and external service calls. At the same time, modules enable us to group features within an application in a way that allows us to load them on demand. This is called lazy loading, a topic we will cover in `Chapter 4`, *Building Personal Trainer*.

Each Angular application will have one or more modules that contain its components. Angular has introduced `NgModule` as a way to conveniently specify the components that make up a module. Every Angular application must have at least one of these modules—the root module.

Angular itself is built as modules that we import into our application. So you will see the use of modules all over as you build Angular apps.

# The basic steps to building Angular applications

To sum up: at a basic level, you will see that to develop applications in Angular, you will do the following:

1. Create components.
2. Bundle them into modules.
3. Bootstrap your application.

The best way to understand Angular and the component design pattern is by seeing it in action. Hence, we are going to build our first Hello World app in Angular. This app will help you become familiar with the Angular framework and see the component design pattern in action.

Let's get started doing that.

# The customary Hello Angular app – Guess the Number!

As our first exercise, we want to keep things simple but still showcase the framework's capabilities. Therefore, we are going to build a very simple game called *Guess the Number!*. The objective of the game is to guess a random computer-generated number in as few tries as possible.

This is how the game looks:

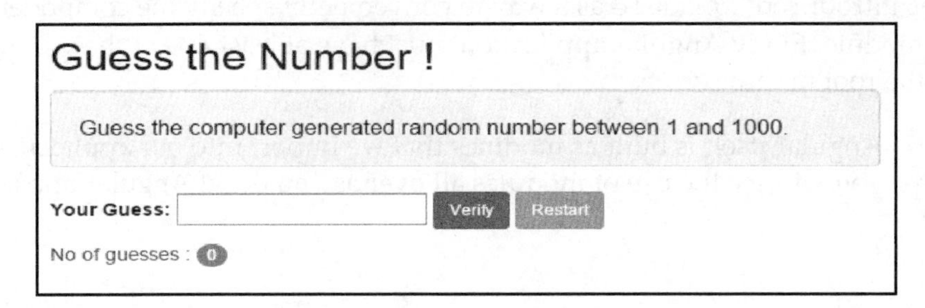


# Setting up a development server

The development web server that we choose greatly depends on the platform we work on and the backend we support. However, since apps in this book target purely client-side development, any web server will do.

Our recommendation is to use `live-server`, a simple HTTP server with live reload capability. You can install it using Node.js. Since Node.js is available cross-platform, you can install Node.js from `http://nodejs.org/`.

For this book, we are using Node.js version 4.4.2 and npm version 3.8.6. You can find more information about installing Node and updating npm to the latest version at `https://docs.npmjs.com/getting-started/installing-node`.

Once Node.js is installed, installing the `live-server` module and running the HTTP server is easy. Open the command line and type the following command:

```
npm install -g live-server
```

This installs `live-server` at the global level.

To run the server, we just navigate to the folder where the app code resides, or open the folder from where we want to serve static files. Then type this:

```
live-server
```

And that's it!

We have an HTTP server running at `http://localhost:8080`. It can serve files from the current directory.

The `live-server` module does support some startup configurations. Check out the documentation at `https://github.com/tapio/live-server`.

Depending on the platform we are on, we can also try Python's `SimpleHTTPServer` module, Mongoose, or any such web server.

Let's now build *Guess the Number!*.

# Building Guess the Number!

The standard practice while building user interfaces is to build them top-down. Start by designing the UI and then plug in the data and behavior according to your needs. With such an approach, the UI, data, and behavioral aspects of the app are all tightly coupled, which is a less than ideal situation!

With component-based design, we work differently. We start by looking at the UI and expected behavior, and then we encapsulate all of this into a building block that we call a **component**. This component is then hosted on our page. Within the component, we separate the UI into a view and the behavior into a class, with the appropriate properties and methods needed to support the behavior. If you are not familiar with classes, don't worry. We'll be discussing what they are in detail as we move through the example.

Okay, so let's identify the UI and behavior that we will need for our application.

# Designing our first component

To determine what needs to go into our component, we will start by detailing the features that we want the app to support:

- Generating random numbers (`original`)
- Providing input for a user to guess the value (`guess`)
- Tracking the number of guesses already made (`noOfTries`)
- Giving the user hints to improve their guess based on their input (`deviation`)
- Giving a success message if the user guesses the number correctly (`deviation`)

Now that we have our features, we can determine what we need to display to the user and what data we need to track. For the preceding feature set, the elements in parentheses denote the properties that will support those features and will need to be included in our component.

Designing the component is a very crucial process. If it is done right, we can logically organize our application in a way that makes it understandable and easy to maintain.

While building any app, we urge you to first think about the functionality you want to offer, and then the data and behavior that can support the functionality. Lastly, think about how to build a user interface for it. This is a good practice irrespective of the library or framework you use to build your app.

# The host file

Let's start by creating files for our component. We'll start by creating a directory for our application. We'll name it `guessthenumber` (but you can name it anything that suits you). We'll use that to add our files.

## An HTML page

First, open your favorite editor and create an HTML page with the following `html` and `script` tags:

```
<!DOCTYPE html>
<html>
  <head>
    <title>Guess the Number!</title>
    <link href="http://netdna.bootstrapcdn.com/bootstrap/
              3.1.1/css/bootstrap.min.css" rel="stylesheet">
    <script src="https://unpkg.com/core-js/client/shim.min.js"></script>
    <script src="https://unpkg.com/reflect-metadata@0.1.3"></script>
    <script src="https://unpkg.com/zone.js@0.6.23?main=browser"></script>
    <script src="https://unpkg.com/typescript@2.0/lib/typescript.js">
    </script>
    <script src="https://unpkg.com/systemjs@0.19.27/dist/system.js">
    </script>
    <script src="systemjs.config.js"></script>
    <script>
      System.import('app').catch(function(err){ console.error(err); });
```

```
    </script>
  </head>
  <body>
    <my-app>Loading...</my-app>
  </body>
</html>
```

**Downloading the example code**

The code in this book is available on GitHub at `https://github.com/chandermani/angular2byexample`. It is organized in checkpoints that allow you to follow along step by step as we build our sample projects in this book. The branch to download for this chapter is GitHub's **Branch: checkpoint1.1**. Look in the `guessthenumber` folder for the code we are covering here. If you are not using Git, download the snapshot of Checkpoint 1.1 (a ZIP file) from the following GitHub location: `https://github.com/chandermani/angular2byexample/tree/checkpoint1.1`. Refer to the `readme.md` file in the `guessthenumber` folder when setting up the snapshot for the first time.

## Script tags

There are several things to note in this file:

- The first five `<script>` tags reference the following:
  - `shim`: This provides ES2015 features for older browsers
  - `reflect-metadata`: This adds decorator support
  - `zone.js`: This manages change detection
  - `typescript.js`: This is the current version of TypeScript
  - `system.js`: This loads our modules
- The next `<script>` tag references a JavaScript file within our application: `systemjs.config.js`. Be sure to add that file from the example code files. We will be discussing this file a little later. Essentially, it provides instructions to SystemJS as to which modules it should load and directs it to dynamically transpile our TypeScript files to ES5 at runtime.
- The final `<script>` tag calls SystemJS to import our component directory: `app`.

We will discuss how the latter two script tags work together to enable module loading a little later in the chapter.

## Custom elements

There is one more important tag on the page:

```
<my-app>Loading...</my-app>
```

This tag is a **custom element**. It instructs Angular where to inject the component that we will be building.

The rest of the app's HTML code is self-explanatory. We reference the Twitter Bootstrap CSS in the `<head>` section and add a title, which is `Guess the Number!`, for our page.

*Guess the Number!* and all the other apps that are part of this book have been tested against the Angular final release.

# The component file

Now let's create a file for our component.

1. Add a subdirectory to our application called `app`.
2. Then, using your editor, create a file with the name `guess-the-number.component.ts` and place it in that subdirectory. The `.ts` extension identifies our file as a TypeScript file that will be compiled into ES5 at runtime.

## The import statement

At the top of the page, place the following line:

```
import { Component }from '@angular/core';
```

This is an import statement. It tells us what modules we will be loading and using in our component. In this case, we are selecting one module that we need to load from Angular: `Component`. Angular has many other modules, but we load only what we need.

You'll notice that the location from which we are importing is not identified as a path or directory within our application. Instead, it is identified as `@angular/core`. Angular has been divided into barrel modules that are prefixed with `@angular`.

These barrels combine several modules that are logically related. In this case, we are indicating that we want to import the `core` barrel module, which in turn brings in the `Component` module. This naming convention ties into the loading of our modules, which we will discuss in more detail later in this chapter.

## Decorators

Next, add the following block of script to your `guess-the-number.component.ts` file:

```
@Component({
  selector: 'my-app',
  template: `
    <div class="container">
      <h2>Guess the Number !</h2>
      <p class="well lead">Guess the computer generated random
         number between 1 and 1000.</p>
      <label>Your Guess: </label>
      <input type="number" [value]="guess" (input)="guess =
       $event.target.value" />
      <button (click)="verifyGuess()" class="btn btn-primary btn-sm">
      Verify</button>
      <button (click)="initializeGame()" class="btn btn-warning btn-sm">
      Restart</button>
    <div>
      <p *ngIf="deviation<0" class="alert alert-warning">
      Your guess is higher.</p>
      <p *ngIf="deviation>0" class="alert alert-warning">
      Your guess is lower.</p>
      <p *ngIf="deviation===0" class="alert alert-success">
      Yes! That's it.</p>
    </div>
      <p class="text-info">No of guesses :
        <span class="badge">{{noOfTries}}</span>
      </p>
    </div>
})
```

This is the decorator for our component and is placed directly above the class definition, which we will discuss soon. The `@` symbol is used to identify a decorator. The `@Component` decorator has a property called selector, and you may not be surprised to see that it is set to the `<my-app>` tag in our HTML page. This setting tells Angular to inject this component into that tag on the HTML page.

The decorator also has a property called template, and this property identifies the HTML markup for our component. Notice the use of back ticks (introduced by ES2015) for rendering the template string over multiple lines. Alternatively, we can set a `templateUrl` property that would point to a separate file.

## Defining the class

Now, add the following block of code to your `guess-the-number.component.ts` file:

```
export class GuessTheNumberComponent {
  deviation: number;
  noOfTries: number;
  original: number;
  guess: number;
  constructor() {
    this.initializeGame();
  }
  initializeGame() {
    this.noOfTries = 0;
    this.original = Math.floor((Math.random() * 1000) + 1);
    this.guess = null;
    this.deviation = null;
  }
  verifyGuess() {
    this.deviation = this.original - this.guess;
    this.noOfTries = this.noOfTries + 1;
  }
}
```

If you have been developing in ES5, the version of JavaScript that is supported in all current browsers, you may not be familiar with the use of classes here. So, we will take a few moments to walk through what makes up a class (for those of you who have developed using an object-oriented programming language, such as C# or Java, this should be familiar territory).

The class file holds the code that we will use to run our component. At the top, we give the class a name, which is `GuessTheNumberComponent`. Then, inside the curly braces, we have four lines that declare the properties for our class. These are similar to ES5 variables, and we will use them to hold the values that we will need to run the application (you'll notice that these are the four values that we identified when we designed our component).

What makes these properties different from standard JavaScript variables is that each property name is followed by : and number. These set the type of the property. In this case, we are indicating that each of these four properties will be set to the number type, which means we are expecting the values of all of these properties to be numbers. The ability to specify types for our properties is provided by TypeScript and is not available in standard JavaScript.

As we move down, we will see three blocks of script that have names, followed by parentheses, and then curly braces with several lines of script inside them. These are the methods for our class, and they contain the operations that our component will support. They are a lot like standard JavaScript functions.

The first of these methods is `constructor()`, which is a special method that will run when an instance of our component is first created. In our example, the constructor does only one thing when the class is created; it calls another method in our class, called `initializeGame()`.

The `initializeGame()` method sets the starting values of the four properties in the class using the assignment operator =. We set these values to `null` or `zero`, except for `original`, in which we use a random number generator to create the number to be guessed.

The class holds one more method called `verifyGuess()`, which updates the `deviation` and `noOfTries` properties. This method is not being called from within the component class; instead, it will be called from the view, as we will see when we examine the view more closely later. You'll also notice that our methods refer to properties in the same class by prepending `this` to them.

# The module file

As we mentioned earlier, every Angular component must be contained within an Angular module. This means that at a minimum we must add at least one Angular module file to the root of our application. We call this the **root module**. For a simple application like Guess the Number!, the root module may be the only module we will need. However, as an Angular application increases in size, it will often make sense to have multiple Angular module files broken down by features. We will cover that situation as we move into building more complex applications in later chapters in this book.

Let's go ahead and create our Angular module file. First, create a new file named `app.module.ts` in the same directory as `guess-the-number.component.ts` and add the following code to it:

```
import { NgModule }      from '@angular/core';
import { BrowserModule } from '@angular/platform-browser';

import { GuessTheNumberComponent }
from './guess-the-number.component';

@NgModule({
    imports:      [ BrowserModule ],
    declarations: [ GuessTheNumberComponent ],
    bootstrap:    [ GuessTheNumberComponent ]
})
export class AppModule { }
```

The first two statements import `NgModule` and `BrowserModule`. Notice that, while `NgModule` is being imported from `@angular/core`, `BrowserModule` is being imported from a different module: `@angular/platform-browser`. What's significant here is that the import is not coming from `@angular/core` but from a separate module that is specific to browser-based applications. This is a reminder that Angular can support devices other than browsers, such as mobile devices, hence the need to place `BrowserModule` into a separate module.

The other import in this file is our component `GuessTheNumberComponent`, which we just built. If you go back to that component you will notice that we added `export` in front of the class definition, which means we are using module loading within our own application.

We next define a new component AppModule. This component looks different from GuessTheNumberComponent, which we just defined. There is nothing in the class itself other than a few imports and a decorator: @ngModule. We can use this decorator to configure the module in our application. We first add imports, which in this case include the BrowserModule. As the name suggests, this module will provide the functionality needed to run our application in a browser. The next property is declarations and with that property we provide an array of the components that will be used in our application. In this case, we have just one component: GuessTheNumberComponent.

Finally, we set the bootstrap property. This indicates the first component that will be loaded when our application starts up. Again this is the GuessTheNumberComponent.

With this configuration in place, we are now ready to bootstrap our component.

# Bootstrapping

The class definition for GuessTheNumberComponent operates as a blueprint for the component, but the script inside it does not run until we have created an instance of the component. In order to run our application then, we need something in our application that creates this instance. The process of doing that requires us to add code that bootstraps our component.

In the app subdirectory, create another file named main.ts and add the following code to it:

```
import { platformBrowserDynamic }
from '@angular/platform-browser-dynamic';
import { AppModule } from './app.module';
const platform = platformBrowserDynamic();
platform.bootstrapModule(AppModule);
```

As you can see, we are first importing the platformBrowserDynamic module from @angular/platform-browser-dynamic. Like the import of BrowseModule in the appModule file, this import is specifically for browser-based applications.

Next we add an import of our AppModule, which we have just defined.

Finally, we assign `platformBrowserDynamic()`, which we just imported, to a constant: platform. Then we call its `bootstrapModule` method with our `AppModule` as a parameter. The `bootstrapModule` method then creates a new instance of our `AppModule` component, which in turn initializes our `GuessTheNumberComponent`, which we have marked as the component to bootstrap. It does that by calling the component's constructor method and setting the starting values for our game.

We will discuss in greater detail how this `bootstrap` method fits into the entire process of app initialization a little later in this chapter.

# We're up-and-running!

Well, the app is complete and ready to be tested! Navigate to the directory where the file is located and type this:

```
live-server
```

The app should appear on your browser.

If you are having trouble running the app, you can check out a working version available on GitHub at `https://github.com/chandermani/angular2byexample`. If you are not using Git, download the snapshot of Checkpoint 1.1 (a ZIP file) from the following GitHub location: `https://github.com/chandermani/angular2byexample/tree/checkpoint1.1`. Refer to the `readme.md` file in the `guessthenumber` folder when setting up the snapshot for the first time.

If we glance at our component file now, we should be mightily impressed with what we have achieved with these 43 lines. We are not writing any code to update the UI when the application is running. Still, everything works perfectly.

# Digging deeper

To understand how this app functions in the Angular context, we need to delve a little deeper into our component. While the class definition in the component is pretty simple and straightforward, we need to look more closely at the HTML in the template that is inside the `@Component` decorator to understand how Angular is working here. It looks like standard HTML with some new symbols, such as `[ ]`, `( )`, `{{`, and `}}`.

In the Angular world, these symbols mean the following:

- `{{` and `}}` are interpolation symbols
- `[ ]` represents property bindings
- `( )` represents event bindings

Clearly, these symbols have some behavior attached to them and seem to be linking the view HTML and component code. Let's try to understand what these symbols actually do.

## Interpolation

Look at this HTML fragment from the Guess the Number! code:

```
<p class="text-info">No of guesses :
  <span class="badge">{{noOfTries}}</span>
</p>
```

The term `noOfTries` is sandwiched between two interpolation symbols. Interpolation works by replacing the content of the interpolation markup with the value of the expression (`noOfTries`) inside the interpolation symbol. In this case, `noOfTries` is the name of a component property. So the value of the component property will be displayed as the contents inside the interpolation tags.

Interpolations are declared using this syntax: `{{expression}}`. This expression looks similar to a JavaScript expression but is always evaluated in the context of the component. Notice that we did not do anything to pass the value of the property to the view. Instead, the interpolation tags read the value of the property directly from the component without any need for additional code.

## Tracking changes in the number of tries

Another interesting aspect of interpolation is that changes made to component properties are automatically synchronized with the view. Run the app and make some guesses; the `noOfTries` value changes after every guess and so does the view content:

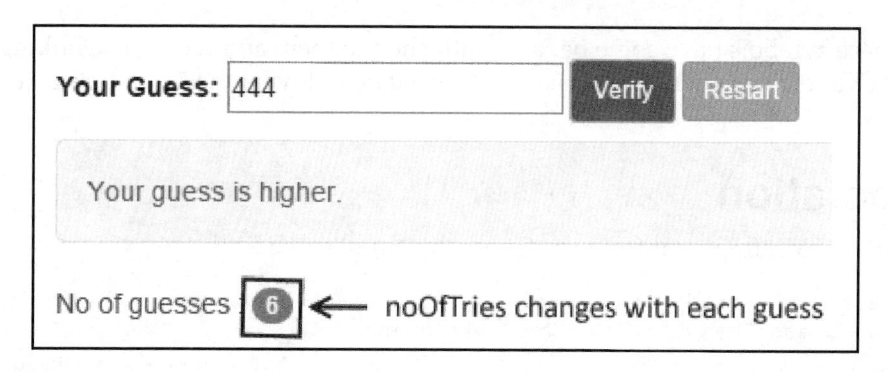


Interpolation is an excellent debugging tool in scenarios where we need to see the state of the model. With interpolation, we don't have to put a breakpoint in code just to know the value of a component property. Since interpolation can take an expression, we can pass a component's method call or a property and see its value.

## Expressions

Before going any further, we need to spend a few moments understanding what template expressions are in Angular.

Template expressions in Angular are nothing but pieces of plain JavaScript code that are evaluated in the context of the component instance associated with the template instance in which they are used. But as the documentation at `https://angular.io/docs/ts/latest/guide/template-syntax.html#template-expressions` makes it clear, there are some differences:

- Assignment is prohibited, except in event bindings
- The new operator is prohibited
- The bitwise operators `|` and `&` are not supported
- Increment and decrement operators, `++` and `--`, aren't supported
- Template expression operators, such as `|` and `?.` add new meanings

In the light of our discussion on component-based design, you probably won't be surprised to learn that the documentation also makes some more things clear; template expressions cannot:

- Refer to anything in the global namespace
- Refer to a window or document
- Call `console.log`

Instead, these expressions are confined to the expression context, which is typically the component instance supporting a particular template instance.

However, these limitations do not stop us from doing some nifty stuff with expressions. As we can see in the following examples, these all are valid expressions:

```
// outputs the value of a component property
{{property}}

// adds two values
{{ 7 + 9 }}

//outputs the result of boolean comparison. Ternary operator
{{property1 >=0?'positive': 'negative'}}

//call a component's testMethod and outputs the return value
{{testMethod()}}
```

Having looked into expressions, we strongly advise you to keep your expressions simple, thus keeping the HTML readable.
The `*ngIf="formHasErrors()"` expression is always better than `*ng-if="name==null || email==null || emailformatInValid(email) || age < 18"`.
So, when an expression starts to become complex, move it into a method in your component.

## The safe navigation operator

Before we move on there is one other expression that we should touch on: the Angular safe navigation operator (`?.`). This operator provides a convenient way to check for null values in lengthy property paths like so:

```
{{customer?.firstName }}
```

If the safe navigation operator finds a null value (here the customer), it stops processing the path but lets the application continue running. Without it, the application will crash when it reaches anything after the first null (here the customer name) and the view will not display. The safe navigation operator is especially helpful in situations where you are loading data asynchronously and it might not be immediately available to the view. The safe navigation operator will prevent the application from crashing and then load the data when it is available.

# Data binding

Learning interpolation and expressions was easy; now let's look at another framework construct that is being used by our sample app-data binding. We will be covering data binding in far more detail in the upcoming chapters. At this point, we will just touch briefly on the bindings that are used in the sample app we are building.

## Property binding

If we look through the HTML for the view, we will see several places where square brackets `[ ]` are used. These are **property bindings**.

Let's look at the first of the bindings that we created:

```
<input type="number" [value]="guess" (input)="guess =
$event.target.value" />
```

This binding works by linking the value of the `guess` property in our component class to the `value` of the input field in the view. The binding is dynamic; so, as the value of the `guess` property changes, the `value` of the input field will be synchronized to the same value. And, we do not have to write any code to do that.

At the outset, when we initialize the game, this property is set to null in the initialization method of the component class, so we will not see anything in the input field. However, as the game progresses, this number will be updated with the value of the guess as it changes.

## Event binding

Looking again at the HTML view, we find several places where parentheses `( )` appear. These are **event bindings**.

Let's look at the HTML code line that we created for the first of these event bindings. It should be familiar since the event binding is on the same tag that we first looked at for property binding: the `input` tag:

```
<input type="number" [value]="guess" (input)="guess =
$event.target.value" />
```

In this case, the `input` event of the input element is bound to an expression. The expression sets the `guess` property in our component class to `$event.target.value`, which is the value being entered by the user. Behind the scenes, when we use this syntax, Angular sets up an event handler for the event that we are binding to. In this case, the handler updates the `guess` property in our component class whenever the user enters a number in the `input` field.

There are a couple of other places in your code where the `( )` parentheses appear:

```
<button (click)="verifyGuess()" class="btn btn-primary btn-
sm">Verify</button>
<button (click)="initializeGame()" class="btn btn-warning    btn-
sm">Restart</button>
```

These two event bindings tie the `click` events for the buttons on the screen to methods in our component. So in this case, behind the scenes, Angular sets up event handlers that bind directly to the methods in our component. When the **Verify** button is clicked, the `verifyGuess` method is called, and when the **Restart** button is clicked, the `initializeGame` method is called.

As you work through the samples in this book, you will see many places where the `[]` tags for property bindings are combined with the `()` tags for events. In fact, this pairing is so common that, as we will see later, Angular has come up with a shorthand syntax to combine these tags into one.

# Structural directives

Next, we'll examine something that looks similar to data binding but incorporates an Angular feature we have haven't seen before: **structural directives**:

```
<div>
  <p *ngIf="deviation<0" class="alert alert-warning"> Your guess is
higher.</p>
  <p *ngIf="deviation>0" class="alert alert-warning"> Your guess is
lower.</p>
  <p *ngIf="deviation===0" class="alert alert-success"> Yes! That"s it.</p>
```

```
</div>
```

`*ngIf` inside the `<p>` tags is the `NgIf` structural directive. Structural directives allow us to manipulate the structure of DOM elements. The `NgIf` directive removes or adds DOM elements based on the result of an expression that is assigned to it.

The asterisk `*` in front of `ngIf` is a simplified syntax that Angular, under the hood, expands into an HTML5 `<template>`. We'll be learning a lot more about this syntax and about structural directives in the next chapter.

In this case we are using `NgIf` with a simple expression, similar to the types of expression we saw with interpolation. The expression resolves to either `true` or `false` based on the value of the guess being made and its relation (higher, lower, or equal) to the correct number. It then assigns that result to `NgIf`, which will either add the DOM element if the result is `true` or remove it if it is `false`.

# Revisiting our app

So now that we have looked more closely at what makes up our view, let's take another look at our app when it is up-and-running. When we run our app, Angular binding starts up once the browser has rendered the raw HTML in our view. The framework then compiles this view template and, in the process, sets up the necessary binding. Next, it does the necessary synchronization between our component class and the view template that produces the final rendered output. The following screenshot depicts the transformations that happen to the view template after data binding is done for our app:

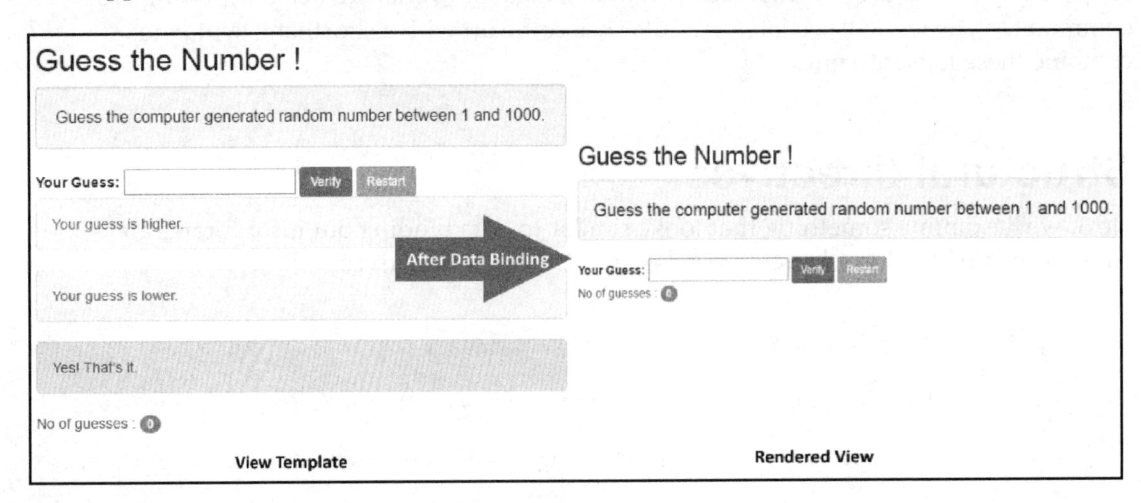

We can ourselves see the untransformed view template of the app (what is shown on the left-hand side of the preceding screenshot) by removing the three `*ngIf` directives and the expressions assigned to them from the paragraphs below the input box and refreshing the app in the browser.

Angular differs from other template frameworks, in that these bindings between a component and its view are live. Changes made to the properties of the component update the view. Angular never regenerates the HTML; it just works on the relevant part of the HTML and updates only the HTML elements that need to change as component properties change. This data binding capability makes Angular an exceptional view templating engine too.

# Looking at how our code handles updates

If we go back and look at the code for our class, we will see that the properties and methods in the class do not directly reference the view. Instead, the methods simply operate on the properties in the class. As a consequence, the code for our class is more readable, hence more maintainable (and of course, testable).

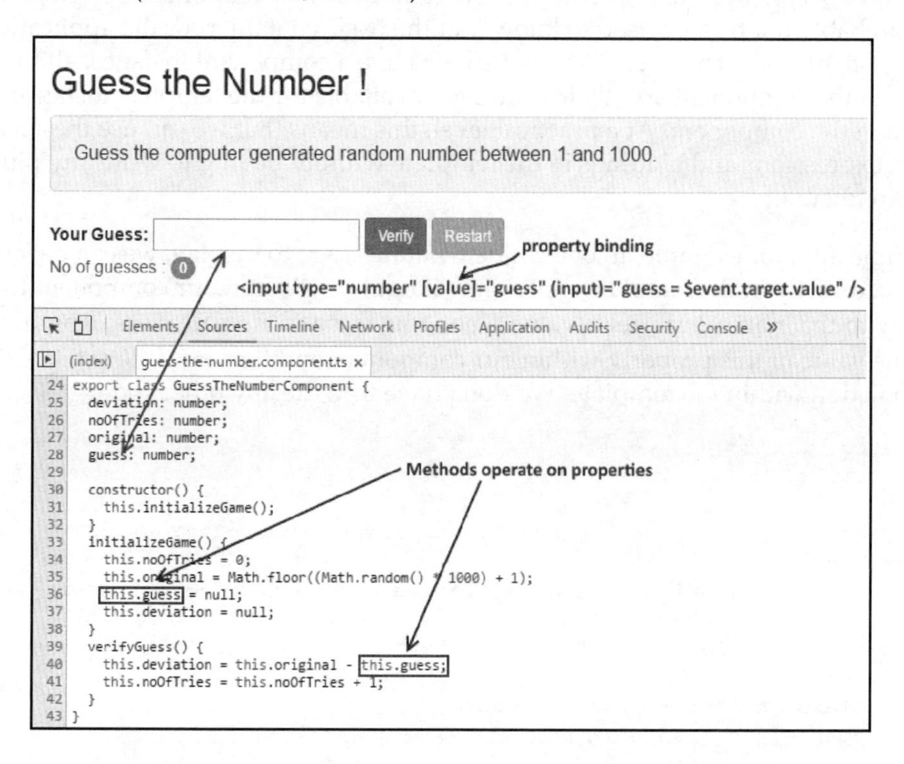

So far, we have discussed how Angular updates the view based on changes in a component's properties. This is an important concept to understand as it can save us from countless hours of debugging and frustration. The next section is dedicated to learning about change detection and how these updates are managed.

# Maintaining the state

First, let's look at how we maintain the state in our Angular application. Since Angular apps are dynamic and not static, we need to understand the mechanisms that are used to make sure that these dynamic values are kept up-to-date as the data in an application gets updated. For example, in our application, how does the number of guesses get updated on the screen? How does the application decide to display the correct message about whether the guess is correct based on the user input?

# Component as the container for the state

Since we have been emphasizing so far that Angular uses the component design pattern, you will probably not be surprised to know that the basic container for the application state is the component itself. This means that when we have a component instance, all the properties in the component and their values are available for the template instance that is referenced in the component. At a practical level, this means that we can use these values directly in expressions and bindings in the template without having to write any plumbing code to wire them up.

In the sample app, for example, to determine what message to display, we can use `deviation` directly in the template expression. Angular will scan our component to find a property with that name and use its value. The same is true for `noOfTries`; Angular will look for the value of this property within our component and then use it to set its value in the interpolation within the template. We don't have to write any other code:

```
template: `
.............
  <div>
    <p *ngIf="deviation<0" class="alert alert-warning"> Your guess is
higher.</p>
    <p *ngIf="deviation>0" class="alert alert-warning"> Your guess is
lower.</p>
    <p *ngIf="deviation===0" class="alert alert-success"> Yes! That's
it.</p></div>
    <p class="text-info">No of guesses :
      <span class="badge">{{noOfTries}}</span>
```

```
    </p>
  </div>
`
```

# Change detection

So how does Angular keep track of changes in our component as it runs? So far, it appears as if this is all done by magic. We just set up our component properties and methods, and then we bind them to the view using interpolation along with property and event binding. Angular does the rest!

But this does not happen by magic, of course, and in order to make effective use of Angular, you need to understand how it updates these values as they change. This is called **change detection**, and Angular has a very different approach to doing this than what previously existed.

If you use the debugger tool in your browser to walk through the application, you will see how change detection works. Here, we are using Chrome's developer tools and setting a watch for the `noOfTries` property. If you place a breakpoint at the end of the `verifyGuess()` method, you will see that when you enter a guess, the `noOfTries` property is first updated as soon as you hit the breakpoint, like this:

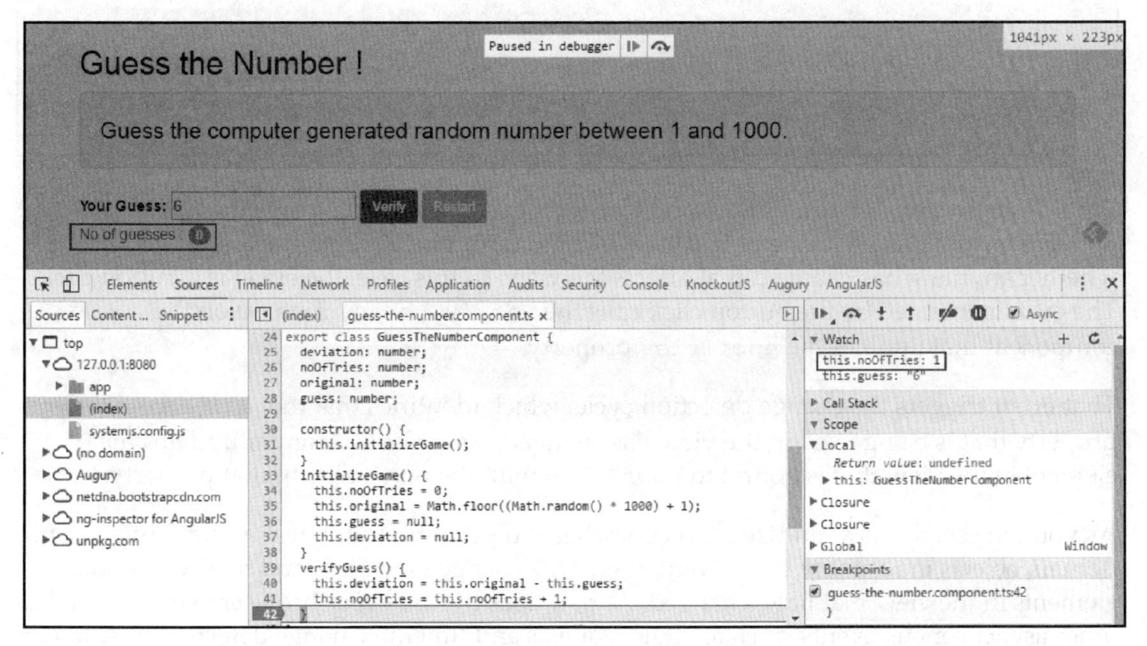

Once you move past the breakpoint, the display on the screen updates with the correct number of guesses, as seen in the following screenshot:

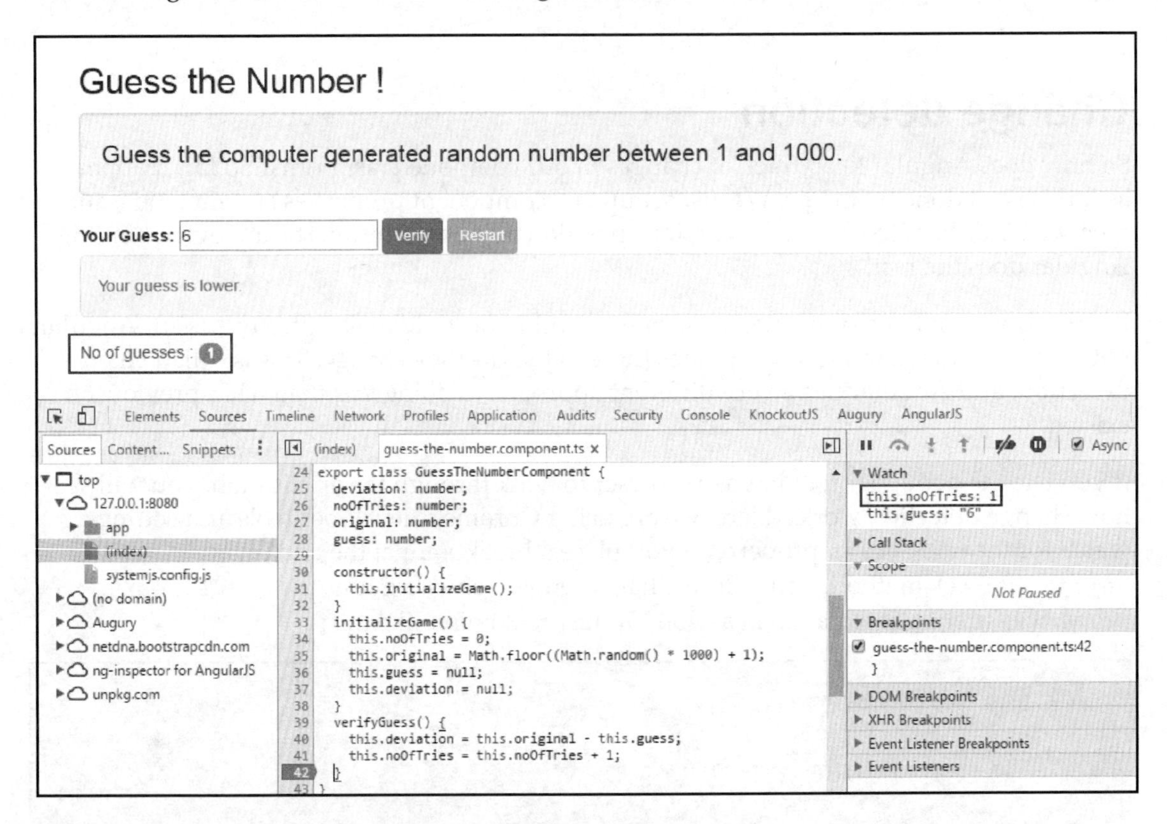


What is really going on under the hood is that Angular is reacting to events in the application and using change detectors, which go through every component to determine whether anything has changed that affects the view. In this case, the event is a button click. The event generated by the button click calls the `verifyGuess()` method on the component that updates the `noOfTries` property.

That event triggers the change detection cycle, which identifies that the `noOfTries` property that is being used in the view has changed. As a result, Angular updates the element in the view that is bound to `noOfTries` with the new value of that property.

As you can see, this is a multistep process where Angular first updates the components and domain objects in response to an event, then runs change detection, and finally rerenders elements in the view that have changed. And, it does this on every browser event (as well as other asynchronous events, such as XHR requests and timers). Change detection in Angular is reactive and one way.

This approach allows Angular to make just one pass through the change detection graph. It is called **one-way data binding**, and it vastly improves the performance of Angular.

We'll be covering Angular change detection in depth in `Chapter 8`, *Some Practical Scenarios*. For a description of this process by the Angular team, visit `https://vsavkin.com/two-phases-of-angular-2-applications-fda2517604be#.fabhc0ynb`.

# Initializing the app

We touched on app initialization earlier when we created the `app.module.ts` and `main.ts` files. The first file wraps our application in a module and the second file bootstraps that module. Now we will take a more detailed look at the initialization process.

# Loading the modules needed by our application

Let's go back to the `index.html` page again and take a look at the following script tags on this page:

```
<script src="https://unpkg.com/systemjs@0.19.27/dist/system.js"></script>
<script src="systemjs.config.js"></script>
```

The first tag indicates that we will be using a JavaScript library called SystemJS in our application. The second tag references a file within our application that sets the configuration for SystemJS.

As we mentioned earlier, ES2015 introduced a new syntax for module loading. One attractive feature of ES2015 module loading syntax is that it allows for modules to be loaded selectively and asynchronously. We will be using module loading throughout our application.

TypeScript supports the ES2015 module loading syntax, and SystemJS allows us to add module loading to applications that run ES5. Putting the two together:

- TypeScript will transpile (compile) our application's components to ES5
- SystemJS will load them as ES5 modules

As part of this process, each of our TypeScript files will be compiled to a SystemJS module. SystemJS will then load all the related dependencies and when requested the module itself.

Other module loaders exist, including Webpack, so you are not limited in your choice of module loaders to SystemJS. For more information on using Webpack with Angular refer to the following documentation on the Angular site: `https://angular.io/docs/ts/latest/guide/webpack.html`.

Open `systemjs.config.js`. You will see that it contains mappings that indicate where SystemJS should look for the files that are needed in our application:

```
map : {
    'app': 'app',
    'rxjs': 'https://unpkg.com/rxjs@5.0.0-beta.12',
    '@angular/common': 'https://unpkg.com/@angular/common@2.0.0',
    '@angular/compiler': 'https://unpkg.com/@angular/compiler@2.0.0',
    '@angular/core': 'https://unpkg.com/@angular/core@2.0.0',
    '@angular/platform-browser': 'https://unpkg.com/@angular/
                                 platform-browser@2.0.0',
    '@angular/platform-browser-dynamic': 'https://unpkg.com/@angular/
    platform-browser-dynamic@2.0.0'
},
```

First, we map our own application, app, to the directory in which it resides. Then we add mappings for RxJS and several Angular modules. They are being downloaded from a **content delivery network** (**CDN**), the **Node Package Manager** (`https://unpkg.com/#/`) CDN. As you can see, we are using the `@` notation with the Angular modules. As we mentioned before, this means we will be importing Angular barrels containing multiple modules.

The module bundles that Angular identifies as barrels are referred to as scoped packages in NPM. For a description of how Angular is using NPM scoped packages, visit `https://www.reddit.com/r/Angular2/comments/4hk0f8/the_angular2_team_is_changing_the_repository_on/`.

Next, we identify the packages that we will be importing and their main entry points. In the case of our `app` package, we indicate that `main.ts` will be the entry point and `ts` will be the default extension since we are using TypeScript for it. Then we identify the specific Angular `@barrels` that we will be using in our application:

```
packages:{
    'app':  { main: 'main.ts',  defaultExtension: 'ts' },
    '@angular/common': { main: 'bundles/common.umd.js',
                        defaultExtension: 'js' },
    '@angular/compiler': { main: 'bundles/compiler.umd.js',
                          defaultExtension: 'js' },
    '@angular/core': { main: 'bundles/core.umd.js',
```

```
                        defaultExtension: 'js' },
    '@angular/platform-browser': { main: 'bundles/platform-browser.umd.js',
                               defaultExtension: 'js' },
    '@angular/platform-browser-dynamic': { main:
    'bundles/platform-browser-dynamic.umd.js', defaultExtension: 'js' },
},
```

These packages will be all we need to run our application. As you move to more sophisticated applications, more packages will be required. But you only need to add what is required to get the application running and this is what we are doing here. The flexibility to select just the packages that you need helps to minimize the size of the download, always a good thing for performance.

The file also contains the following script that directs SystemJS to transpile our TypeScript files:

```
transpiler: 'typescript',
typescriptOptions: {
    emitDecoratorMetadata: true
}
```

This script will transpile our TypeScript files in the browser, which we would typically not do in a production app. Also, in order to keep the setup of our application as simple as possible, we are downloading the Angular modules and other libraries on-the-fly as the browser loads. In a production application, we would move both these steps into a build process that takes place before the application is launched in the browser. This will reduce the download size of the application and improve performance. We will be exploring setting up such a build process in `Chapter 2`, *Building Our First App – 7 Minute Workout*.

We then implement the module loading for our app by adding the following script to our `index.html` file:

```
system.import('app').catch(function(err){ console.error(err); });
```

The parameter being passed to our import statement `app` is the location of our application, which in this case is the `app` directory. Because we have indicated in `systemjs.config.js` that `main.ts` will be the entry point for our app, SystemJS will look for that file in the `app` directory when the application first starts.

# Bootstrapping our app

The next step in the initialization process is to bootstrap our application. As the entry point for our app, `main.ts` does that by executing the following code:

```
platform.bootstrapModule(AppModule);
```

Not surprisingly, we are bootstrapping `AppModule`. As we already know, bootstrapping instantiates `GuessTheNumberComponent` because it has been assigned to the bootstrap property within `AppModule`.

This makes `GuessTheNumberComponent` the root component for our application. While our app has only one component, every Angular application typically starts off with one root component.

We identify where this component will appear using the following syntax in our `index.html` file:

```
<my-app>Loading...</my-app>
```

As you might recall, the `@Component` decorator for our component identifies the component's selector:

```
@Component({
  selector: 'my-app'
```

Taken together, these indicate to Angular that when our page is loaded, it needs to bind our component to the `my-app` tag in `index.html`. Note that here we are using a custom element based on the Web Components standard.

So, this starts the bootstrapping process, which continues as follows:

- Angular scans our component definition and imports the modules that we have identified in the `import` statement of our component along with its related dependencies. These are the modules that we discussed earlier.
- It compiles the HTML view, starting from where the `my-app` tag is declared. In this compilation process, the framework traverses the template, looks for all interpolations, and sets up the binding between the view and our class.
- Post compilation, it links the view and our component class where changes are synced across the model and viewed in real time as we interact with the app.

# Tools

Tools make our lives easy, and we are going to share some tools that will help you with different aspects of Angular development, from code writing to debugging:

- **Augury** (`https://augury.angular.io/`)**:** This is a Chrome Dev Tools extension for debugging Angular applications. While the developers of the tool, rangle.io, indicate that it should be treated as a preview, it does support the Angular final release.
- **Browser developer console**: All current browsers have excellent capabilities when it comes to JavaScript debugging. Since we are working with JavaScript, we can put in breakpoints, add a watch, and do everything that is otherwise possible with JavaScript. Remember that a lot of errors with code can be detected just by looking at the browser's console window.
- **JSFiddle and Plunker**: JSFiddle (`http://jsfiddle.net/`) and Plunker (`http://plnkr.co/`) are excellent playgrounds for trying out HTML, CSS, and JavaScript code. These tools also have great versioning and sharing capabilities that can come in handy if we want someone's help.
- **IDE extensions**: Many of the popular IDEs on the market have plugins/extensions that make Angular development easy for us. Examples include:
    - **JetBrains WebStorm 2016:2.3**: `https://www.jetbrains.com/webstorm/`.
    - **Angular2 Snippets for Sublime Text**: `https://github.com/evanplaice/angular2-snippets`.
    - **Atom Angular 2 Snippets and Completions**: `https://github.com/d3viant0ne/angular2-atom-snippets`.
- **Visual Studio Code**: This is a brand new IDE that Microsoft has come up with (`https://code.visualstudio.com/`). It provides excellent IntelliSense and code completion support for Angular and TypeScript. Visual Studio 2015 (`https://www.visualstudio.com/`) also includes support for Angular and TypeScript.
- The Angular community is also developing CLI tools (`https://cli.angular.io/`) with the goal of taking the developer from the initial project setup all the way to the final deployment. We'll cover using Angular CLI in `Chapter 8`, *Some Practical Scenarios*.

- Component vendors are starting to offer support for Angular as well. For example, Telerik has released Kendo UI for Angular: `http://www.telerik.com/kendo-angular-ui/`.

# Resources

Angular is a new framework, but already a vibrant community is starting to emerge around it. Together with this book, there are also blogs, articles, support forums, and plenty of help. Some of the prominent resources that will be useful are explained as follows:

- **Framework code and documentation**: The Angular documentation can be found at `https://angular.io/docs/js/latest/`. The documentation is a work-in-progress and not fully finalized. Then, there is always the Angular source code, a great source of learning. It can be found at `https://github.com/angular/angular`.
- **The Angular team's blog**: You can refer to the Angular team's blog for more information about Angular 2 at `http://angularjs.blogspot.com/`.
- **Awesome Angular: A curated list of awesome Angular resources**: This is a new community resource at `https://angularclass.github.io/awesome-angular2/`.
- **The Angular Slack channel (**`https://angularchat.co/`**) and the gitter chat room (**`https://gitter.im/angular/angular`**)**: These are other great resources. Also check out Angular on Reddit: `https://www.reddit.com/r/Angular2`.
- **The Angular Google group (**`https://groups.google.com/forum/#!forum/angular`**) and the Stack Overflow channel (**`http://stackoverflow.com/questions/tagged/Angular2`**)**: Head over here if you have any questions or are stuck with some issue.
- **Built with Angular (**`http://builtwithangular2.com/`**)**: People have already created some amazing apps using Angular. This site showcases such apps, and most of them have source code available for us to take a look at.

That's it! The chapter is complete and it's time to summarize what you've learned.

# Summary

The journey has started and we have reached the first milestone. Despite this chapter being named *Getting Started*, we have covered a lot of concepts that you will need to know in order to understand the bigger picture. Your learning was derived from our Guess the Number! app, which we built and dissected throughout the chapter.

You learned how Angular implements the component design pattern using the emerging standards for Web Components, along with the latest versions of JavaScript and TypeScript. We also reviewed some of the constructs that are used in Angular, such as interpolation, expressions, and the data binding syntax. Finally, we took a look at change detection and app initialization.

The groundwork has been laid, and now we are ready for some serious app development on the Angular framework. In the next chapter, we will start working on a more complex exercise and expose ourselves to a number of new Angular constructs.

# 2

# Building Our First App - 7 Minute Workout

I hope the first chapter was intriguing enough and you want to learn more about Angular. Believe me, we have just scratched the surface! The framework has a lot to offer, and it strives to make frontend development using JavaScript more organized and hence manageable.

Keeping up with the theme of this book, we will be building a new app in Angular, and in the process, developing a better understanding of the framework. This app will also help us explore some new capabilities of the framework.

The topics that we will cover in this chapter include the following:

- **7 Minute Workout problem description**: We detail the functionality of the app that we build in this chapter.
- **Code organization**: For our first real app, we will try to explain how to organize code, specifically Angular code.
- **Designing the model**: One of the building blocks of our app is its model. We design the app model based on the app's requirements.
- **Understanding the data binding infrastructure**: While building the *7 Minute Workout* view, we will look at the data binding capabilities of the framework, which include *property*, *attribute*, *class*, *style*, and *event* bindings.
- **Exploring the Angular platform directives**: Some of the directives that we will cover are `ngFor`, `ngIf`, `ngClass`, `ngStyle`, and `ngSwitch`.

- **Cross-component communication with input properties**: Having built some child components, we learn how *input properties* can be used to pass data from the parent to its child components.
- **Cross-component communication with events**: Angular components can subscribe to and raise events. We get introduced to event binding support in Angular.
- **Angular pipes**: Angular pipes provide a mechanism to format view content. We explore some standard Angular pipes and build our own pipe too to support conversions from seconds to hh:mm:ss.

Let's get started! The first thing we will do is define theof our *7 Minute Workout* app.

# What is 7 Minute Workout?

We want everyone reading this book to be physically fit. Therefore, this book should serve a dual purpose; it should not only simulate your grey matter but also urge you to look after your physical fitness. What better way to do it than to build an app that targets physical fitness!

*7 Minute Workout* is an exercise/workout plan that requires us to perform a set of twelve exercises in quick succession within the seven minute time span. *7 Minute Workout* has become quite popular due to its benefits and the short duration of the workout. We cannot confirm or refute the claims but doing any form of strenuous physical activity is better than doing nothing at all. If you are interested to know more about the workout, then check out `http://well.blogs.nytimes.com/2013/05/09/the-scientific-7-minute-workout/`.

The technicalities of the app include performing a set of 12 exercises, dedicating 30 seconds for each of the exercises. This is followed by a brief rest period before starting the next exercise. For the app that we are building, we will be taking rest periods of 10 seconds each. So, the total duration comes out be a little more than 7 minutes.

At the end of the chapter, we will have the *7 Minute Workout* app ready, which will look something like this:

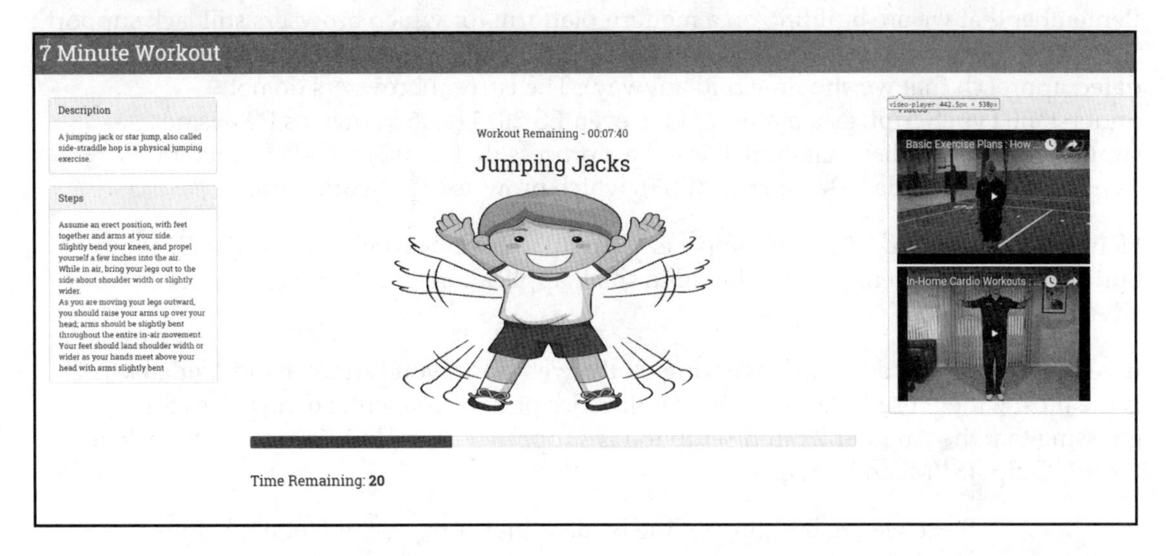


# Downloading the code base

The code for this app can be downloaded from the GitHub site `https://github.com/chandermani/angular2byexample` dedicated to this book. Since we are building the app incrementally, we have created multiple checkpoints that map to **GitHub branches** such as `checkpoint2.1`, `checkpoint2.2`, and so on. During the narration, we will highlight the branch for reference. These branches will contain the work done on the app up to that point in time.

The *7 Minute Workout* code is available inside the repository folder named `trainer`.

So let's get started!

# Setting up the build

Remember that we are building on a modern platform for which browsers still lack support. Therefore, directly referencing script files in HTML is out of question (while common, it's a dated approach that we should avoid anyway). The current browsers do not understand**TypeScript**; as a matter of fact, even **ES 2015** (also known as ES6) is not supported. This implies that there has to be a process that converts code written in TypeScript into standard **JavaScript (ES5)**, which browsers can work with.

Hence, having a build setup for almost any Angular 2 app becomes imperative. Having a build process may seem like overkill for a small application, but it has some other advantages as well.

If you are a frontend developer working on the web stack, you cannot avoid **Node.js**. This is the most widely used platform for Web/JavaScript development. So, no prizes for guessing that the Angular 2 build setup too is supported over Node.js with tools such as **Grunt**, **Gulp**, **JSPM**, and **webpack**.

Since we are building on the Node.js platform, install Node.js before starting.

While there are quite elaborate build setup options available online, we go for a minimal setup using **Gulp**. The reason is that there is no one size fits all solution out there. Also, the primary aim here is to learn about Angular 2 and not to worry too much about the intricacies of setting up and running a build.

Some of the notable starter sites plus build setups created by the community are as follows:

| **Start site** | **Location** |
|---|---|
| `angular2-webpack-starter` | `http://bit.ly/ng2webpack` |
| `angular2-seed` | `http://bit.ly/ng2seed` |
| `angular-cli`-We touched upon this tool in `Chapter 1`, *Getting Started*. It allows us to generate the initial code setup, including the build configurations, and has good scaffolding capabilities too. | `http://bit.ly/ng2-cli` |

A natural question arises if you are very new to **Node.js** or the overall build process: what does a typical Angular build involve? It depends! To get an idea about this process, it would be beneficial if we look at the build setup defined for our app. Let's set up the app's build locally then. Follow these steps to have the boilerplate Angular 2 app up and running:

1. Download the base version of this app from `http://bit.ly/ng2be-base` and unzip it to a location on your machine. If you are familiar with how Git works, you can just clone the repository and checkout the branch `base`:

   ```
   git checkout base
   ```

   This code serves as the starting point for our app.

2. Navigate to the `trainer` folder from the command line and execute these commands:

   ```
   npm i -g gulp typings
   npm install
   ```

   The first command installs Gulp (and typings) globally so that you can invoke the **Gulp** command line tool from anywhere and execute *Gulp tasks*. A **Gulp task** is an activity that Gulp performs during the build execution. If we look at the Gulp build script (which we will do shortly), we realize that it is nothing but a sequence of tasks performed whenever a build occurs. The second command installs the app's dependencies (in the form of `npm` packages). **Packages** in the Node.js world are third-party libraries that are either used by the app or support the app's building process. For example, Gulp itself is a Node.js package. The **npm** is a command-line tool for pulling these packages from a central repository.

3. Once Gulp is installed and npm pulls dependencies from the npm store, we are ready to build and run the application. From the command line, enter the following command:

   ```
   gulp play
   ```

   This compiles and runs the app. If the build process goes fine, the default browser window/tab will open with a rudimentary Hello World page (`http://localhost:9000/index.html`). We are all set to begin developing our app in Angular 2!

But before we do that, it would be interesting to know what has happened under the hood.

# The build internals

Even if you are new to Gulp, looking at `gulpfile.js` gives you a fair idea about what the build process is doing. A Gulp build is a set of tasks performed in a predefined order. The end result of such a process is some form of package code that is ready to be run. And if we are building our apps using *TypeScript/ES2015* or some other similar language that browsers do not understand natively, then we need an additional build step, called **transpilation**.

## Code transpiling

As it stands in 2016, browsers still cannot run *ES2015* code. While we are quick to embrace languages that hide the not-so-good parts of JavaScript (ES5), we are still limited by the browser's capabilities. When it comes to language features, **ES5** is still the safest bet as all browsers support it. Clearly, we need a mechanism to convert our *TypeScript* code into plain JavaScript (**ES5**). Microsoft has a TypeScript compiler that does this job.

The TypeScript compiler takes the TypeScript code and converts it into ES5-format code that can run in all browsers. This process is commonly referred to as **transpiling**, and since the TypeScript compiler does it, it's called a **transpiler**.

Interestingly, transpilation can happen at both build/compile time and runtime:

- **Build-time transpilation**: Transpilation as part of the build process takes the script files (in our case, TypeScript `.ts` files) and compiles them into plain JavaScript. Our build setup uses build-time transpilation.
- **Runtime transpilation**: This happens in the browser at runtime. We include the raw language-specific script files (`.ts` in our case), and the TypeScript compiler, which is loaded in the browser beforehand, compiles these script files on the fly. While runtime transpilation simplifies the build setup process, as a recommendation, it should be limited to development workflows only, considering the additional performance overhead involved in loading the transpiler and transpiling the code on the fly. The example in `Chapter 1`, *Getting Started*, used runtime transpilation, and hence it did not require any build setup. Go check it out again!

The process of transpiling is not limited to TypeScript. Every language targeted towards the Web, such as **CoffeeScript**, **ES2015**, or any other language that is not inherently understood by a browser, needs transpilation. There are transpilers for most languages, and the prominent ones (other than TypeScript) are **tracuer** and **babel**.

To compile TypeScript files, we can install the TypeScript compiler manually from the command line using this:

```
npm install -g typescript
```

Once installed, we can compile any TypeScript file into ES5 format using the compiler (`tsc.exe`).

But for our build setup, this process is automated using the `ts2js` Gulp task (check out `gulpfile.js`). And if you are wondering when we installed TypeScript... well, we did it as part of the `npm install` step, when setting up the code for the first time. The `gulp-typescript` package downloads the TypeScript compiler as a dependency.

With this basic understanding of transpilation, we can summarize what happens with our build setup:

- The `gulp play` command kicks off the build process. This command tells Gulp to start the build process by invoking the `play` task.
- Since the `play` task has a dependency on the `ts2js` task, `ts2js` is executed first. The `ts2js` compiles the TypeScript files (`.ts`) located in `src` folder and outputs them to the `dist` folder at the root.
- Post build, a static file server is started that serves all the app files, including static files (images, videos, and HTML) and script files (check `gulp.play` task).
- Thenceforth, the build process keeps a watch on any script file changes (the `gulp.watch` task) you make and recompiles the code on the fly.

The **livereload** has also been set up for the app. Any changes to the code refresh the browser running the app automatically. In case automatic browser refresh fails, we can always do a manual refresh.

This is a rudimentary build setup required to run an Angular app. For complex build requirements, we can always look at the starter/seed projects that have a more complete and robust build setup, or build something of our own.

Next let's look at the boilerplate app code already there and the overall code organization.

# Organizing code

This is how we are going to organize our code and other assets for the app:

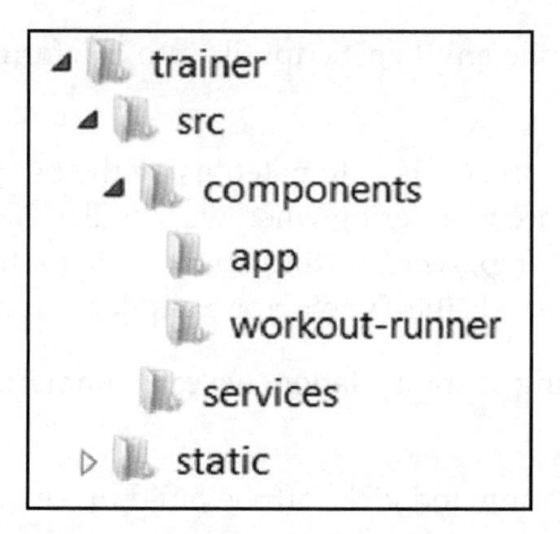


The `trainer` folder is the root folder for the app and it has a folder (`static`) for the static content (such as images, CSS, audio files, and others) and a folder (`src`) for the app's source code.

The organization of the app's source code is heavily influenced by the design of Angular and the Angular style guide (`http://bit.ly/ng2-style-guide`) released by the Angular team. As we saw in `Chapter 1`, *Getting Started*, among the primary building blocks of Angular 2 are **components**. The `components` folder hosts all the components that we create. We will be creating subfolders in this folder for every major component of the application. Each component folder will contain artifacts related to that component, which includes its template, its implementation, and other related items. We will also keep adding more top-level folders (inside the `src` folder) as we build the application.

If we look at the code now, the `components/app` folder has defined a *root level component* `TrainerAppComponent` and root level module `AppModule`. The `bootstrap.ts` file contains code to bootstrap/load the application module (`AppModule`).

*7 Minute Workout* uses **Just-in-time (JIT) compilation** to compile Angular views. This implies that views are compiled just before they are rendered in the browser. Angular has a compiler running in the browser that compiles these views.
Angular also supports the **Ahead-of-time (AoT) compilation** model. With AoT, the views are compiled on the server side using a server version of the Angular compiler. The views returned to the browser are precompiled and ready to be used.
For *7 Minute Workout*, we stick to the JIT compilation model just because it is easy to set up as compared to AoT, which requires server-side tweaks and package installation.
We highly recommend that you use AoT compilation for production apps due the numerous benefits it offers. AoT can improve the application's initial load time and reduce its size too. Look at the AoT platform documentation (*cookbook*) at `http://bit.ly/ng2-aot` to understand how AoT compilation can benefit you.

Time to start working on our first focus area, which is the app's model!

# The 7 Minute Workout model

Designing the model for this app requires us to first detail the functional aspects of the *7 Minute Workout* app, and then derive a model that satisfies those requirements. Based on the problem statement defined earlier, some of the obvious requirements are as follows:

- Being able to start the workout.
- Providing a visual clue about the current exercise and its progress. This includes the following:
    - Providing a visual depiction of the current exercise
    - Providing step-by-step instructions on how to do a specific exercise
    - The time left for the current exercise
- Notifying the user when the workout ends.

Some other valuable features that we will add to this app are as follows:

- The ability to pause the current workout.
- Providing information about the next exercise to follow.
- Providing audio clues so that the user can perform the workout without constantly looking at the screen. This includes:
    - A timer click sound
    - Details about the next exercise
    - Signaling that the exercise is about to start
- Showing related videos for the exercise in progress and the ability to play them.

As we can see, the central theme for this app is **workout** and **exercise**. Here, a workout is a set of exercises performed in a specific order for a particular duration. So, let's go ahead and define the model for our workout and exercise.

Based on the requirements just mentioned, we will need the following details about an exercise:

- The name. This should be unique.
- The title. This is shown to the user.
- The description of the exercise.
- Instructions on how to perform the exercise.
- Images for the exercise.
- The name of the audio clip for the exercise.
- Related videos.

With TypeScript, we can define the classes for our model.

Create a folder called `workout-runner` inside the `src/components` folder and copy the `model.ts` file from the checkpoint2.1 branch folder `workout-runner`(`http://bit.ly/ng2be-2-1-model-ts`) to the corresponding local folder. `model.ts` contains the model definition for our app.

The `Exercise` class looks like this:

```
export class Exercise {
  constructor(
    public name: string,
    public title: string,
    public description: string,
    public image: string,
    public nameSound?: string,
    public procedure?: string,
    public videos?: Array<string>) { }
}
```

**TypeScript tips**

Passing constructor parameters with `public` or `private` is a shorthand for creating and initializing class members at one go.

The `?` suffix after `nameSound`, `procedure`, and `videos` implies that these are optional parameters.

For the workout, we need to track the following properties:

- The name. This should be unique.
- The title. This is shown to the user.
- The exercises that are part of the workout.
- The duration for each exercise.
- The rest duration between two exercises.

So, the model class (`WorkoutPlan`) looks like this:

```
export class WorkoutPlan {
  constructor(
    public name: string,
    public title: string,
    public restBetweenExercise: number,
    public exercises: ExercisePlan[],
    public description?: string) { }

  totalWorkoutDuration(): number { ... }
}
```

The `totalWorkoutDuration` function returns the total duration of the workout in seconds.

`WorkoutPlan` has a reference to another class in the preceding definition: `ExercisePlan`. It tracks the exercise and the duration of the exercise in a workout, which is quite apparent once we look at the definition of `ExercisePlan`:

```
export class ExercisePlan {
  constructor(
    public exercise: Exercise,
    public duration: number) { }
}
```

These three classes constitute our base model, and we will decide in the future whether or not we need to extend this model as we start implementing the app's functionality.

Since we have started with a preconfigured and basic Angular app, you just need to understand how this app bootstrapping is occurring.

# App bootstrapping

`Chapter 1`, *Getting Started*, had a good introduction to the app bootstrapping process. The app bootstrapping process for *7 Minute Workout* remains the same, look at the `src` folder. There is a `bootstrap.ts` file with only the execution bit (other than `imports`):

```
platformBrowserDynamic().bootstrapModule(AppModule);
```

The `boostrapModule` function call actually bootstraps the application by loading the root module, `AppModule`. The process is triggered by this call in `index.html`:

```
System.import('app').catch(console.log.bind(console));
```

The `System.import` statement sets off the app bootstrapping process by loading the first module from `bootstrap.ts`.

Modules defined in the context of Angular 2, (using `@NgModule` decorator) are different from modules SystemJS loads. SystemJS modules are JavaScript modules, which can be in different formats adhering to *CommonJS*, *AMD*, or *ES2015* specifications.
Angular modules are constructs used by Angular to segregate and organize its artifacts.
Unless the context of discussion is SystemJS, any reference to *module* implies *Angular module.*

The next section details how SystemJS loads our Angular app. Remember all references to module in the next section are JavaScript module. If you want you can skip this section as it does not directly relate to Angular development.

# App loading with SystemJS

SystemJS starts loading the JavaScript module with the call to `System.import('app')` in `index.html`.

SystemJS starts by loading `bootstrap.ts` first. The imports defined inside `bootstrap.ts` cause SystemJS to then load the imported modules. If these module imports have further `import` statements, SystemJS loads them too, recursively.

And finally the `platformBrowserDynamic().bootstrapModule(AppModule);` function gets executed once all the imported modules are loaded.

For the SystemJS `import` function to work, it needs to know where the modules are located. We define this in the file, `systemjs.config.js`, and reference it in `index.html`, before the `System.import` script:

```
<script src="systemjs.config.js"></script>
```

This configuration file contains all of the necessary configuration for SystemJS to work correctly.

Open `systemjs.config.js`, the `app` parameter to `System.import` function points to a folder `dist` as defined on the `map` object:

```
var map = {
    'app': 'dist',
    ...
 }
```

And the next variable, `packages`, contains settings that hint to SystemJS how to load a module from a package when no filename/extension is specified. For `app`, the default module is `bootstrap.js`:

```
var packages = {
    'app': { main: 'bootstrap.js',  defaultExtension: 'js' },
    ...
};
```

Are you wondering what the `dist` folder has to do with our application? Well, this is where our transpiled scripts end up. As we build our app in TypeScript, the TypeScript compiler converts these `.ts` script files in the `src` folder to JavaScript modules and deposits them into the `dist` folder. SystemJS then loads these compiled JavaScript modules. The transpiled code location has been configured as part of the build definition in `gulpfile.js`. Look for this excerpt in `gulpfile.ts`:

```
return tsResult.js
    .pipe(sourcemaps.write())
    .pipe(gulp.dest('dist'))
```

The module specification used by our app can again be verified in `gulpfile.js`. Take a look at this line:

```
noImplicitAny: true,
module: 'system',
target: 'ES5',
```

These are TypeScript compiler options, with one being `module`, that is, the target module definition format.

The `system` module type is a new module format designed to support the exact semantics of ES2015 modules within ES5.

Once the scripts are *transpiled* and the module definitions created (in the target format), SystemJS can load these modules and their dependencies.

It's time to get into the thick of action; let's build our first component.

# Our first component – WorkoutRunnerComponent

To implement the `WorkoutRunnerComponent`, we need to outline the behavior of the application.

What we are going to do in the `WorkoutRunnerComponent` implementation is as follows:

1. Start the workout.
2. Show the workout in progress and show the progress indicator.
3. After the time elapses for an exercise, show the next exercise.
4. Repeat this process until all the exercises are over.

Let's start with the implementation. The first thing that we will create is the `WorkoutRunnerComponent` implementation.

Open the `workout-runner` folder in the `src/components` folder and add a new code file called `workout-runner.component.ts` to it. Add this chunk of code to the file:

```
import {WorkoutPlan, ExercisePlan, Exercise, ExerciseProgressEvent,
ExerciseChangedEvent} from '../../services/model';
export class WorkoutRunnerComponent { }
```

The `import` module declaration allows us to reference the classes defined in the `model.ts` file in `WorkoutRunnerComponent`.

We first need to set up the workout data. Let's do that by adding a constructor and related class properties to the `WorkoutRunnerComponent` class:

```
workoutPlan: WorkoutPlan;
restExercise: ExercisePlan;
constructor() {
   this.workoutPlan = this.buildWorkout();
   this.restExercise = new ExercisePlan(
     new Exercise("rest", "Relax!", "Relax a bit", "rest.png"),
     this.workoutPlan.restBetweenExercise);
}
```

The `buildWorkout` on `WorkoutRunnerComponent` sets up the complete workout, as we will see shortly. We also initialize a `restExercise` variable to track even the rest periods as exercise (note that `restExercise` is an object of type `ExercisePlan`).

The `buildWorkout` function is a lengthy function, so it's better if we copy the implementation from the workout runner's implementation available in Git branch checkpoint2.1 (`http://bit.ly/ng2be-2-1-workout-runner-component-ts`). The `buildWorkout` code looks like this:

```
buildWorkout(): WorkoutPlan {
let workout = new WorkoutPlan("7MinWorkout",
"7 Minute Workout", 10, []);
   workout.exercises.push(
      new ExercisePlan(
        new Exercise(
          "jumpingJacks",
          "Jumping Jacks",
          "A jumping jack or star jump, also called side-straddle hop
           is a physical jumping exercise.",
          "JumpingJacks.png",
          "jumpingjacks.wav",
          `Assume an erect position, with feet together and
           arms at your side. ...`,
          ["dmYwZH_BNd0", "BABOdJ-2Z6o", "c4DAnQ6DtF8"]),
        30));
   // (TRUNCATED) Other 11 workout exercise data.
   return workout;
}
```

This code builds the `WorkoutPlan` object and pushes the exercise data into the exercises array (an array of `ExercisePlan` objects), returning the newly built workout.

The initialization is complete; now, it's time to actually implement the start workout. Add a `start` function to the `WorkoutRunnerComponent` implementation, as follows:

```
start() {
   this.workoutTimeRemaining =
   this.workoutPlan.totalWorkoutDuration();
   this.currentExerciseIndex = 0;
this.startExercise(this.workoutPlan.exercises[this.currentExerciseIndex]);
}
```

Then declare the new variables used in the function at the top, with other variable declarations:

```
workoutTimeRemaining: number;
currentExerciseIndex: number;
```

The `workoutTimeRemaining` variable tracks the total time remaining for the workout, and `currentExerciseIndex` tracks the currently executing exercise index. The call to `startExercise` actually starts an exercise. This is how the code for `startExercise` looks:

```
startExercise(exercisePlan: ExercisePlan) {
    this.currentExercise = exercisePlan;
    this.exerciseRunningDuration = 0;
    let intervalId = setInterval(() => {
      if (this.exerciseRunningDuration >=
        this.currentExercise.duration) {
          clearInterval(intervalId);
      }
      else { this.exerciseRunningDuration++;     }
    }, 1000);
}
```

We start by initializing `currentExercise` and `exerciseRunningDuration`. The `currentExercise` variable tracks the exercise in progress and `exerciseRunningDuration` tracks its duration. These two variables also need to be declared at the top:

```
currentExercise: ExercisePlan;
exerciseRunningDuration: number;
```

We use the `setInterval` JavaScript function with a delay of 1 second (1,000 milliseconds) to track the exercise progress by incrementing`exerciseRunningDuration`. The `setInterval` invokes the callback every second. The `clearInterval` call stops the timer once the exercise duration lapses.

**TypeScript Arrow functions**
The callback parameter passed to `setInterval` (`()=>{...}`) is a **lambda function** (or an **arrow function** in ES 2015). Lambda functions are short-form representations of anonymous functions, with added benefits. You can learn more about them at `https://basarat.gitbooks.io/typescript/content/docs/arrow-functions.html`.

As of now, we have a `WorkoutRunnerComponent` class. We need to convert it into an *Angular component* and define the component view. We did something similar in `Chapter 1`, *Getting Started*, too.

Add the import for `Component` and a component decorator (highlighted code):

```
import {WorkoutPlan, ExercisePlan, Exercise} from './model'
import {Component} from '@angular/core';

@Component({
  selector: 'workout-runner',
  template: `
    <pre>Current Exercise: {{currentExercise | json}}</pre>
    <pre>Time Left: {{currentExercise?.duration-
exerciseRunningDuration}}</pre>`
})
export class WorkoutRunnerComponent {
```

There is not much to explain here, as you already know how to create an Angular component. You understand the role of the `@Component` decorator, what `selector` does, and how the `template` is used.

The JavaScript generated for the `@Component` decorator contains enough metadata about the component. This allows Angular framework to instantiate the correct component at runtime.
Strings enclosed in **backticks** (` `) are a new addition to ES2015. Also called **template literals**, such string literals can be multiline and allow expressions to be embedded inside (not to be confused with Angular expressions). Look at the MDN article here at `http://bit.ly/template-literals` for more details.

The preceding template HTML will render the raw `ExercisePlan` object and the exercise time remaining. It has an interesting expression inside the first interpolation: `currentExercise | json`. The `currentExercise` property is defined in `WorkoutRunnerComponent`, but what about the `|` symbol and what follows it (`json`)? In the Angular 2 world, it is called a **pipe**. *The sole purpose of a pipe is to transform/format template data.* The `json` pipe here does JSON data formatting. You will learn more about pipes later in this chapter, but to get a general sense of what the `json` pipe does, we can remove the `json` pipe plus the `|` symbol and render the template; we are going to do this next.

As we saw in `Chapter 1`, *Getting Started*, before we can use `WorkoutRunnerComponent`, we need to declare it inside a module. Since our app currently has only one module (`AppModule`), we add the `WorkoutRunnerComponent` declaration to it. Update `app.module.ts` by adding the highlighted code:

```
import {WorkoutRunnerComponent} from '../workout-runner/workout-
runner.component';
@NgModule({
```

```
    imports: [BrowserModule],
    declarations: [TrainerAppComponent, WorkoutRunnerComponent],
```

Now `WorkoutRunnerComponent` can be referenced in the root component so that it can be rendered. Modify `src/components/app/app.component.ts` as highlighted in the following code:

```
@Component({
...
    template: `
<div class="navbar ...> ...
</div>
<div class="container ...>
<workout-runner></workout-runner>
</div>`
})
```

We have changed the root component template and added the `workout-runner` element to it. This will render the `WorkoutRunnerComponent` inside our root component.

While the implementation may look complete, there is a crucial piece missing. Nowhere in the code do we actually start the workout. The workout should start as soon as we load the page.

**Component life cycle hooks** are going to rescue us!

# Component life cycle hooks

The life of an Angular component is eventful. Components get created, change state during their lifetime, and finally they are destroyed. Angular provides some **life cycle hooks/functions** that the framework invokes (on the component) when such an event occurs. Consider these examples:

- When component is initialized, Angular invokes `ngOnInit`
- When a component's input properties change, Angular invokes `ngOnChanges`
- When a component is destroyed, Angular invokes `ngOnDestroy`

As developers, we can tap into these key moments and perform some custom logic inside the respective component.

Angular has TypeScript interfaces for each of these hooks that can be applied to the component class to clearly communicate the intent. For example:

```
class WorkoutRunnerComponent implements OnInit {
  ngOnInit (){
    ...
}
...
```

The interface name can be derived by removing the prefix `ng` from the function names.

The hook we are going to utilize here is `ngOnInit`. The `ngOnInit` function gets fired when the component's data-bound properties are initialized but before the view initialization starts.

Add the `ngOnInit` function to the `WorkoutRunnerComponent` class with a call to start the workout:

```
ngOnInit() {
    this.start();
}
```

And implement the `OnInit` interface on `WorkoutRunnerComponent`; it defines the `ngOnInit` method:

```
import {Component,OnInit} from '@angular/core';
...
export class WorkoutRunnerComponent implements OnInit {
```

There are a number of other life cycle hooks, including `ngOnDestroy`, `ngOnChanges`, and `ngAfterViewInit`, that components support; but we are not going to dwell into any of them here. Look at the developer guide ( `http://bit.ly/ng2-lifecycle`) on *Life Cycle Hooks* to learn more about other such hooks.

Time to run our app! Open the command line, navigate to the `trainer` folder, and type this line:

```
gulp play
```

If there are no compilation errors and the browser automatically loads the app (`http://localhost:9000/index.html`), we should see the following output:

```
Current Exercise: {
  "exercise": {
    "name": "jumpingJacks",
    "title": "Jumping Jacks",
    "description": "A jumping jack or star jump, also called side-
    "image": "JumpingJacks.png",
    "nameSound": "",
    "procedure": "Assume an erect position, with feet together and
inches into the air.\n                            While in air, br
 you are moving your legs outward, you should raise your arms up o
          Your feet should land shoulder width or wider as your h
    "videos": [
      "dmYwZH_BNd0",
      "BABOdJ-2Z6o",
      "c4DAnQ6DtF8"
    ]
  },
  "duration": 30
}

Time Left: 28
```

The model data updates with every passing second! Now you'll understand why interpolations (`{{ }}`) are a great debugging tool.

This will also be a good time to try rendering `currentExercise` without the `json` pipe (use `{{currentExercise}}`), and see what gets rendered.

We are not done yet! Wait long enough on the `index.html` page and you will realize that the timer stops after 30 seconds. The app does not load the next exercise data. Time to fix it!

Update the code inside the `setIntervalif` condition:

```
if (this.exerciseRunningDuration >=
this.currentExercise.duration) {
    clearInterval(intervalId);
    let next: ExercisePlan = this.getNextExercise();
    if (next) {
      if (next !== this.restExercise) {
        this.currentExerciseIndex++;
        }
      this.startExercise(next);}
    else { console.log("Workout complete!"); }
}
```

The `if` condition `if (this.exerciseRunningDuration >= this.currentExercise.duration)` is used to transition to the next exercise once the time duration of the current exercise lapses. We use `getNextExercise` to get the next exercise and call `startExercise` again to repeat the process. If no exercise is returned by the `getNextExercise` call, the workout is considered complete.

During exercise transitioning, we increment `currentExerciseIndex` only if the next exercise is not a rest exercise. Remember that the original workout plan does not have a rest exercise. For the sake of consistency, we have created a rest exercise and are now swapping between rest and the standard exercises that are part of the workout plan. Therefore, `currentExerciseIndex` does not change when the next exercise is rest.

Let's quickly add the `getNextExercise` function too. Add the function to the `WorkoutRunnerComponent` class:

```
getNextExercise(): ExercisePlan {
    let nextExercise: ExercisePlan = null;
    if (this.currentExercise === this.restExercise) {
      nextExercise =
this.workoutPlan.exercises[this.currentExerciseIndex + 1];
    }
    else if (this.currentExerciseIndex <
    this.workoutPlan.exercises.length - 1) {
      nextExercise = this.restExercise;
    }
    return nextExercise;
}
```

The `WorkoutRunnerComponent.getNextExercise` returns the next exercise that needs to be performed.

Note that the returned object for `getNextExercise` is an `ExercisePlan` object that internally contains the exercise details and the duration for which the exercise runs.

The implementation is quite self-explanatory. If the current exercise is rest, take the next exercise from the `workoutPlan.exercises` array (based on `currentExerciseIndex`); otherwise, the next exercise is rest, given that we are not on the last exercise (the `else if` condition check).

With this, we are ready to test our implementation. So go ahead and refresh `index.html`. Exercises should flip after every 10 or 30 seconds. Great!

The current build setup automatically compiles any changes made to the script files when the files are saved; it also refreshes the browser post these changes. But just in case the UI does not update or things do not work as expected, refresh the browser window.
If you are having a problem with running the code, look at the Git branch `checkpoint2.1` for a working version of what we have done thus far. Or if you are not using Git, download the snapshot of Checkpoint 2.1 (a ZIP file) from `http://bit.ly/ng2be-checkpoint2-1`. Refer to the `README.md` file in the `trainer` folder when setting up the snapshot for the first time.

We have done enough work on the controller for now, let's build the view.

# Building the 7 Minute Workout view

Most of the hard work has already been done while defining the model and implementing the component. Now we just need to skin the HTML using the super-awesome data binding capabilities of Angular. It's going to be simple, sweet, and elegant!

For the *7 Minute Workout* view, we need to show the exercise name, the exercise image, a progress indicator, and the time remaining. Copy the `workout-runner.html` file from the Git branch `checkpoint2.2`, the `workout-runner` folder (or download it from `http://bit.ly/ng2be-2-2-workout-runner-html`), to the corresponding folder locally. The view HTML looks like this:

```
<div class="row">
  <div id="exercise-pane" class="col-sm-8 col-sm-offset-2">
    <div class="row workout-content">
      <div class="workout-display-div">
```

```
        <h1>{{currentExercise.exercise.title}}</h1>
        <img class="img-responsive" [src]="'/static/images/' +
         currentExercise.exercise.image" />
        <div class="progress time-progress">
          <div class="progress-bar" role="progressbar"
          [attr.aria-valuenow]="exerciseRunningDuration"
          aria-valuemin="0"
          [attr.aria-valuemax]="currentExercise.duration"
          [ngStyle] = "{'width':(exerciseRunningDuration/
                      currentExercise.duration) * 100 + '%'}"></div>
        </div>
        <h1>Time Remaining:{{currentExercise.duration -
           exerciseRunningDuration}}</h1>
      </div>
    </div>
  </div>
</div>
```

Before we take a deep dive into the view HTML, we still need to reference the view. The `WorkoutRunnerComponent` currently uses an inline template. We need to externalize it by pointing to the above HTML file. Update the `workout-runner.component.ts` file by replacing the `template` property with property `templateUrl`:

```
templateUrl: '/src/components/workout-runner/workout-runner.html'
```

The decorator property, `templateUrl`, allows us to reference an external file for the view HTML.

**Inline versus External templates**

We just saw two ways of defining a view template: using `template` (inline) and using the `templateUrl` (external) property. Which one is preferred?

Well, considering the way in which components are envisioned in Angular2, as self-contained building blocks, for our app, inline templating makes sense. Everything that is required by the component is available in a single file. However, inline templates have a disadvantage; formatting HTML becomes difficult and IDE support is very limited as the content is treated as a string literal. When we externalize HTML, we can develop a template as a normal HTML document.

One possible solution that allows us to have the best of both worlds would be to define our HTML templates in separate files during development and reference them using the standard `templateUrl` property. But as part of a production release, configure the build process to inline the template HTML into the component definition.

Before we understand the Angular pieces in the view, let's just run the app again. Save the changes in `workout-runner.component.ts`. If `gulp play` is already running, just refresh the page and see the workout app in its full glory!

We have our basic app running. The exercise image and title show up, the progress indicator shows the progress, and exercise transitioning occurs when the exercise time lapses. This surely feels great!

If you are having a problem with running the code, look at the Git branch `checkpoint2.2` for a working version of what we have done thus far. You can also download the snapshot of `checkpoint2.2` (a zip file) from this GitHub location: `http://bit.ly/ng2be-checkpoint2-2`. Refer to the `README.md` file in the `trainer` folder when setting up the snapshot for the first time.

If we look back at the view HTML, we see that other than the styling done using *bootstrap* CSS and some custom CSS, there are some interesting Angular pieces that need our attention. Also, since everything ties into the Angular binding infrastructure, it's time to dig deeper into this infrastructure and explore its capabilities.

# The Angular 2 binding infrastructure

Any modern JavaScript framework comes with strong model-view binding support, and Angular is no different. The primary aim of any binding infrastructure is to reduce the boilerplate code that we need to write to keep the view and the model in sync. A robust binding infrastructure is always declarative and terse.

The Angular binding infrastructure allows us to transform a template (raw) HTML into a live view that is bound to model data. Based on the binding constructs used, data can flow and be synced in both directions: from model to view and view to model.

Angular established a link between the model data on the component and the view using the `template` or `templateUrl` property of the `@Component` decorator. With the exception of the `script` tag, almost any piece of HTML can act as a template for the Angular binding infrastructure.

To make this binding magic work, Angular needs to take the view template, compile it, link it to the model data, and keep it in sync with model updates without the need for any boilerplate synchronization code.

Based on the data flow direction, these bindings can be of three categories:

- *One-way binding from model to view*: In model-to-view binding, changes to model are kept in sync with the view. Interpolations, property, attribute, class, and style bindings fall in this category.
- *One-way binding from view to model*: In this category, view changes flow towards the model. Event bindings fall in this category.
- *Two-way/bidirectional binding*: Two-way binding, as the name suggests, keeps the view and model in sync. There is a special binding construct used for two-way binding, `ngModel`.

Let's understand how to utilize the binding capabilities of Angular to support view templatization. Angular provides these binding constructs:

- Interpolations
- Property binding
- Attribute binding
- Class binding
- Style binding
- Event binding

The *7 Minute Workout* view uses some of these constructs, so this is a good time to learn about them. **Interpolation** is the first one.

# Interpolations

**Interpolations** are quite simple. The expression inside the interpolation symbols (`{{ }}`) is evaluated in the context of the model (or the component class members), and the outcome of the evaluation is embedded in HTML. We have seen these all along in `Chapter 1`, *Getting Started*, and in the workout runner view. We render the exercise title and the exercise time remaining using interpolation:

```
<h1>{{currentExercise.exercise.title}}</h1>
<h1>Time Remaining: {{currentExercise.duration?-
exerciseRunningDuration}}</h1>
```

Remember that interpolations synchronize model changes with the view. This is one way of binding from a model to a view.

Interpolations in fact are a special case of property binding, which allows us to bind HTML element/custom component properties to a model. We shortly discuss how an interpolation can be written using property binding syntax. We can consider interpolation as syntactical sugar over property binding.

# Property binding

Look at this view excerpt:

```
<img class="img-responsive" [src]="'/static/images/' +
currentExercise.exercise.image" />
```

It seems that we are setting the `src` attribute of `img` to an expression that gets evaluated at runtime when the app runs. Not true!

What seems to be an **attribute binding** is, in fact, a **property binding**. Moreover, since many of us are not aware of the difference between an HTML element's *property* and its *attribute*, this statement is very confusing. Therefore, before we look at how property bindings work, let's try to grasp the difference between an element's *property* and its *attribute*.

# Property versus attribute

Take any DOM element API and you will find *attributes, properties, functions,* and *events.* While events and functions are self-explanatory, it is difficult to understand the difference between *properties* and *attributes.* We use these words interchangeably, which does not help much either. Take, for example, this line of code:

```
<input type="text" value="Awesome Angular2">
```

When the browser creates a DOM element (`HTMLInputElement` to be precise) for this input textbox, it uses the `value` attribute on `input` to set the initial state of the `input`'s `value` property to `Awesome Angular2`.

Post this initialization, any changes to `input value` property do not reflect on the `value` attribute; the attribute always has `Awesome Angular2` (unless set explicitly again). This can be confirmed by querying the `input` state.

Suppose we change the `input` data to `Angular2 rocks!` and query the `input` element state:

```
input.value // value property
```

The `value` property always returns the current input content, which is `"Angular2 rocks!"`. Whereas this DOM API function:

```
input.getAttribute('value')  // value attribute
```

Returns the `value` attribute, and is always `Awesome Angular2` set initially.

The primary role of an element attribute is to initialize the state of the element when the corresponding DOM object is created.

There are a number of other nuances that add to this confusion. These include the following:

- *Attribute* and *property* synchronization is not consistent across properties. As we saw in the preceding example, changes to the `value` *property* on `input` do not affect the `value` *attribute,* but this is not true for all property-value pairs. The `src` property of an image element is a prime example of this; changes to property or attribute value are always kept in sync.

- It's surprising to learn that the mapping between attributes and properties is also not one-to-one. There are a number of properties that do not have any backing attribute (such as `innerHTML`), and there are also attributes that do not have a corresponding property defined on the DOM (such as `colspan`).
- Attribute and property mapping too adds to this confusion as it does not follow a consistent pattern. An excellent example of this is available in the Angular 2 developer's guide, which we are going to reproduce here verbatim:

*The* `disabled` *attribute is another peculiar example. A button's* `disabled` *property is* `false` *by default so the button is enabled. When we add the disabled attribute, its presence alone initializes the button's* `disabled` *property to true so the button is disabled. Adding and removing the disabled attribute disables and enables the button. The value of the attribute is irrelevant which is why we cannot enable a button by writing* `<button disabled="false">Still Disabled</button>`.

The aim of this discussion is to make sure that you understand the difference between the properties and attributes of a DOM element. This new mental model will help you as we continue to explore the framework's property and attribute binding capabilities. Let's get back to our discussion on property binding.

## Property binding continued...

Now that we understand the difference between a property and an attribute, let's look at the binding example again:

```
<img class="img-responsive" [src]="'/static/images/' +
currentExercise.exercise.image" />
```

The `[propertName]` square bracket syntax is used to bind the `img.src` property to an Angular expression.

The general syntax for property binding looks like this:

```
[target]="sourceExpression";
```

In the case of property binding, the **target** is a property on the DOM element or component. The target can also be an event, as we will see shortly when we perform event binding.

Binding source and target

It is important to understand the difference between source and target in an Angular binding.
The property appearing inside [] is a **target**, sometimes called **binding target**. The target is the consumer of the data and always refers to a property on the component/element. The **source** expression constitutes the data source that provides data to the target.

At runtime, the expression is evaluated in the context of the component's/element's property (the `WorkoutRunnerComponent.currentExercise.exercise.image` property in the preceding case).

Property binding, event binding, and attribute binding do not use the interpolation symbol. The following is invalid:
`[src]="{{'/static/images/' + currentExercise.exercise.image}}"`
If you have worked on Angular 1, property binding to any DOM property allows Angular 2 to get rid of a number of directives, such as `ng-disable`, `ng-src`, `ng-key*`, `ng-mouse*`, and a few others.

Property binding works on component properties too! Components can define input and output properties that can be bound to the view, such as this:

```
<workout-runner [exerciseRestDuration]="restDuration"></workout-runner>
```

This hypothetical snippet binds the `exerciseRestDuration` property on the `WorkoutRunnerComponent` class to the `restDuration` property defined on the container component (parent), allowing us to pass the rest duration as a parameter to the `WorkoutRunnerComponent`. As we enhance our app and develop new components, you will learn how to define custom properties and events on a component.

We can enable property binding using the `bind-` syntax, which is a *canonical form* of property binding. This implies that:
`[src]="'/static/images/' + currentExercise.exercise.image"`
Is equivalent to the following:
`bind-src="'/static/images/' + currentExercise.exercise.image"`

Property binding, like interpolation, is unidirectional, from the component/element source to the view. Changes to the model data are kept in sync with the view.

When we concluded the last section by describing interpolation as syntactical sugar over property binding, the intent was to highlight how both can be used interchangeably.
The interpolation syntax is terser than property binding and hence very useful. This is how Angular interprets an interpolation:

```
<h3>Main heading - {{heading}}</h3>
<h3 [text-content]="' Main heading - '+heading"></h3>
```

Angular translates the interpolation in the first statement into the `textContent` property binding (second statement).

While property binding makes it easy for us to bind any expression to the target property, we should be careful with the expression we employ. This is also because of the fact that Angular's change detection system will evaluate your expression binding multiple times during the life cycle of the application, while your component is alive. Therefore, while binding an expression to a property target, keep these two guidelines in mind.

## Quick expression evaluation

A property binding expression should evaluate quickly. Slow evaluation can happen when a function is used as an expression. Consider this binding:

```
<div>{{doLotsOfWork()}}</div>
```

This interpolation binds the return value of `doLotsOfWork` to the `div`. This function then gets called every time Angular performs a change detection run, which Angular does quite often based on some internal heuristics. Hence it becomes imperative that the expressions we use evaluate quickly.

## Side-effect-free binding expressions

If a function is used in a binding expression, it should be side-effect-free. Consider yet another binding:

```
<div [innerHTML]="getContent()"></div>
```

And the underlying function, `getContent`:

```
getContent() {
  var content=buildContent();
  this.timesContentRequested +=1;
  return content;
}
```

The `getContent` call changes the state of the component by updating the `timesContentRequested` property every time it is called. If this property is used in view such as:

```
<div>{{timesContentRequested}}</div>
```

Angular throws errors such as:

```
Expression '{{getContent()}}' in AppComponent@0:4' has changed after it was
checked. Previous value: '1'. Current value: '2'
```

Angular framework works in two modes, *dev* and *production*. If we enable production mode in the application, the preceding error does not show up. Look at the framework documentation at http://bit.ly/enableProdMode for more details.

The bottom line is that your expression used inside property binding should be side-effect-free.

Let's now look at some interesting behavior related to the square bracket syntax used for property binding. The target specified in `[]` is not limited to a component/element property. To understand *target selection*, we need to introduce a new concept: **directives**.

# Angular directives

As a framework, Angular tries to enhance the HTML **DSL** (short for **Domain-Specific Language**).

*Components* are referenced in HTML using custom tags such as `<workout-runner></workout-runner>` (not part of standard HTML constructs). This highlights the first extension point.

The use of `[]` and `()` for property and event binding defines the second.

And then there is the third one, called **attribute directives**.

While components come with their own view, attribute directives are there to enhance the appearance and/or behavior of existing elements/components. The `ngStyle` directive used in workout-runner's view is a good example of an attribute directive:

```
<div class="progress-bar" role="progressbar"
[ngStyle] = "{'width':(exerciseRunningDuration/
currentExercise.duration) * 100 + '%'}"></div>
```

The `ngStyle` directive does not have its own view; instead it allows us to set multiple styles on an HTML element using binding expressions. We will be covering a number of framework attribute directives later in this book.

There is also another class of directives, called **structural directives**. Again, structural directives do not have their own view; they change the DOM layout of the elements on which they are applied. The `ngFor` and `ngIf` directives fall into this category. We dedicate a complete section later in the chapter to understanding these structural directives.

**Directive nomenclature**
"Directives" is an umbrella term used for component directives (also known as components), attribute directives, and structural directives. Throughout the book, when we use the term "directive," we will be referring to either an attribute directive or a structural directive depending on the context. Component directives are always referred to as *components*.

With this understanding of the different types of directives that Angular has, we can comprehend the process of target selection for binding.

# Target selection for binding

As described earlier, the target specified in `[]` is not limited to a component/element property. While the property name is a common target, the Angular templating engine actually does a heuristics to decide the target type. Angular first searches the registered known directives (*attribute* or *structural*) that have matching selectors before looking for a property that matches the target expression. Consider this view fragment:

```
<div [ngStyle]='expression'></div>
```

The search starts for a directive having selector `ngStyle` first. Since Angular already has an `ngStyle` directive, it becomes the target. If Angular did not have a built-in `ngStyle` directive, the binding engine would have looked for a property called `ngStyle` on the underlying component.

If the nothing matches the target expression, an *unknown directive* error is thrown.

That covers most of Angular's property binding capabilities. Next, let's look at attribute binding and understand what role it plays.

# Attribute binding

The only reason attribute binding exists in Angular is that there are HTML attributes that do not have a backing DOM property. The `colspan` and `aria` attributes are some good examples of attributes without backing properties. The progress bar div in our view uses attribute binding.

It may seem that we can use standard interpolation syntax to set an attribute, but that does not work! Open `workout-runner.html` and replace the two aria attributes `attr.aria-valuenow` and `attr.aria-valuemax` enclosed in [], with this highlighted code:

```
<div class="progress-bar" role="progressbar"
aria-valuenow = "{{exerciseRunningDuration}}"
aria-valuemin="0"
aria-valuemax= "{{currentExercise.duration}}"
[ngStyle]= "{'width':(exerciseRunningDuration/currentExercise.duration) *
100 + '%'}"> </div>
```

Save and refresh the page. Then, Angular will throw an interesting error:

```
Can't bind to 'ariaValuenow' since it isn't a known native property in
WorkoutRunnerComponent ...
```

Angular is trying to search for a property called `ariaValuenow` in the `div` that does not exist! Remember, interpolations are actually property binding.

We hope that this gets the point across: *to bind to an attribute, use attribute binding*.

Angular binds to properties by default and not to attributes.

To support attribute binding, Angular uses a prefix notation, `attr`, within `[]`. An attribute binding looks like this:

```
[attr.attribute-name]="expression"
```

Revert to the original aria setup to make attribute binding work:

```
<div ... [attr.aria-valuenow]="exerciseRunningDuration" [attr.aria-
valuemax]="currentExercise.duration" ...>
```

Remember that unless an explicit `attr.` prefix is attached, attribute binding does not work.

While we have not used style- and class-based binding in our workout view, these are some binding capabilities that can come in handy. Hence, they are worth exploring.

# Style and class binding

We use **class binding** to set and remove a specific class based on the component state, as follows:

```
[class.class-name]="expression"
```

This adds `class-name` when `expression` is `true` and removes it when it is `false`. A simple example can look like this:

```
<div [class.highlight]="isPreferred">Jim</div> // Toggles the highlight
class
```

Use **style bindings** to set inline styles based on the component state:

```
[style.style-name]="expression";
```

While we have used the `ngStyle` While we have used the ngStyle directive for the workout view, we could have easily used *style binding* as well, as we are dealing with a single style. With style binding, the same `ngStyle` expression would become the following:

```
[style.width.%]="(exerciseRunningDuration/currentExercise.duration) * 100"
```

`width` is a style, and since it takes units too, we extend our target expression to include the `%` symbol.

Remember that `style.` and `class.` are convenient bindings for setting a single class or style. For more flexibility, there are corresponding attribute directives: `ngClass` and `ngStyle`. It's time now to formally introduce you to attribute directives.

# Attribute directives

Attribute directives are HTML extensions that change the behavior of a component/element. As described in the section on *Angular directives,* these directives do not define their own view.

Other than `ngStyle` and `ngClass` directives, there are a few more attribute directives that are part of the core framework. `ngValue`, `ngModel`, `ngSelectOptions`, `ngControl`, and `ngFormControl` are some of the attribute directives Angular provides.

While the next section is dedicated to learning how to use the `ngClass` and `ngStyle` attribute directives, it is not until `Chapter 6`, *Angular 2 Directives in Depth,* that we learn how to create our own attribute directives.

# Styling HTML with ngClass and ngStyle

Angular has two excellent directives that allow us to dynamically set styles on any element and toggle CSS classes. For the bootstrap progress bar, we use the **ngStyle** directive to dynamically set the element's style, `width`, as the exercise progresses:

```
<div class="progress-bar" role="progressbar" ...
[ngStyle]="{'width':(exerciseRunningDuration/currentExercise.duration) *
100 + '%'}"> </div>
```

`ngStyle` allows us to bind one or more styles to component properties at once. It takes an object as a parameter. Each property name on the object is the style name, and the value is the Angular expression bound to that property, such as the following:

```
<div [ngStyle]= "{
'width':componentWidth,
'height':componentHeight,
'font-size': 'larger',
'font-weight': ifRequired ? 'bold': 'normal' }"></div>
```

The styles can not only bind to component properties (`componentWidth` and `componentHeight` above) but also be set to a constant value (`larger`). The expression parser also allows the use of the ternary operator (`?:`); check out `isRequired`.

If styles become unwieldy in HTML, we also have the option of writing in our component a function that returns the object hash, and setting that as an expression:

```
<div [ngStyle]= "getStyles()"></div>
```

Moreover, `getStyles` on the component looks like this:

```
getStyles () {
    return {
      'width':componentWidth,
      ...
    }
}
```

The `ngClass` too works on the same lines, except that it is used to toggle one or multiple classes. For example, check out the following code:

```
<div [ngClass]= "{'required':inputRequired, 'email':whenEmail}"></div>
```

The `required` class is applied when `inputRequired` is true, and is removed when it evaluates to `false`.

Directives (custom or platform), like components, have to be registered on a module before they can be used.

Well! That covers everything we had to explore for our newly developed view.

And as described earlier, if you are having a problem with running the code, look at the Git branch `checkpoint2.2`.
If not using Git, download the snapshot of `checkpoint2.2` (a zip file) from `http://bit.ly/ng2be-checkpoint2-2`. Refer to the `README.md` file in the `trainer` folder when setting up the snapshot for the first time.

Time to add some enhancement and learn a bit more about the framework!

To start with, we are going to create a new module dedicated to workout runner. Everything that related to workout runner, including `WorkoutRunnerComponent`, goes into this module. This gives us a great opportunity to revisit Angular modules in great details.

# Exploring Angular modules

As the *7 Minute Workout* app grows and we add new components/directives/pipes/other artifacts to it, a need arises to organize these items. Each of these items needs to be part of an Angular module.

A naïve approach would be to declare everything in our app's root module (`AppModule`), as we did with `WorkoutRunnerComponent`, but this defeats the whole purpose of Angular modules.

To understand why a single-module approach is never a good idea, let's revisit Angular modules.

## Comprehending Angular modules

In Angular, **modules** are a way to organize code into chunks that belong together and work as a cohesive unit. Modules are Angular's way of grouping and organizing code.

An Angular module primarily defines:

- The components/directives/pipes it owns
- The components/directives/pipes it makes public for other modules to consume
- Other modules that it depends on
- Services that the module wants to make available application wide

Any decent-sized Angular app will have modules interlinked with each other: some modules consuming artifacts from other, some providing artifacts to others, and some modules doing both.

As a standard practice, module segregation is *feature-based*. One divides the app into features or sub features (for large features) and modules are created for each of the features. Even the framework adheres to this guideline as all of the framework constructs are divided across modules:

- There is the omnipresent `BrowserModule` that aggregates the standard framework constructs used in every browser-based Angular app
- There is the `RouterModule` if we want to use the Angular routing framework
- There is the `HtppModule` if our app needs to communicate with the server over HTTP

Angular modules are created by applying the `@NgModule` decorator to a TypeScript class, something we learned in `Chapter 1`, *Getting Started*. The decorator definition exposes enough metadata allowing Angular to load everything the module refers to.

The decorator has multiple attributes that allow us to define:

- External dependencies (using `imports`)
- Module artifacts (using `declarations`)
- Module exports (using `exports`)
- The services defined inside the module that need to be registered globally (using `providers`)

This diagram highlights the internals of a module and how they link to each other:

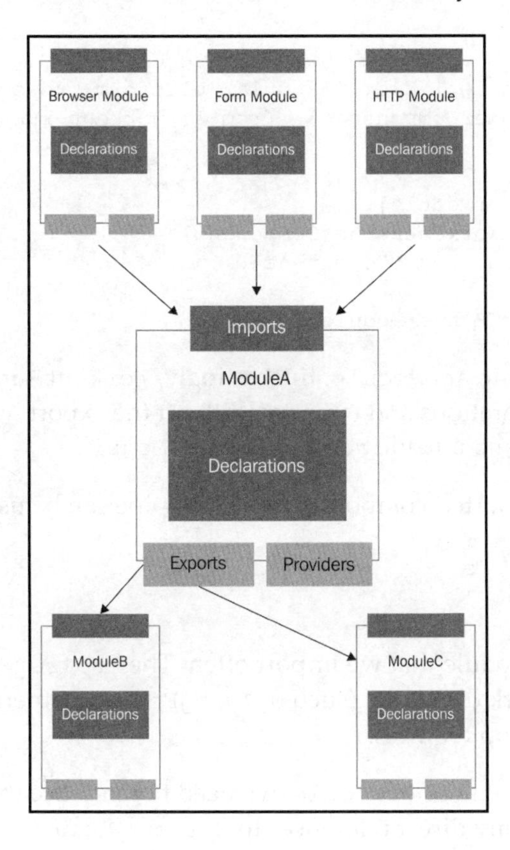

We hope one thing is clear from all this discussion: creating a single application-wide module is not the right use of Angular modules, unless you are building something rudimentary.

# Adding a new module to 7 Minute Workout

We are going to add more modules to *7 Minute Workout* too (hey, we do what we preach!). To start with, we are going to create a new module, `WorkoutRunnerModule`, and declare the `WorkoutRunnerComponent` in it. Henceforth, every component/directive/pipe we create specifically for workout execution goes into `WorkoutRunnerModule`.

Create a new file, `workout-runner.module.ts`, inside the `workout-runner` folder, and add this module definition:

```
import { NgModule }      from '@angular/core';
import { BrowserModule } from '@angular/platform-browser';
import { WorkoutRunnerComponent }  from './workout-runner.component';

@NgModule({
    imports: [BrowserModule],
    declarations: [WorkoutRunnerComponent],
    exports: [WorkoutRunnerComponent],
})
export class WorkoutRunnerModule { }
```

The module looks similar to `AppModule`. Interestingly, `WorkoutRunnerComponent` is a part of both module declarations and exports. Without the export, we cannot use `WorkoutRunnerComponent` outside `WorkoutRunnerModule`.

Only an exported component/directive/pipe can be used across modules.

`BrowserModule` is the module that we import often. The `BrowserModule` module declares all the common framework directives (such `NgIf`, `NgFor`, and others), pipes, and services that any browser-based app requires.

Now that we have `WorkoutRunnerModule`, we need to reference this module in `AppModule` and remove any direct references to `WorkoutRunnerComponent` in `AppModule`.

Open app.module.ts and remove the import and declaration specific to WorkoutRunnerComponent. Next, import the WorkoutRunnerModule by appending it to the existing module imports and add the necessary import references. See the highlighted code here:

```
import {WorkoutRunnerModule} from '../workout-runner/workout-
runner.module';
@NgModule({
  imports: [BrowserModule, WorkoutRunnerModule],
```

And we are good to go! Behaviorally nothing has changed, but we are now a bit more organized.

Are you wondering what happens if we leave the WorkoutRunnerComponent declaration in AppModule too? Remember, WorkoutRunnerModule also declares WorkoutRunnerComponent. Let's try. Go ahead; add WorkoutRunnerComponent back to the AppModuledeclarations section and run the application.

The application throws this error in the browser console:

> *Type WorkoutRunnerComponent is part of the declarations of 2 modules: WorkoutRunnerModule and AppModule! Please consider moving WorkoutRunnerComponent to a higher module that imports WorkoutRunnerModule and AppModule. You can also create a new NgModule that exports and includes WorkoutRunnerComponent then import that NgModule in WorkoutRunnerModule and AppModule.*

This leads to very important conclusion:

*A component/directive/pipe can only belong to a single module.*

We are not allowed to re-declare a component once it is a part of a module. To use a component that is a part of a specific module, we should import the respective module instead of re-declaring the component.

Importing dependencies through modules presents another challenge too. Circular dependency is not allowed. ModuleA cannot import ModuleB if ModuleB already imports ModuleA. This can happen if ModuleA wants to use some component from ModuleB and at the same time ModuleB is dependent on a component from ModuleA.

In such a case, as the preceding error describes, the common dependency should be moved into another Angular module, ModuleC, and both the modules, ModuleA and ModuleB, should reference it.

That is enough on Angular modules for now. Let's add some more enhancements to the app.

# Learning more about an exercise

For people who are doing this workout for the first time, it will be good to detail the steps involved in each exercise. We can also add references to some YouTube videos for each exercise to help the user understand the exercise better.

We are going to add the exercise description and instructions in the left panel and call it the **description panel**. We will add a reference to YouTube videos in the right panel, which is **the video panel**. To make things more modular and learn some new concepts, we are going to create independent components for each description panel and video panel.

The model data for this is already available. The `description` and `procedure` properties in the `Exercise` class (see `model.ts`) provide the necessary details about the exercise. The `videos` array contains some related YouTube video IDs, that will be used to fetch these videos.

# Adding descriptions and video panels

An Angular app is nothing but a hierarchy of components, similar to a tree structure. As of now, *7 Minute Workout* has two components, the root component, `TrainerAppComponent`, and its child, `WorkoutRunnerComponent`, in line with the HTML component layout, which now looks like this:

```
<trainer-app>
   <workout-runner></workout-runner>
</trainer-app>
```

We are going to add two new components to `WorkoutRunnerComponent`, one each to support the exercise description and exercise videos. While we could have added some HTML directly to the `WorkoutRunnerComponent` view, what we are hoping here is to learn a bit more about cross-component communication. Let's start with adding the description panel on the left and understand how a component can accept inputs.

# Providing component inputs

Create a folder called `exercise-description` (inside the `components/workout-runner` folder) and add a new file, `exercise-description.component.ts` to it. Add this piece of code to the file:

```
import {Component, Input} from '@angular/core';

@Component({
  selector: 'exercise-description',
  templateUrl: '/src/components/workout-runner/
  exercise-description/exercise-description.html',
})
export class ExerciseDescriptionComponent {
  @Input() description: string;
  @Input() steps: string;
}
```

Before we discuss the role of the `@Input` decorator, let's integrate the component into `WorkoutRunnerComponent`.

Copy the corresponding view HTML, `exercise-description.html`, from the Git branch `checkpoint2.3`, folder `workout-runner/exercise-description` (GitHub location: `http://bit.ly/ng2be-2-3-exercise-description-html`). To understand the role of `@Input`, let's highlight the relevant parts of the `exercise-description` HTML:

```
<div class="panel-body">
  {{description}}
</div>
 ...
 <div class="panel-body">
   {{steps}}
 </div>
```

The preceding interpolation refers to the input properties of the `ExerciseDescriptionComponent`: `description` and `steps`. The `@Input` decorator signifies that the component property is available for view binding.

The component definition is complete. Now, we just need to reference `ExerciseDescriptionComponent` in `WorkoutRunnerComponent` and provide values for `description` and `steps` for the `ExerciseDescriptionComponent` view to render correctly.

`WorkoutRunnerComponent` can use `ExerciseDescriptionComponent` only if:

- Either `ExerciseDescriptionComponent` is registered with the module that `WorkoutRunnerComponent` belongs to
- Or `ExerciseDescriptionComponent` has been imported (using `imports`) from another module into the module that `WorkoutRunnerComponent` belongs to

For this case, we are going to register `ExerciseDescriptionComponent` with `WorkoutRunnerModule`, which already has `WorkoutRunnerComponent`.

Open `workout-runner.module.ts` and append `ExerciseDescriptionComponent` to the `declarations` array. Check out the highlighted code:

```
import {ExerciseDescriptionComponent} from './exercise-
description/exercise-description.component';
...
declarations: [WorkoutRunnerComponent,
 ExerciseDescriptionComponent],
```

We can now use `ExerciseDescriptionComponent`. Open `workout-runner.html` and update the HTML fragments as highlighted in the following code. Add a new div called `description-panel` before the `exercise-pane` div and adjust some styles on `exercise-pane` div, as follows:

```
<div class="row">
  <div id="description-panel" class="col-sm-2">    <exercise-description
[description]="currentExercise.exercise.description"
[steps]="currentExercise.exercise.procedure"></exercise-description>
</div>
  <div id="exercise-pane" class="col-sm-7">
...
```

Make sure that `gulp play` is running and refresh the page. The description panel should show up on the left with the relevant exercise details.

Look at the `exercise-description` declaration in the preceding view. We are referring to the `description` and `steps` properties in the same manner as we did with the HTML element properties earlier in the chapter (`<img [src]='expression' ...`). Simple, intuitive, and very elegant!

The Angular data binding infrastructure makes sure that whenever the `currentExercise.exercise.description` and `currentExercise.exercise.procedure` properties on `WorkoutRunnerComponent` change, the bound properties on `ExerciseDescriptionComponent`, `description` and `steps`, are also updated .

The `@Input` decoration can take a **property alias** as a parameter, which means the following; consider a property declaration such as this: `@Input("myAwesomeProperty") myProperty:string;`
It can be referenced in the view as follows: `<my-component [myAwesomeProperty]="expression"....`

The power of the Angular binding infrastructure allows us to use any component property as a bindable property by attaching the `@Input` decorator (and `@Output` too) to it. We are not limited to basic data types such as `string`, `number`, and `boolean`; there can be complex objects too, which we will see next as we add the video panel.

Copy `video-player.component.ts` and `video-player.html` from the Git branch `checkpoint2.3` folder in `trainer/src/components/workout-runner/video-player` (GitHub location: `http://bit.ly/ng2be-2-3-video-player`).

Let's look at the implementation for a video player. Open `video-player.component.ts` and check out the `VideoPlayerComponent` class:

```
export class VideoPlayerComponent implements OnChanges {
  private youtubeUrlPrefix = '//www.youtube.com/embed/';

  @Input() videos: Array<string>;
  safeVideoUrls: Array<SafeResourceUrl>;

  constructor(private sanitizer: DomSanitizationService) { }

  ngOnChanges() {
    this.safeVideoUrls = this.videos ?
    this.videos
    .map(v => this.sanitizer.bypassSecurityTrustResourceUrl(
    this.youtubeUrlPrefix + v))
    : this.videos;
  }
}
```

The `videos` input property here takes an array of strings (YouTube video codes). While we take the `videos` array as input, we do not use this array directly in video player view; instead we transform the input array into a new array of `safeVideoUrls` and bind it. This can be confirmed by looking at the view implementation:

```
<div *ngFor="let video of safeVideoUrls">
   <iframe width="330" height="220" [src]="video"
frameborder="0" allowfullscreen></iframe>
</div>
```

The view uses a new Angular directive called `ngFor` to bind to the `safeVideoUrls` array. The `ngFor` directive belongs to a class of directives called *structural directives*. The directive's job is to take an HTML fragment and regenerate it based on the number of elements in the bound collection.

If you are confused about how the `ngFor` directive works with `safeVideoUrls` and why we need to generate `safeVideoUrls` instead of using the `videos` input array, wait for a while as we are shortly going to address these queries. But let's first complete the integration of `VideoPlayerComponent` with `WorkoutRunnerComponent` to see the final outcome.

As we did with the `ExerciseDescriptionComponent`, we need to add `VideoPlayerComponent` to `WorkoutRunnerModule` as well. We will leave this exercise to the reader.

Next, update the `WorkoutRunnerComponent` view by adding the component declaration after the `exercise-pane` div:

```
<div id="video-panel" class="col-sm-3">
    <video-player
[videos]="currentExercise.exercise.videos">
</video-player>
</div>
```

The `VideoPlayerComponent`'s `videos` property binds to the exercise's videos collection.

Start/refresh the app and the video thumbnails should show up on the right.

If you are having a problem with running the code, look at the Git branch `checkpoint2.3` for a working version of what we have done thus far. You can also download the snapshot of `checkpoint2.3` (a zip file) from `http://bit.ly/ng2be-checkpoint2-3`. Refer to the `README.md` file in the `trainer` folder when setting up the snapshot for the first time.

Now it's time to go back and look at the parts of the `VideoPlayerComponent` implementation that we skipped earlier. We specifically need to understand:

- How the `ngFor` directive works
- Why there is a need to transform the input `videos` array into `safeVideoUrls`
- The significance of the Angular component life cycle event `OnChanges`

To start with it's time to formally introduce the `ngFor` and the class of directives it belongs to: *structural directives.*

## Structural directives

The third categorization of directives, **structural directives**, work on the components/elements to manipulate their layout.

The Angular documentation describes structural directives in a succinct manner:

> *Instead of defining and controlling a view like a Component Directive, or modifying the appearance and behavior of an element like an Attribute Directive, the Structural Directive manipulates the layout by adding and removing entire element sub-trees.*

Since we have already touched upon *component directives* (such as `workout-runner` and `exercise-description`) and *attribute directives* (such as `ngClass` and `ngStyle`), we can very well contrast their behaviors against *structural directives*.

The `ngFor` directive belongs to this class. We can identify such directives by the `*` prefix. Other than `ngFor`, Angular comes with some other structural directives such as `ngIf` and `ngSwitch`.

## The ever-so-useful NgFor

Every templating language has constructs that allow the templating engine to generate HTML (by repetition). Angular has **ngFor**. The ngFor directive is a super useful directive used to duplicate a piece of an HTML fragment *n* number of times.

The preceding code repeats the div fragment for each exercise video (using the `videos` array). The `let video of videos` string expression is interpreted as follows: take each video in the videos array and assign it to a **template input variable**, `video`. This input variable can now be referenced inside the `ngFor` template HTML, as we do when we set the `src` property binding.

To provide more details about the iteration context, the `ngFor` directive provides an optional `index` that increases from 0 to the length of the array for each iteration, something similar to a `for` loop, which we all are familiar with. This `index` too can be captured into a *template input variable* and used inside the template:

```
<div *ngFor="let video of videos; let i=index">
     <div>This is video - {{i}}</div>
</div>
```

Other than `index`, there is some more iteration context information available, which includes `first`, `last`, `even`, and `odd`. This information can come in handy as we can do some nifty stuff with it. Consider this example:

```
<div *ngFor="let video of videos; let i=index; let f=first">
     <div [class.special]="f">This is video - {{i}}</div>
</div>
```

It applies a `special` class to the first video `div`.

The `ngFor` directive can be applied to HTML elements as well as our custom components. This is a valid use of ngFor:

```
<user-profile *ngFor="let userDetail of users"
[user]= "userDetail"></user-profile>
```

> The `*` prefix is a terser format to represent a structural directive. In reality, the `ngFor` directive used with the preceding `videos` array expands to:
>
> ```
> <template ngFor let-video [ngForOf]="videos">
>   
>     <iframe width="330" height="220"
>     [src]="'//www.youtube.com/embed/' + video" ...>
>     </iframe>
>   
> </template>
> ```
>
> The `template` tag has a declaration for `ngFor`, a *template input variable* (`video`), and a property (`ngForOf`) that points to the `videos` array.

## NgFor performance

Since `NgFor` generates HTML-based on collection elements, it is notorious for causing performance issues. But we cannot blame the directive. It does what it is supposed to do: iterate and generate elements! UI rendering performance can take a hit if the backing collection is huge or there is repeated re-rendering of DOM due to the bound collection changing often.

One of the performance tweaks for `NgFor` allows us to alter the behavior of this directive when it comes to creating and destroying DOM elements (when the underlying collection elements are added or removed).

Imagine a scenario where we frequently get an array of objects from the server and bind it to the view using `NgFor`. The default behavior of `NgFor` is to regenerate the DOM every time we refresh the list (since Angular does a standard object equality check). However, as developers, we may very well know not much has changed. Some new objects may have been added, some removed, and maybe some modified. But Angular just regenerates the complete DOM.

To alleviate this situation, Angular allows us to specify a custom **tracking function**, which lets Angular know when two objects being compared are equal. Have a look at the following function:

```
trackByUserId(index: number, hero: User) { return user.id; }
```

A function such as this can be used in the `NgFor` template to tell Angular to compare the *user* object based on its `id` property instead of doing a reference equality check.

This is how we then use the preceding function in the `NgFor` template:

```
<div *ngFor="let user of users;
trackBy: trackByUserId">{{user.name}}</div>
```

`NgFor`

*will now avoid recreating DOM for users with IDs already rendered.*

*Remember, Angular may still update the existing DOM elements if the bound properties of a user have changed.*

That's enough on the `ngFor` directive; let's move ahead.

We still need to understand the role of the `safeVideoUrls` and the `OnChange` life cycle event in the `VideoPlayerComponent` implementation. Let's tackle the former first and understand the need for `safeVideoUrls`.

## Angular 2 security

The easiest way to understand why we need to bind to `safeVideoUrls` instead of the `videos` input property is by trying it out. Replace the existing `ngFor` fragment HTML with this:

```
<div *ngFor="let video of videos">
<iframe width="330" height="220"
[src]="'//www.youtube.com/embed/' + video"
frameborder="0" allowfullscreen></iframe>
</div>
```

And look at the browsers console log (a page refresh may be required). There are a bunch of errors thrown by the framework, such as:

**Error: unsafe value used in a resource URL context (see http://g.co/ng/security#xss)**

No prize for guessing what is happening! Angular is trying to safeguard our application against a *Cross-Site Scripting (XSS)* attack.

Such an attack enables the attacker to inject malicious code into our web pages. Once injected, the malicious code can read data from the current site context. This allows it to steal confidential data and also impersonate the logged-in user, hence gaining access to privileged resources.

Angular has been designed to block these attacks by sanitizing any external code/script that is injected into an Angular view. Remember, content can be injected into a view through number of mechanisms, including *property/attribute/style bindings* or *interpolation*.

Interpolations escape any content that we bind to them.

When we use the `innerHTML` property of an HTML element (*property binding*), while the HTML content is emitted, any unsafe content (*script*) embedded in the HTML is stripped. We will shortly see an example of it when we format the exercise steps as HTML instead of plain text.

But what about *Iframes*? In our preceding example, Angular is blocking property binding to Iframe's `src` property too. This is a warning against third-party content being embedded in our own site using Iframe. Angular prevents this too.

All in all, the framework defines four security contexts around content sanitization. These include:

1. HTML content sanitization, when HTML content is bound using the `innerHTML` property.
2. Style sanitization, when binding CSS into the `style` property.
3. URL sanitization, when URLs are used with tags such as `anchor` and `img`.
4. Resource sanitization when using `Iframes` or `script` tag. In this case, content cannot be sanitized and hence it is blocked by default.

Angular is trying its best to keep us out of danger. But at times, we know that the content is safe to render and hence want to circumvent the default sanitization behavior.

## Trusting safe content

To let Angular know that the content being bound is safe, we use the **DomSanitizer** and call the appropriate method based on the security contexts just described. The available functions are:

- `bypassSecurityTrustHtml`
- `bypassSecurityTrustScript`
- `bypassSecurityTrustStyle`
- `bypassSecurityTrustUrl`
- `bypassSecurityTrustResourceUrl`

In our video player implementation, we use `bypassSecurityTrustResourceUrl`; it converts the video URL into a trusted `SafeResourceUrl` object:

```
this.sanitizer.bypassSecurityTrustResourceUrl(this.youtubeUrlPrefix + v)
```

The `map` method transforms the videos array into a collection of `SafeResourceUrl` objects and assigns it to `safeVideoUrls`.

Each of the methods listed previously takes a *string parameter*. This is the content we want Angular to know is safe. The return object, which could be any of `SafeStyle`, `SafeHtml`, `SafeScript`, `SafeUrl`, or `SafeResourceUrl`, can then be bound to the view.

A comprehensive treatment of this topic is available in the framework security guide available at http://bit.ly/ng2-security. A highly recommended read!

The last question to answer is: why do this in the `OnChanges` Angular life cycle event?

*The* `OnChanges` *life cyle event is triggered whenever the component's inputs change.* In the case of `VideoPlayerComponent`, it is the `videos` array input property. We use this life cycle event to recreate the `safeVideoUrls` array and re-bind it to the view. Simple!

Video panel implementation is now complete. Let's add a few more minor enhancements and explore a bit more in Angular.

# Formatting exercise steps with innerHTML binding

One of the sore points in the current app is the formatting of the exercise steps. It's a bit difficult to read these steps.

The steps should either have a line break (`<br>`) or be formatted as an HTML `list` for easy readability. This seems to be a straightforward task, and we can just go ahead and change the data that is bounded to the step interpolation, or write a pipe that can add some HTML formatting using the line delimiting convention (`.`). For a quick verification, let's update the first exercise steps in `workout-runner.component.ts` by adding a break (`<br>`) after each line:

```
`Assume an erect position, with feet together and arms at your side. <br>
 Slightly bend your knees, and propel yourself a few inches into the air.
<br>
 While in air, bring your legs out to the side about shoulder width or
slightly wider. <br>
 ...
```

Now refresh the workout page. The output does not match our expectations, as shown here:

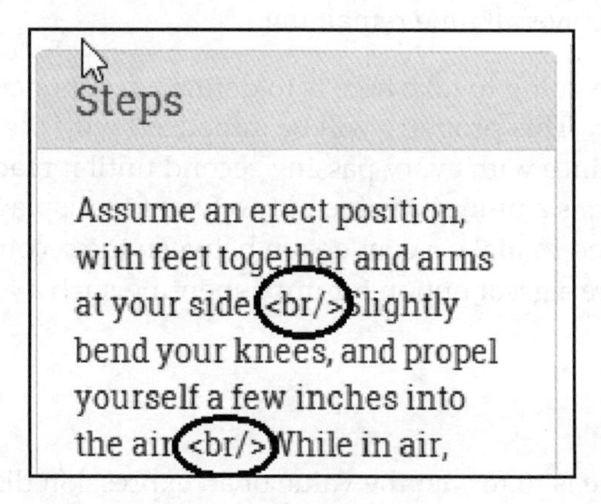


The break tags were literally rendered in the browser. Angular did not render the interpolation as HTML; instead, it escaped the HTML characters, and we know why!

How to fix it? Easy! Replace the interpolation with the property binding to bind step data to the element's `innerHTML` property (in `exercise-description.html`), and you are done!

```
<div class="panel-body" [innerHTML]="steps"> </div>
```

Refresh the workout page to confirm.

**Preventing Cross-Site Scripting Security (XSS) issues**
As discussed earlier, Angular by default sanitizes the input HTML when injected dynamically. This safeguards our app against XSS attacks. Still if you want to dynamically inject styles/script into HTML, use the **DomSanitizer** to bypass this sanitization check.

Time for another enhancement! It's time to learn about Angular pipes.

# Displaying the remaining workout duration using pipes

It will be nice if we can tell the user the time left to complete the workout and not just the duration of the exercise in progress. We can add a countdown timer somewhere in the exercise pane to show the overall time remaining.

The approach that we are going to take here is to define a component property called `workoutTimeRemaining`. This property will be initialized with the total time at the start of the workout, and will reduce with every passing second until it reaches zero. Since `workoutTimeRemaining` is a numeric value but we want to display a timer in the `hh:mm:ss` format, we need to make a conversion between the seconds' data and the time format. **Angular pipes** are a great option for implementing such a feature.

# Angular pipes

The primary aim of a pipe is to format the value of an expression displayed in the view. The framework comes with multiple predefined pipes, such as `date`, `currency`, `lowercase`, `uppercase`, and others. This is how we use a pipe in a view:

```
{{expression | pipeName:inputParam1}}
```

An expression is followed by the *pipe symbol* (`|`), which is followed by the pipe name and then an optional parameter (`inputParam1`) separated by a colon (`:`). If the pipe takes multiple inputs, they can be placed one after another separated by a colon, like the inbuilt `slice` pipe, which can slice an array or string:

```
{{fullName | slice:0:20}} //renders first 20 characters
```

The parameter passed to the filter can be a constant or a property from the underlying component, such as this:

```
{{fullName | slice:0:truncateAt}} //renders based on value truncateAt
```

Here are some examples of the use of the `date` pipe, as described in the Angular `date` documentation. Assume that `dateObj` is initialized to June 15, 2015 21:43:11 and locale is *en-US*:

```
{{ dateObj | date }}                // output is 'Jun 15, 2015'
{{ dateObj | date:'medium' }}       // output is 'Jun 15, 2015, 9:43:11 PM'
{{ dateObj | date:'shortTime' }}    // output is '9:43 PM'
{{ dateObj | date:'mmss' }}         // output is '43:11'
```

Some of the most commonly used pipes are the following:

- **date**: As we just saw, the date filter is used to format the date in a specific manner. This filter supports quite a number of formats and is locale-aware too. To know about the other formats supported by the date pipe, check out the framework documentation at `http://bit.ly/ng2-date`.
- **uppercase** and **lowercase**: These two pipes, as the name suggests, change the case of the string input.
- **decimal** and **percent**: `decimal` and `percent` pipes are there to format decimal and percentage values based on the current browser locale.
- **currency**: This is used to format numeric values as a currency based on the current browser locale.

```
{{14.22|currency:"USD" }} <!-Renders USD 14.22 -->
{{14.22|currency:"USD":true}}  <!-Renders $14.22 -->
```

- **json**: This is a handy pipe for debugging that can transform any input into a string using `JSON.stringify`. We made good use of it at the start of this chapter to render the `WorkoutPlan` object (see the Checkpoint 2.1 code).

Another really powerful feature of pipes is that they can be chained, wherein the output from one pipe can serve as the input to another pipe. Consider this example:

```
{{fullName | slice:0:20 | uppercase}}
```

The first pipe slices the first 20 characters of `fullName` and the second pipe transforms them to uppercase.

Now that we have seen what pipes are and how to use them, why not implement one for *7 Minute Workout* app: a **seconds to time** pipe?

# Implementing a custom pipe – SecondsToTimePipe

The `SecondsToTimePipe` converts a numeric value into the `hh:mm:ss` format.

Create a file called `seconds-to-time.pipe.ts` in the `workout-runner` folder and add the following pipe definition (the definition can also be downloaded from the Git branch `checkpoint.2.4` on the GitHub site at `http://bit.ly/ng2be-2-4-seconds-to-time-pipe-ts`):

```
export class SecondsToTimePipe implements PipeTransform {
  transform(value: number): any {
    if (!isNaN(value)) {
      var hours = Math.floor(value / 3600);
      var minutes = Math.floor((value - (hours * 3600)) / 60);
      var seconds = value - (hours * 3600) - (minutes * 60);

      return ("0" + hours).substr(-2) + ':'
        + ("0" + minutes).substr(-2) + ':'
        + ("0" + seconds).substr(-2);
    }
    return;
  }
}
```

In an Angular pipe, the implementation logic goes into the `transform` function. Defined as part of the `PipeTransform` interface, the preceding `transform` function transforms the input seconds' value into an *hh:mm:ss* string. The first parameter to the `transform` function is the pipe input. The subsequent parameters, if provided, are the arguments to pipe, passed using colon separator (`pipe:argument1:arugment2..`) from the view. We do not make use of any pipe argument as the implementation does not require it.

The implementation is quite straightforward, as we convert seconds into hours, minutes, and seconds. Then we concatenate the result into a string value and return the value. The addition of 0 on the left for each of the `hours`, `minutes`, and `seconds` variables is done to format the value with a leading 0 in case the calculated value for hours, minutes, or seconds is less than 10.

The pipe that we just created is a standard TypeScript class. Unless we apply the `@Pipe` decorator to it, it does not qualify as an Angular pipe.

Add the necessary `import` statement and the `@Pipe` decorator to the `seconds-to-time.pipe.ts` file just before the class definition:

```
import {Pipe, PipeTransform} from '@angular/core';

@Pipe({
  name: 'secondsToTime'
})
```

The pipe definition is complete, but before we can use `SecondsToTimePipe` in `WorkoutRunnerComponent`, the pipe needs to be declared in `WorkoutRunnerModule`. This is something we did for `ExerciseDescriptionComponent` and `VideoPlayerComponent` earlier.

Open `workout-runner.module.ts`, and add the highlighted code:

```
import {VideoPlayerComponent} from
'./video-player/video-player.component';
import {SecondsToTimePipe} from './seconds-to-time.pipe';

...
    declarations: [WorkoutRunnerComponent, ...
SecondsToTimePipe],
```

Finally, we just need to add the pipe in the view. Update `workout-runner.html` by adding the highlighted fragment:

```
<div class="workout-display-div">
  <h4>Workout Remaining - {{workoutTimeRemaining | secondsToTime}}</h4>
  <h1>{{currentExercise.exercise.title}}</h1>
```

Surprisingly the implementation is still not complete! We have a pipe definition, and we have referenced it in the view, but `workoutTimeRemaining` needs to update with each passing second for `SecondsToTimePipe` to be of any use.

We have already initialized the `WorkoutRunnerComponent's` `workoutTimeRemaining` property in the `start` function with the total workout time:

```
start() {
this.workoutTimeRemaining = this.workoutPlan.totalWorkoutDuration();
...
}
```

Now the question is: how to update the `workoutTimeRemaining` variable with each passing second? Remember that we already have a `setInterval` setup that updates `exerciseRunningDuration`. While we can write another `setInterval` implementation for `workoutTimeRemaining`, it will be better if a single `setInterval` setup can take care of both the requirements.

Add a function called `startExerciseTimeTracking` to `WorkoutRunnerComponent`; it looks like this:

```
startExerciseTimeTracking() {
    this.exerciseTrackingInterval = setInterval(() => {
      if (this.exerciseRunningDuration >=
          this.currentExercise.duration) {
        clearInterval(this.exerciseTrackingInterval);
        let next: ExercisePlan = this.getNextExercise();
        if (next) {
          if (next !== this.restExercise) {
            this.currentExerciseIndex++;
          }
          this.startExercise(next);
        }
        else {
          console.log("Workout complete!");
        }
        return;
      }
      ++this.exerciseRunningDuration;
      --this.workoutTimeRemaining;
    }, 1000);
  }
```

As you can see, the primary purpose of the function is to track the exercise progress and flip the exercise once it is complete. However, it also tracks `workoutTimeRemaining` (it decrements this counter). The first `if` condition setup just makes sure that we clear the timer once all the exercises are done. The inner `if` conditions are used to keep `currentExerciseIndex` in sync with the running exercise.

This function uses an instance variable called `exerciseTrackingInterval`. Add it to the class declaration section. We are going to use this variable later to implement an exercise pausing behavior.

Remove the complete `setInterval` setup from `startExercise` and replace it with a call to `this.startExerciseTimeTracking();`. We are all set to test our implementation. Refresh the browser and verify the implementation.

The next section is about another inbuilt Angular directive, `ngIf`, and another small enhancement.

# Adding the next exercise indicator using ngIf

It will be nice for the user to be told what the next exercise is during the short rest period between exercises. This will help them prepare for the next exercise. So let's add it.

To implement this feature, we can simply output the title of the next exercise in line from the `workoutPlan.exercises` array. We show the title next to the `Time Remaining` countdown section. Change the workout div (`class="workout-display-div"`) to include the highlighted content, and remove `Time Remainingh1`:

```
<div class="workout-display-div">
<!-- Exiting html -->
   <div class="progress time-progress">
   <!-- Exiting html -->
   </div>
   <div class="row"> <h3 class="col-sm-6 text-left">Time Remaining:
   <strong>{{currentExercise.duration-exerciseRunningDuration}}</strong>
   </h3><h3 class="col-sm-6 text-right" *ngIf= "currentExercise.exercise
   .name=='rest'">Next up: <strong>{{workoutPlan.exercises[
   currentExerciseIndex + 1].exercise.title}}</strong></h3> </div>
</div>
```

We wrap the existing Time Remaining h1 and add another h3 tag to show the next exercise inside a new div with some style updates. Also, there is a new directive, ngIf, in the second h3. The * prefix implies that it belongs to the same set of directives that ngFor belongs: structural directives.

The **ngIf** directive is used to add or remove a specific section of the DOM based on whether the expression provided to it returns true or false. The DOM element is added when the expression evaluates to true and destroyed otherwise. Isolate the ngIf declaration from the preceding view:

```
ngIf="currentExercise.details.name=='rest'"
```

The expression then checks whether we are currently at the rest phase and the directive accordingly shows or hides the linked h3.

Other than this, in the same h3, we have an interpolation that shows the name of the exercise from the workoutPlan.exercises array.

A word of caution here: ngIf adds and destroys the DOM element, and hence it is not similar to the visibility constructs that we employed to show and hide elements. While the end result of style, display:none is the same as that of ngIf, the mechanism is entirely different:

```
<div [style.display]="isAdmin" ? 'block' : 'none'">Welcome Admin</div>
```

Versus this line:

```
<div *ngIf="isAdmin" ? 'block' : 'none'">Welcome Admin</div>
```

With ngIf, whenever the expression changes from false to true, a complete re-initialization of the content occurs. Recursively, new elements/components are created and data binding is set up, starting from the parent down to the children. The reverse happens when the expression changes from true to false: all of this is destroyed. Therefore, using ngIf can sometimes become an expensive operation if it wraps a large chunk of content and the expression attached to it changes very often. But otherwise, wrapping a view in ngIf is more performant than using css/style-based show or hide, as neither the DOM is created nor the data binding expressions are set up when the ngIf expression evaluates to false.

There is another directive that belongs in this league: **ngSwitch**. When defined on the parent HTML, it can swap the child HTML elements based on the ngSwitch expression. Consider this example:

```
<div id="parent" [ngSwitch] ="userType">
<div *ngSwitchCase="'admin'">I am the Admin!</div>
<div *ngSwitchCase="'powerUser'">I am the Power User!</div>
<div *ngSwitchDefault>I am a normal user!</div>
</div>
```

We bind the `userType` expression to `ngSwitch`. Based on the value of `userType` (`admin`, `powerUser`, or any other `userType`), one of the inner div elements will be rendered. The `ngSwitchDefault` directive is a wildcard match/fallback match, and it gets rendered when `userType` is neither `admin` nor `powerUser`.

If you have not realized it yet, note that there are three directives working together here to achieve switch-case-like behavior:

- `ngSwitch`
- `ngSwitchCase`
- `ngSwitchDefault`

Coming back to our next exercise implementation, we are ready to verify the implementation, so we refresh the page and wait for the rest period. There should be a mention of the next exercise during the rest phase, as shown here:

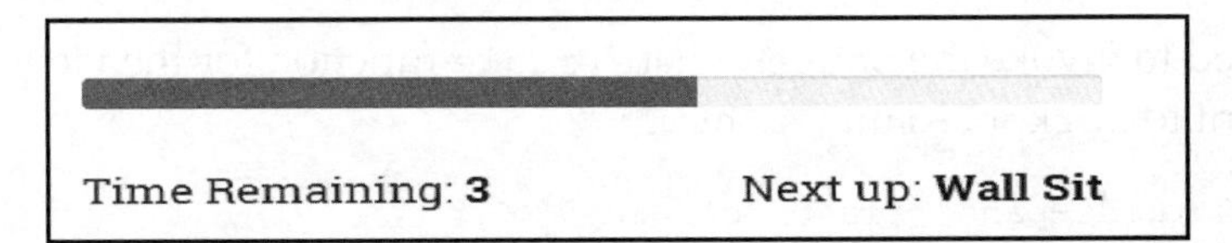


The app is shaping up well. If you have used the app and done some physical workouts along with it, you will be missing the exercise pause functionality badly. The workout just does not stop until it reaches the end. We need to fix this behavior.

# Pausing an exercise

To pause an exercise, we need to stop the timer. We also need to add a button somewhere in the view that allows us to pause and resume the workout. We plan to do this by drawing a button overlay over the exercise area in the center of the page. When clicked on, it will toggle the exercise state between paused and running. We will also add keyboard support to pause and resume the workout using the key binding p or P. Let's update the component.

Update the `WorkoutRunnerComponent` class, add these three functions, and add a declaration for the `workoutPaused` variable:

```
workoutPaused: boolean;
pause() {
    clearInterval(this.exerciseTrackingInterval);
    this.workoutPaused = true;
  }

resume() {
    this.startExerciseTimeTracking();
    this.workoutPaused = false;
  }

pauseResumeToggle() {
    if (this.workoutPaused) { this.resume();    }
    else {      this.pause();    }
  }
```

The implementation for pausing is simple. The first thing we do is cancel the existing `setInterval` setup by calling `clearInterval(this.exerciseTrackingInterval);`. While resuming, we again call `startExerciseTimeTracking`, which again starts tracking the time from where we left off.

Now we just need to invoke the `pauseResumeToggle` function for the view. Add the following content to `workout-runner.html`:

```
<div id="exercise-pane" class="col-sm-7">
    <div id="pause-overlay" (click)="pauseResumeToggle()">
    <span class="glyphicon pause absolute-center"
    [ngClass]="{'glyphicon-pause' : !workoutPaused,
    'glyphicon-play' : workoutPaused}"></span>    </div>
    <div class="row workout-content">
```

The `click` event handler on the div toggles the workout running state, and the `ngClass` directive is used to toggle the class between `glyphicon-pause` and `glyphicon-play`-standard Angular stuff. What is missing now is the ability to pause and resume on a p or P key press.

We can apply a `keyup` event handler on the div:

```
<div id="pause-overlay" (keyup)= "onKeyPressed($event)">
```

But the `div` element does not have a concept of focus, so we also need to add the `tabIndex` attribute on the div to make it work. Even then, it works only when we have clicked on the div at least once. There is a better way to implement this; attach the event handler to the global `window` event `keyup`. This is how the event binding will look now:

```
<div id="pause-overlay" (window:keyup)= "onKeyPressed($event)">
```

Make note of the special `window:` prefix before the `keyup` event. We can use this syntax to attach events to any global object, such as `document`. The `onKeyPressed` event handler needs to be added to `WorkoutRunnerComponent`. Add this function to the class:

```
onKeyPressed = function(event:KeyboardEvent) {
    if (event.which == 80 || event.which == 112) {
      this.pauseResumeToggle();
    }
}
```

The `$event` object is the standard **DOM event object** that Angular makes available for manipulation. Since this is a keyboard event, the specialized class is `KeyboardEvent`. The `which` property is matched to ASCII values of `p` or `P`. Refresh the page and you should see the play/pause icon when your mouse hovers over the exercise image, like this:

While we are on event binding, it would be a good opportunity to explore Angular's event binding infrastructure

# The Angular event binding infrastructure

Angular event binding allows a component to communicate with its parent through events.

If we look back at the app implementation, what we have encountered thus far are the property/attribute bindings. Such bindings allow a component/element to take inputs from the outside world. The data flows into the component.

Event bindings are the reverse of property bindings. They allow a component/element to inform the outside world about any state change.

As we saw in the pause/resume implementation, event binding employs *round brackets* (`()`) to specify the target event:

```
<div id="pause-overlay" (click)="pauseResumeToggle()">
```

This attaches a `click` event handler to the `div` that invokes the expression `pauseResumeToggle()` when the `div` is clicked.

Like properties, there is a canonical form for events too. Instead of using round brackets, the `on-` prefix can be used:
`on-click="pauseResumeToggle()"`

Angular supports all types of events. Events related to keyboard inputs, mouse movements, button clicks, and touches. The framework even allows us to define our own event for the components we create, such as:

```
<workout-runner (paused)= "stopAudio()"></workout-runner>
```

We will be covering custom component events in the next chapter, where we will add audio support to *7 Minute Workout*.

It is expected that events have side effects; in other words, an event handler may change the state of the component, which in turn may trigger a chain reaction in which multiple components react to the state change and change their own state. This is unlike a property binding expression, which should be side-effect-free. Even in our implementation, clicking on the `div` element toggles the exercise run state.

## Event bubbling

When Angular attaches event handlers to standard HTML element events, the event propagation works in the same way as standard DOM event propagation works. This is also called **event bubbling**. Events on child elements are propagated upwards, and hence event binding is also possible on a parent element, like this:

```
<div id="parent " (click)="doWork($event)"> Try
  <div id="child ">me!</div>
</div>
```

Clicking on either of the divs results in the invocation of the `doWork` function. Moreover, `$event.target` contains the reference to the `div` that dispatched the event.

Custom events created on Angular components do not support event bubbling.

Event bubbling stops if the expression assigned to the target evaluates to a `falsey` value (such as `void`, `false`). Therefore, to continue propagation, the expression should evaluate to `true`:

```
<div id="parent" (click)="doWork($event) || true">
```

Here too, the `$event` object deserves some special attention.

## Event binding an $event object

Angular makes an `$event` object available whenever the target event is triggered. This `$event` contains the details of the event that occurred.

The important thing to note here is that the shape of the `$event` object is decided based on the event type. For HTML elements, it is a DOM event object (`https://developer.mozilla.org/en-US/docs/Web/Events`), which may vary based on the actual event.

But if it is a custom component event, what is passed in the `$event` object is decided by the component implementation. We will return to this discussion again, in the next chapter.

We have now covered most of the data binding capabilities of Angular, with the exception of two-way binding. A quick introduction to the two-way binding constructs is warranted before we conclude the chapter.

# Two-way binding with ngModel

**Two-way binding** helps us keep the model and view in sync. Changes to the model update the view and changes to the view update the model. The obvious area where two-way binding is applicable is *form input*. Let's look at a simple example:

```
<input [(ngModel)]="workout.name">
```

The `ngModel` directive here sets a two-way binding between the `input`'s `value` property and the `workout.name` property on the underlying component. Anything that the user enters in the above `input` is synced with `workout.name`, and any changes to `workout.name` are reflected back on the preceding `input`.

Interestingly we can achieve the same result without using the `ngModel` directive too, by combining both *property and event binding syntax*. Consider the next example; it works in the same way as `input` before:

```
<input [value]="workout.name"
(input)="workout.name=$event.target.value" >
```

There is a property binding set up on the `value` property and an event binding set up on the `input` event that make the bidirectional sync work.

We get into more details on two-way binding in `Chapter 4`, *Building Personal Trainer*, where we build our own custom workouts.

We have created a diagram that summarizes the data flow patterns for all the bindings that we have discussed thus far. A handy illustration to help you memorize each of the binding constructs and how data flows is as follows:

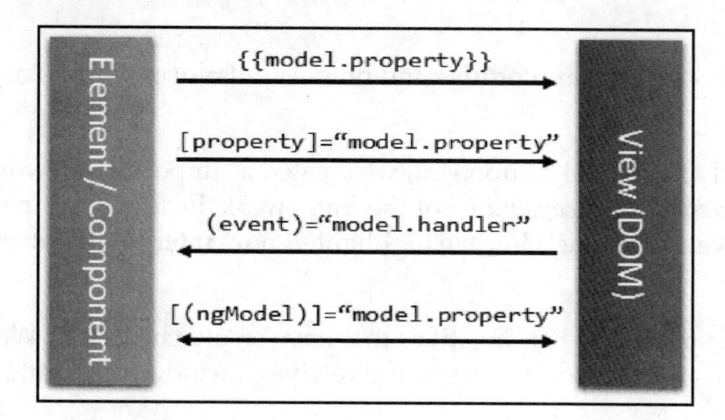


We now have a fully functional *7 Minute Workout*, with some bells and whistles too, and hopefully it was fun creating the app. It's time to conclude the chapter and summarize the lessons.

If you are having a problem with running the code, look at the Git branch `checkpoint2.4` for a working version of what we have done thus far. You can also download a snapshot of `checkpoint2.4` (a zip file) from this GitHub location: `http://bit.ly/ng2be-checkpoint2-4`. Refer to the `README.md` file in the `trainer` folder when setting up the snapshot for the first time.

# Summary

We started this chapter with the aim of creating an Angular app that is more complex than the sample we created in the first chapter. The *7 Minute Workout* app fitted the bill, and you learned a lot about the Angular framework while building this app.

We started by defining the functional specifications of the *7 Minute Workout* app. We then focused our efforts on defining the code structure for the app.

To build the app, we started off by defining the model of the app. Once the model was in place, we started the actual implementation, by building an **Angular component**. Angular components are nothing but classes that are decorated with a framework-specific decorator, `@Component`.

We also learned about **Angular modules** and how Angular uses them to organize code artifacts.

Once we had a fully functional component, we created a supporting view for the app. We also explored the *data binding capabilities* of the framework, including **property**, **attribute**, **class**, **style,** and **event binding**. Plus, we highlighted how **interpolations** are a special case of *property binding*.

Components are a special class of directives that have an attached view. We touched upon what directives are and the special classes of directives, including **attribute** and **structural directives**.

We learned how to perform cross-component communication using **input properties**. The two child components that we put together (`ExerciseDescriptionComponent` and `VideoPlayerComponent`) derived their inputs from the parent `WorkoutRunnerComponent` using input properties.

We then covered another core construct of Angular, **pipes**. We saw how to use pipes such as the *date pipe* and how to create one of our own.

Throughout the chapter, we touched upon a number of Angular directives, including the following:

- `ngClass/ngStyle`: For applying multiple styles and classes using Angular binding capabilities
- `ngFor`: For generating dynamic HTML content using a looping construct
- `ngIf`: For conditionally creating/destroying DOM elements.
- `ngSwitch`: For creating/destroying DOM elements using the switch-case construct

We now have a basic 7 Minute Workout app. For a better user experience, we have added a number of small enhancements to it too, but we are still missing some good-to-have features that would make our app more usable. From the framework perspective, we have purposefully ignored some core/advanced concepts such as **change detection**, **dependency injection**, **component routing**, as well as data flow patterns, which we plan to cover in the next chapter.

# 3

# More Angular 2 – SPA, Routing, and Data Flows in Depth

If the previous chapter was about building our first useful app in Angular, then this chapter is about adding a whole lot of Angular goodness to it. Within the learning curve, we have made a start in exploring a technology platform and now we can build some rudimentary apps using Angular. But that's just the start! There is a lot more to learn before we can make effective use of Angular in a decent-sized application. This chapter takes us one step closer to realizing this goal.

The *7 Minute Workout* app still has some rough edges/limitations that we can fix while making the overall app experience better. This chapter is all about adding those enhancements and features. And as always, this app-building process provides us with enough opportunities to enhance our understanding of the framework and learn new things about it.

The topics we cover in this chapter include:

- **Exploring Angular Single Page Applications (SPA)**: We explore Angular's Single Page Application capabilities, which include route navigation, link generation, and routing events.
- **Understanding Dependency Injection**: One of the core platform features. In this chapter, we learn how Angular makes effective use of dependency injection to inject components and services across the application.
- **Angular pure (stateless) and impure (stateful) pipes**: We explore the primary data transformation construct of Angular, pipes, in more detail as we build some new pipes.

- **Cross-component communication**: Since Angular is all about components and their interactions, we look at how to do cross-component communication in a parent-child and sibling component setup. We learn how Angular template variables and events facilitate this communication.
- **Creating and consuming events**: We learn how a component can expose its own events and how to bind to these events from template HTML and from other components.

AS a side note, I expect you are using the *7 Minute Workout* on a regular basis and working on your physical fitness. If not take a 7-minute exercise break and exercise now. I insist!

Hope the workout was fun! Now let's get back to some serious business. Let's start with exploring Angular Single Page Application (SPA) capabilities.

We are starting from where we left off in `Chapter 2`, *Building Our First App – 7 Minute Workout*. The Git branch `checkpoint2.4` can serve as the base for this chapter.
The code is also available on GitHub (`https://github.com/chandermani/angular2byexample`) for everyone to download. Checkpoints are implemented as branches in GitHub.
If you are not using Git, download the snapshot of `checkpoint2.4` (a ZIP file) from the GitHub location `http://bit.ly/ng2be-checkpoint2-4`. Refer to the `README.md` file in the `trainer` folder when setting up the snapshot for the first time.

# Exploring Single Page Application capabilities

The *7 Minute Workout* starts when we load the app, but it ends with the last exercise sticking to the screen permanently. Not a very elegant solution. Why don't we add a start and finish page to the app? This makes the app more professional and allows us to understand the single-page nomenclature of AngularJS.

# The Angular SPA infrastructure

With modern web frameworks such as Angular (Angular 1.x) and Ember, we are now getting used to apps that do not perform full page refreshes. But if you are new to this scene it's worth mentioning what these SPAs are.

Single Page Applications (SPAs) are browser-based apps devoid of any full page refresh. In such apps, once the initial HTML is loaded, any future page navigations are retrieved using AJAX as HTML fragments and injected into the already loaded view. Google Mail is a great example of a SPA. SPAs make for a great user experience as the user gets a desktop app-like feel, with no constant post-backs and page refreshes, which are typically associated with traditional web apps.

Like its predecessor, Angular 2 too provides the necessary constructs for SPA implementation. Let's understand them and add our app pages too.

## Angular routing

Angular supports SPA development using its routing infrastructure. This infrastructure tracks browser URLs, enables hyperlink generation, exposes routing events, and provides a set of directives/components for view.

There are four major framework pieces that work together to support the Angular routing infrastructure:

- **The Router (Router)**: The primary infrastructure piece that actually provides component navigation
- **Routing configuration (Route)**: The component router is dependent upon the routing configuration for setting up routes
- **RouterOutlet component**: The `RouterOutlet` component is the placeholder container (*host*) where route-specific views are loaded
- **RouterLink directive**: This generates hyperlinks that can be embedded in the anchor tags for navigation

The following diagram highlights the roles these components play within the routing setup:

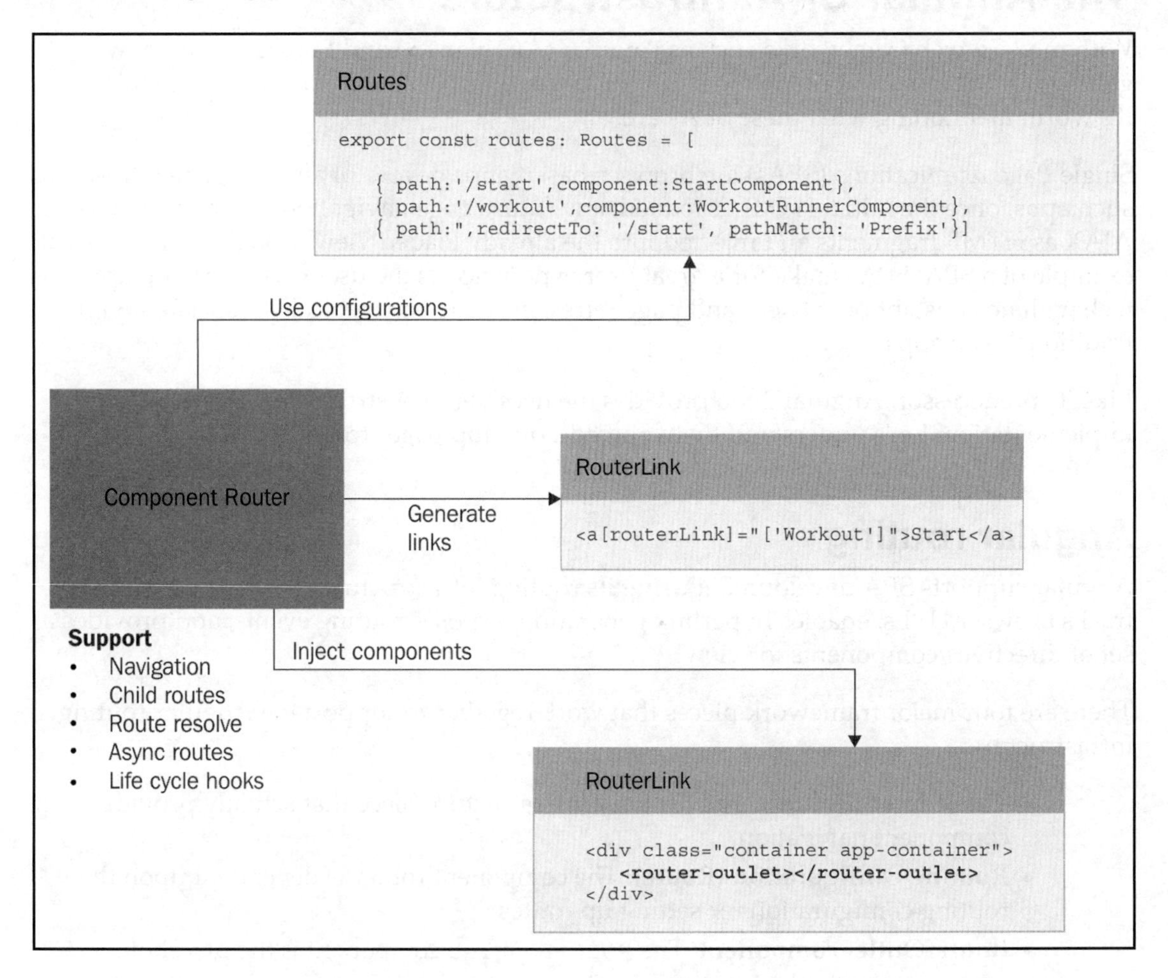


I highly encourage everyone to keep revisiting this diagram as we set up routing for *7 Minute Workout*.

*Router* is the central piece of this complete setup; therefore a quick overview of the router will be helpful.

## Angular router

If you have worked with any JavaScript framework with SPA support, this is how things work. The framework watches the browser URL and serves views based on the URL loaded. There are dedicated framework components for this job. In the Angular world, this tracking is done by a *framework service*, the router.

In Angular, any class, object, or function that provides some generic functionality is termed a **service**. Angular does not provide any special construct to declare a service as it does for components, directives, and pipes. Anything that can be consumed by components/directives/pipes can be termed a service. Router is one such service. And there are many more services that are part of the framework.
If you are from the Angular 1 realm, this is a pleasant surprise-no service, factory, provider, value, or constant!

When building a component try to delegate as much functionality as possible to a service. Components should only act as a mediator that helps in synchronizing the component model and the view state

The Angular router is there to:

- Enable navigation between components on route change
- Pass routing data between component views
- Make the state of the currently route available to active/loaded components
- Provide APIs that allow navigation for component code
- Track the navigation history, allowing us the move back and forward between component views using browser buttons
- Provide life cycle events and guard conditions that allow us to affect navigation based on some external factors

The router also supports some advance routing concepts such as parent-child routes. This gives us the ability to define routes at multiple levels inside the component tree. The parent component can define routes and child components can further add more sub-routes to the parent route definition. This is something that we cover in detail in `Chapter 4`, *Building Personal Trainer*.

The router does not work alone. As highlighted in the preceding diagram, it depends upon other framework pieces to achieve the desired results. Let's add some app pages and work with each piece of the puzzle.

## Routing setup

For the component router to work, we first need to reference it as the router is not part of the core framework.

Open `package.json` and add a package reference to the router as highlighted here:

```
"@angular/platform-browser-dynamic": "2.0.0",
"@angular/router": "3.0.0",
```

Next install the package from the command line using:

```
npm install
```

Finally, reference the package in `systemjs.config.js`. This allows SystemJS to load the `router` module correctly. Add the router package to the `ngPackageNames` array to set up the `packages` configuration:

```
var ngPackageNames = [
...
    'router',
...];
```

Also add the `base` reference (highlighted) to the `head` section of `index.html` if not present:

```
<link rel="stylesheet" href="static/css/app.css" />
<base href="/">
```

The router requires `base href` to be set. The `href` value specifies the base URL to use for all relative URLs within an HTML document, including links to CSS, scripts, images, and any other resource.

The router uses the **pushstate** mechanism for URL navigation. This allows us to use URLs such as:

- `localhost:9000/start`
- `localhost:9000/workout`
- `localhost:9000/finish`

This may not seem a big deal, but remember that we are doing client-side navigation, not the full-page redirects that we are used to. As the **developer guide** states:

> *Modern HTML 5 browsers support* `history.pushState`, *a technique that changes a browser's location and history without triggering a server page request. The router can compose a "natural" URL that is indistinguishable from one that would otherwise require a page load.*

## Pushstate API and server-side url-rewrites

The pushstate API used by the router works only when we click on links embedded in the view (`<a>` tag) or use the router API. The router intercepts any navigation events, loads the appropriate component view, and finally updates the browser URL. The request never goes to the server.

But what if we refresh the browser?

The Angular router cannot intercept the browser's refresh event, and hence a complete page refresh happens. In such a scenario, the server needs to respond to a resource request that only exists on the client side. A typical server response is to send the app host file (such as `index.html`) for any arbitrary request that may result in a **404 (Not Found)** error. This is what we call server **url-rewrite**.

Even our server setup does url-rewrite. Check out the highlighted line in `gulpfile.js`:

```
connect.server({
   ...
   fallback: 'index.html'
 });
```

The last configuration parameter to `connect.server` sets the `fallback` URL for the app server to `index.html`. This implies requests to any non-existent URLs such as `/start`, `/workout`, `/finish` or any other loads the index page.

Each server platform has a different mechanism to support url-rewrite. We suggest you look at the documentation for the server stack you use to enable url-rewrite for your Angular apps.

We can see the server-side rewrites in action once we add some pages to *7 Minute Workout*. Once the new pages are in place, try to refresh the app and see the browser's network log; the server sends `index.html` content every time irrespective of the URL requested.

**Fall back path and debugging**
Setting up a fall back path for all non-existing URLs can be detrimental while debugging the application. Once the fall back mechanism is in place, there are no 404 errors for script/HTML/CSS load failures. This can produce unexpected results for any missing reference as the server always returns the `index.html` file. Watch out for content returned in the browser network log and the browser console for anomalies whenever you add new files to the application.

As part of the preceding router setup, we have learned how router scripts are included, how server-side redirects are set up to support the HTML5 push state and the need to set up `base href`.

Before we proceed any further, we need to add some other pages to our app and configure the routes.

# Adding start and finish pages

The plan here is to have three pages for *7 Minute Workout*:

- **Start page**: This becomes the landing page for the app
- **Workout page**: What we have currently
- **Finish page**: We navigate to this once the workout is complete

The workout component and its view (`workout-runner.component.ts` and `workout-runner.html`) are already there. So let's create `StartComponent` and `FinishComponent`, and their view.

Copy the following files from Git branch `checkpoint3.1`. The files are located in the `start` and `finish` folders, under the `components` folder (the GitHub location to download from is `http://bit.ly/ng2be-3-1-components`):

- `start.component.ts`, `start.html`, and `start.module.ts`: This includes the `StartComponent` implementation and view template. A standard HTML view, and a basic component, which uses the `routerLink` directive to generate a hyperlink.
- `finish.component.ts`, `finish.html`, and `finish.module.ts`: This includes the `FinishComponent` implementation and view template. It follows the same pattern as `StartComponent`.

Both the `Start` and `Finish` components have been defined with their own modules. The convention we will follow is module per top-level view.

All three components are ready. Time to define the route configurations!

# Route configuration

To set up the routes for *7 Minute Workout,* we are going to create a route definition file. Create a file called `app.routes.ts` in the `components/app` folder defining the top-level routes for the app. Add the following routing setup:

```
import { ModuleWithProviders } from '@angular/core';
import { Routes, RouterModule } from '@angular/router';
import {WorkoutRunnerComponent} from '../workout-runner/workout-
runner.component';
import {StartComponent} from '../start/start.component';
import {FinishComponent} from '../finish/finish.component';

export const routes: Routes= [
  { path: 'start', component: StartComponent },
  { path: 'workout', component: WorkoutRunnerComponent },
  { path: 'finish', component: FinishComponent },
  { path: '**', redirectTo:'/start'}
];
export const routing: ModuleWithProviders = RouterModule.forRoot(routes);
```

The `routes` variable is an array of `Route` objects. Each `Route` defines the configuration of a single route, which contains:

- `path`: The target path to match
- `component`: The component to be loaded when the path is hit

Such a route definition can be interpreted as: "When the user navigates to a path (defined in `path`), load the corresponding component defined in the `component` property." Take the first route example; navigating to `http://localhost:9000/start` loads the component view for `StartComponent`.

You may have noticed that the last `Route` definition looks a bit different. The `path` looks odd and it does not have a `component` property either. A path with `**` denotes a catch-all path or the *wildcard route* for our app. Any navigation that does not match one of the first three routes, matches the catch-all route, causing the app to navigate to the start page (defined in the `redirectTo` property).

We can try this once the routing setup is complete. Type any random route such as `http://localhost:9000/abcd` and the app automatically redirects to `http://localhost:9000/start`.

The final call to `RouterModule.forRoot` is used to export this route setup as a module. We use the setup (exported as `routing`) inside AppModule to complete the route setup. Open `app.module.ts` and import the routing setup as well the modules we have created with respect to the `Start` and `Finish` pages:

```
import {StartModule} from '../start/start.module';
import {FinishModule} from '../finish/finish.module';
import {routing} from './app.routes';
@NgModule({
  imports: [..., StartModule, FinishModule, routing],
```

Now that we have all the required components and all the routes defined, where do we inject these components on route change? We just need to define a placeholder for that in the host view.

## Rendering component views with router-outlet

If we check the current `TrainerAppComponent` template, it has an embedded `WorkoutRunnerComponent`:

```
<workout-runner></workout-runner>
```

This needs to change. Remove the preceding declaration and replace it with:

```
<router-outlet></router-outlet>
```

`RouterOutlet` is an Angular component directive that acts as a placeholder for a child component to load on route change. It integrates with the router to load the appropriate component based on the current browser URL and the route definition.

The following diagram helps us to easily visualize what is happening with the `router-outlet` setup:

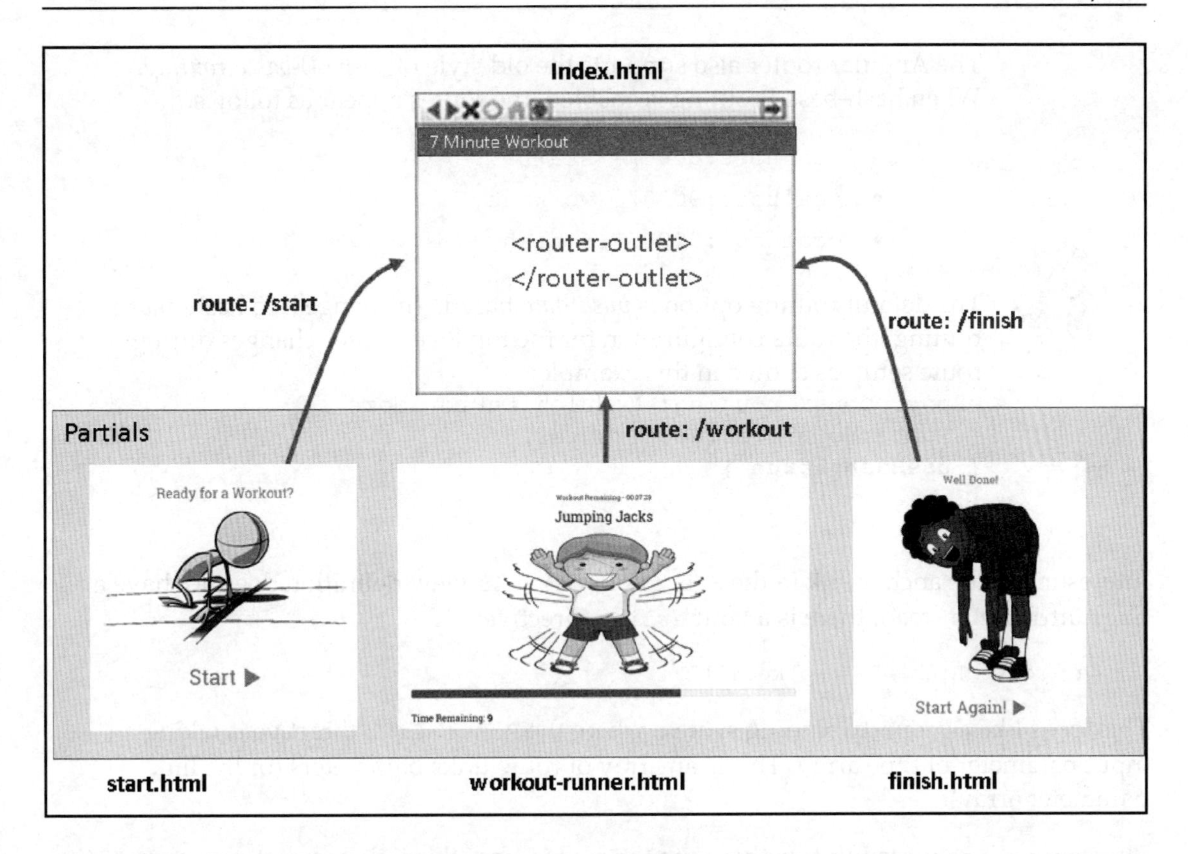


We are almost done now; it's time to trigger navigation.

## Route navigation

Like standard browser navigation, Angular navigation can happen:

- When a user enters a URL directly in the browser
- On clicking on a link on the anchor tag
- On using a script/code to navigate

If not started, start the app and load `http://localhost:9000` or `http://localhost:9000/start`. The start page should be loaded.

Click on the **Start** button of the page and the workout view should be loaded at `http://localhost:9000/workout`.

The Angular router also supports the old style of *hash (#)-based routing*. When hash-based routing is enabled, the routes appear as follows:

- `localhost:9000/#/start`
- `localhost:9000/#/workout`
- `localhost:9000/#/finish`

The default routing option is *pushState*-based. To change it to hash-based routing, the route configuration for the top level routes changes during route setup as shown in this example:

```
export const routing: ModuleWithProviders
= RouterModule.forRoot(routes,
{ useHash: true }
);
```

Interestingly, the anchor link in the `StartComponent`'s view definition does not have a `href` attribute. Instead, there is a `RouterLink` directive:

```
<a [routerLink]="['/workout']">
```

This looks like a property binding syntax, where the `RouterLink` directive is taking an input parameter of type array. This is an array of route links parameters (or the link parameter array).

The `routerLink` directive, together with the router, uses this link parameter array to resolve the correct URL path. In the preceding case, the only element in the array is the name of the route.

Notice the / prefix in the preceding route path. / is used to specify an absolute path. The Angular router also supports relative paths, which are useful when working with child routes. We will explore the concept of child routes in the next few chapters.

Refresh the app and check the rendered HTML for `StartComponent`; the preceding anchor tag is rendered as:

```
<a href="/workout">
```

**Avoid hardcoding route links**
While you could have directly used `<a href="/workout">`, prefer `routerLink` to avoid hardcoding routes.

### Link parameter array

The link parameter array passed to the `routerLink` directive follows a specific pattern:

```
['routePath', param1, param2, {prop1:val1, prop2:val2} ....]
```

The first element is always the route path, and the next set of parameters is there to replace placeholder tokens defined in a route template.

The route setup for the current *7 Minute Workout* is quite simple, and there isn't a need to pass parameters as part of link generation. But the capability is there for non-trivial routes that require dynamic parameters. See this example:

```
@RouteConfig([
  { path: '/users/:id', component: UserDetail },
  { path: '/users', component: UserList},
])
```

This is how the first route can be generated:

```
<a [routerLink]="['/users', 2]  // generates /users/2
```

The Angular router is quite a beast and supports almost everything that we expect of a modern router library. It supports child routes, async routes, lifecycle hooks, and some other advanced scenarios. We'll delay discussion on these topics until later chapters. This chapter just gets us started with Angular routing, but there is more to come!

The router link parameter can also be an object. Such objects are used to supply *optional parameters* to the route. See this example:

```
<a [routerLink]="['/users', {id:2}]  // generates /users;id=2
```

Note that the generated link contains a semicolon to separate parameters from the route and other parameters.

The last missing part in the implementation is routing to the finish pages once the workout completes.

## Using the router service for component navigation

Navigation from the workout page to the finish page is not triggered manually but on completion of the workout. `WorkoutRunnerComponent` needs to trigger this transition.

For this, `WorkoutRunnerComponent` needs to get hold of the component router and invoke the `navigate` method on it.

How does `WorkoutRunnerComponent` get the router instance? Using Angular's *dependency injection framework*. We have been shying away from this topic for some time now. We have achieved a lot without even knowing that there's a dependency injection framework in play all this while.

Let's wait a tad longer and firstly concentrate on fixing the navigation issue.

For `WorkoutRunnerComponent` to get hold of the router service instance, it just needs to declare the service on the constructor. Update the `WorkoutRunnerComponent` constructor and add the imports:

```
import {Router} from '@angular/router';
constructor(private router: Router) {
```

Angular now magically injects the current router into the `router` private variable when `WorkoutRunnerComponent` is instantiated. The magic it uses is the *dependency injection framework*.

It's now just a matter of replacing the statement `console.log("Workout complete!");` with the call to the `navigation` router:

```
this.router.navigate( ['/finish'] );
```

The `navigate` method takes the same link parameter array as the `RouterLink` directive. We can verify the implementation by patiently waiting for the workout to complete!

If you are having a problem with running the code, look at the Git branch `checkpoint3.1` for a working version of what we have done thus far. Or if you are not using Git, download the snapshot of `checkpoint3.1` (a ZIP file) from `http://bit.ly/ng2be-checkpoint3-1`. Refer to the `README.md` file in the `trainer` folder when setting up the snapshot for the first time.

If you are still wondering how to access route parameters for the current route, we have the `ActivatedRoute` service.

# Using the ActivatedRoute service to access route params

There are times when the current view requires access to the active route state. Information such as the current URL fragment, the current route parameters, and other route-related data may come in handy during component implementation.

The `ActivatedRoute` service is a one-stop shop for all current route-related queries. It has a number of properties, including `url` and `params`, that can utilize the route's current state.

Let's look at an example of a parameterized route and how to access a parameter passed from a component. Given this route:

```
{ path: '/users/:id', component: UserDetail },
```

When the user navigates to `/user/5`, the underlying component can access the `:id` parameter value by first injecting `ActivatedRoute` into its constructor:

```
export class UsersComponent {
  constructor( private route: ActivatedRoute ...
```

It then queries the `id` property from the `params` property of the `ActivatedRoute` service. Look at this example:

```
this.route.params.forEach((params: Params) => {
     let id = +params['id']; // (+) converts string 'id' to a number
     var currentUser=this.getUser(id)
});
```

The `params` property on `ActivatedObject` is actually an *observable*. We will learn more about observables later in the chapter, but for now it is enough to understand that observables are objects that raise events and can be subscribed to.

We use the `forEach` function on the `route.params` observable to get the route's parameters. The callback object (`params:Params`) contains properties corresponding to each route parameter. Check how we retrieve the `id` property and use it.

We have now covered the basic Angular routing infrastructure, but there is more to explore in the later chapters. It's now time to concentrate on a much overdue topic: *Dependency Injection*.

# Angular dependency injection

Angular makes heavy use of dependency injection to manage app and framework dependencies. The surprising part is that we could ignore this topic until we started our discussion on the router without compromising our understanding of how things work. All this while, the Angular dependency injection framework has been supporting our implementation. The hallmark of a good dependency injection framework is that the consumer can use it without bothering about the internals and with little ceremony.

If you are not sure what dependency injection is or just have a vague idea about it, an introduction to DI surely does not hurt anyone.

# Dependency injection 101

For any application, its components (not to be confused with Angular components) do not work in isolation. There are dependencies between them. A component may use other components to achieve its desired functionalities. **Dependency injection** is a pattern for managing such dependencies.

The DI pattern is popular in many programming languages as it allows us to manage dependencies in a loosely coupled manner. With such a framework in place, dependent objects are managed by a DI container. This makes dependencies swappable and the overall code more decoupled and testable.

The idea behind DI is that an object does not create\manage its own dependencies. Instead, the dependencies are provided from outside. These dependencies are made available either through a constructor, which is called **constructor injection** (Angular also does this) or by directly setting the object properties, which is called **property injection**.

Here is a rudimentary example of DI in action. Consider a class called `Tracker` that requires a `Logger` for a logging operation:

```
class Tracker() {
  logger:Logger;
  constructor() {
    this.logger = new Logger();
  }
}
```

The dependency of the class `Logger` is hardwired inside `Tracker`. What if we externalize this dependency? So the class becomes:

```
class Tracker {
  logger:Logger;
  constructor(logger:Logger) {
    this.logger = logger;
  }
}
```

This innocuous-looking change has a major impact. By adding the ability to provide the dependency externally, we can now:

- Decouple components and enable extensibility. The DI pattern allows us to alter the logging behavior of the `Tracker` class without touching the class itself. Here is an example:

  ```
  var trackerWithDBLog=new Tracker(new DBLogger());
  var trackerWithMemoryLog=new Tracker(new MemoryLogger());
  ```

  The two `Tracker` objects we just saw have different logging capabilities for the same `Tracker` class implementation. `trackerWithDBLog` logs to a DB and `trackerWithMemoryLog` to the memory (assuming both `DBLogger` and `MemoryLogger` are derived from `Logger` class). Since Tracker is not dependent of specific implementation on a `Logger` (`DBLogger` or `MemoryLogger`), this implies `Logger` and `Tracker` are loosely coupled. In future we can derive a new `Logger` class implementation and use that for logging without changing the `Tracker` implementation.

- Mock dependencies: The ability to mock dependencies makes our components more testable. The tracker implementation can be tested in isolation (unit testing) by providing a mock implementation for Logger such as MockLogger, or by using a mocking framework that can easily mock the `Logger` interface.

We can now understand how powerful DI is. Once DI is in place, the responsibility for resolving the dependencies falls on the calling/consumer code. In the preceding example, a class that was earlier instantiating `Tracker` now needs to create a `Logger` derivation and inject it into `Tracker` before using it.

Clearly, this flexibility in swapping internal dependencies of a component comes at a price. The calling code implementation can become overly complex as it now has to manage child dependencies too. This may seem simple at first, but given the fact that dependent components may themselves have dependencies, what we are dealing with is a complex dependency tree structure.

To make dependency management less cumbersome for the calling code, there is a need for DI containers/frameworks. These containers are responsible for constructing/managing dependencies and providing it to our client/consumer code.

The Angular DI framework manages dependencies for our Angular components, directives, pipes, and services.

# Exploring dependency injection in Angular

Angular employs its very own DI framework to manage dependencies across the application. The very first example of visible dependency injection was the injection of the component router into `WorkoutRunnerComponent`:

```
constructor(private router: Router) {
```

When the `WorkoutRunnerComponent` class gets instantiated, the DI framework internally locates/creates the correct router instance and injects it into the caller (in our case, `WorkoutRunnerComponent`).

While Angular does a good job at keeping the DI infrastructure hidden, it's imperative that we understand how Angular DI works. Otherwise, everything may seem rather magical.

DI is about creating and managing dependencies, and the framework component that does this is dubbed the **injector**. For the injector to manage dependencies, it needs to understand the following:

- **The what:** What is the dependency? The dependency could be a class, an object, a factory function, or a value. Every dependency needs to be registered with the DI framework before it can be injected.
- **The where/when:** The DI framework needs to know where to inject a dependency and when.
- **The how**: The DI framework also needs to know the recipe for creating the dependency when requested.

Any injected dependency needs to answer these questions irrespective of whether it's a framework construct or artefacts created by us.

Take for example the `Router` instance injection done in `WorkoutRunnerComponent`. To answer the what and how parts, we register the `Router` service in the app module (`app.module.ts`) via the import statement on the module decorator:

```
imports: [..., routing];
```

The `routing` variable is a module that exports multiple routes together with all the Angular-router-related services (technically it re-exports `RouterModule`). We export this variable from `app.routes.ts` with this statement:

```
export const routing: ModuleWithProviders = RouterModule.forRoot(routes);
```

The where and when are decided based on the component that requires the dependencies. The constructor of `WorkoutRunnerComponent` takes a dependency of `Router`. This informs the injector to inject the current `Router` instance when `WorkoutRunnerComponent` is created as part of route navigation.

Internally, the Injector determines the dependencies of a class based on the metadata reflected from it when converting TypeScript to ES5 code (done by the TypeScript compiler). The metadata is generated only if we add a decorator such as `@Component` or `@RouteConfig` on the class.

What happens if we inject `Router` into another class? Is the same `Router` instance used? The short answer is yes. The Angular injector creates and caches dependencies for future reuse, and hence these services are singleton in nature.

While dependencies in an injector are singleton, at any given time, there can be multiple injectors active throughout an Angular app. You'll learn about the injector hierarchy shortly.
With the router, there is another layer of complexity. Since Angular supports the *child route* concept, each of these routes has its own router instance. Wait until we cover child routers in the next chapter so that you can understand the intricacies!

Let's create an Angular service to track workout history. This process will help you understand how dependencies are wired using Angular DI.

# Tracking workout history

It would be a great addition to our app if we could track the workout history. When did we last exercise? Did we complete it? How much time did we spend?

Tracing the workout history requires us to track workout progress. Somehow, we need to track when the workout starts and when it stops. This tracking data then needs to be persisted somewhere.

One way to implement this history tracking is by extending our `WorkoutRunnerComponent` with the desired functionality. But that adds unnecessary complexity to `WorkoutRunnerComponent` and that's not its primary job. We need a dedicated history tracking service for this job, a service that tracks historical data and shares it throughout the app. Let's start building the `WorkoutHistoryTracker` service.

# Building the WorkoutHistoryTracker service

With the `WorkoutHistoryTracker` service, we plan to track the execution of the workout. The service also exposes an interface, allowing `WorkoutRunnerComponent` to start and stop workout tracking.

Create a `services` folder inside the `src` folder if not there and add a file called `workout-history-tracker.ts` with this code:

```
import {ExercisePlan} from '../components/workout-runner/model';
export class WorkoutHistoryTracker {
  private maxHistoryItems: number = 20;   //Tracking last 20 exercises
  private currentWorkoutLog: WorkoutLogEntry = null;
  private workoutHistory: Array<WorkoutLogEntry> = [];
  private  workoutTracked: boolean;
  constructor() { }
  get tracking(): boolean {
    return this. workoutTracked;
  }
}

export class WorkoutLogEntry {
  constructor(
    public startedOn: Date,
    public completed: boolean = false,
    public exercisesDone: number = 0,
    public lastExercise?: string,
    public endedOn?: Date) { }
}
```

There are two classes defined: `WorkoutHistoryTracker` and `WorkoutLogEntry`. As the name suggests, `WorkoutLogEntry` defines log data for one workout execution. `maxHistoryItems` allows us to configure the maximum number of items to store in the `workoutHistory` array, the array that contains the historical data. The `get tracking()` method defines a getter property for `workoutTracked` in TypeScript. `workoutTracked` is set to `true` during workout execution.

Let's add the start tracking, stop tracking, and exercise complete functions:

```
startTracking() {
  this.workoutTracked = true;
  this.currentWorkoutLog = new WorkoutLogEntry(new Date());
  if (this.workoutHistory.length >= this.maxHistoryItems) {
    this.workoutHistory.shift();
  }
    this.workoutHistory.push(this.currentWorkoutLog);
}

exerciseComplete(exercise: ExercisePlan) {
  this.currentWorkoutLog.lastExercise = exercise.exercise.title;
  ++this.currentWorkoutLog.exercisesDone;
}

endTracking(completed: boolean) {
  this.currentWorkoutLog.completed = completed;
  this.currentWorkoutLog.endedOn = new Date();
  this.currentWorkoutLog = null;
  this.workoutTracked = false;
};
```

The `startTracking` function creates a `WorkoutLogEntry` and adds it to the `workoutHistory` array. By setting the `currentWorkoutLog` to the newly created log entry, we can manipulate it later during workout execution. The `endTracking` function and the `exerciseComplete` function just alter `currentWorkoutLog`. The `exerciseComplete` function should be called on completion of each exercise that is part of the workout.

Lastly, add a function that returns the complete historical data:

```
getHistory(): Array<WorkoutLogEntry> {
  return this.workoutHistory;
}
```

That completes the `WorkoutHistoryTracker` implementation; now it's time to integrate it into workout execution.

# Integrating with WorkoutRunnerComponent

The `WorkoutRunnerComponent` requires `WorkoutHistoryTracker` to track workout history; hence there is a dependency to be fulfilled.

To make `WorkoutHistoryTracker` discoverable, it needs to be registered with the framework. At this point, we are spoilt for choices. There are a number of ways to register a dependency and a number of places too! This flexibility makes the DI framework very powerful, albeit it adds to the confusion too.

Let's firstly try to understand the different mechanisms we can use to register a dependency using the `WorkoutHistoryTracker` as an example.

## Registering dependencies

The simplest way to register a dependency is to register it at the root/global level. This can be done by passing the dependency type into the `provides` attribute (array) in the module decorator.

As shown in this example, adding `WorkoutHistoryTracker` to any module's `providers` array registers the service globally:

```
@NgModule({...providers: [WorkoutHistoryTracker],})
```

Technically speaking, when a service is added to the `providers` array it gets registered with the **app's root injector**, irrespective of the Angular module it is declared in. Any Angular artefact in any module henceforth can use the service (`WorkoutHistoryTracker`). No module imports are required at all.

This behavior is different from component/directive/pipe registration. Such artefacts have to be exported from a module for another module to use them.

Providers create dependencies when the Angular injector requests them. These providers have the recipe to create these dependencies. While a class seems to be the obvious dependency that can be registered, we can also register:

- A specific object/value
- A factory function

Directly using the class type to register a dependency (as shown in the `bootstrap` function) may mostly meet our needs, but at times we need some flexibility with our dependency registrations. The expanded version of provider registration syntax gives us that flexibility.

To learn about these variations, we need to explore providers and dependency registration in a little more detail.

## Angular providers

Providers create dependencies that are served by the DI framework.

Look at the `WorkoutHistoryTracker` dependency registration in the previous section:

```
providers: [WorkoutHistoryTracker],
```

This syntax is a short-form notation for the following version:

```
providers:({ provide: WorkoutHistoryTracker, useClass:
WorkoutHistoryTracker })
```

The first property (`provide`) is a token that acts like a key for registering a dependency. This key also allows us to locate the dependency during dependency injection.

The second property (`useClass`) is a provider definition object that defines the recipe for creating the dependency value. The framework provides a number of ways to create these dependencies, as we will see shortly.

With `useClass`, we are registering class `provider`. The class `provider` create dependencies by instantiating the type of object requested for.

## Value providers

The class `provider` create class objects and fulfil the dependency, but at times we want to register a specific object/primitive with the DI provider instead. Value providers solve this use case.

Take an example of `WorkoutHistoryTracker` registered using this technique:

```
{provide: WorkoutHistoryTracker, useValue: new WorkoutHistoryTracker()};
```

What is registered is an instance of the `WorkoutHistoryTracker` object created by us, instead of letting Angular DI create one. Take such hand-crafted dependencies (dependencies created manually) into consideration if there are dependencies further down the lineage that too need to be hand-crafted. Take the example of `WorkoutHistoryTracker` again. If `WorkoutHistoryTracker` has some dependencies, those too need to be fulfilled by manual injection:

```
{provide: WorkoutHistoryTracker, useValue: new WorkoutHistoryTracker(new
LocalStorage())});
```

Value providers come in handy in specific scenarios. For example, we can register a common app configuration using a value provider:

```
{provide: AppConfig, {useValue: {name:'Test App', gridSetting: {...} ...}}
```

Or register a mock dependency while unit-testing:

```
{provide:WorkoutHistoryTracker, {useValue: new MockWorkoutHistoryTracker()}
```

## Factory providers

There are times when injection is not a trivial affair. Injection depends upon external factors. These factors decide what objects or class instances are created and returned. Factory providers do this heavy lifting.

Take an example where we want to have different configurations for dev and production releases. We can very well use a factory implementation to select the right configuration:

```
{provide: AppConfig, useFactory: () => {
  if(PRODUCTION) {
    return {name:'My App', gridSetting: {...} ...}
  }
  else {
    return {name:'Test App', gridSetting: {...} ...}
  }
}
```

A factory function can have its own dependencies too. In such a case, the syntax changes a bit:

```
{provide: WorkoutHistoryTracker, useFactory: (environment:Environment) => {
  if(Environment.isTest) {
```

```
      return new MockWorkoutHistoryTracker();
    }
    else {
      return new WorkoutHistoryTracker();
    },
      deps:[Environment]
  }
```

The dependency is passed as a parameter to the factory function and registered on the provider definition object property, `deps`.

Use the `UseFactory` provide if the construction of the dependency is complex and not everything can be decided during wire-up.

While we have a number of options to declare dependencies, consuming dependencies is far simpler.

> Before continuing further, let's register the `WorkoutHistoryTracker` service in a new service module. This new module (`ServicesModule`) will be used to register all application-wide services.
> Copy the module definition from the Git branch `checkpoint3.2` into the `src/services` folder locally. You can download it from this GitHub location: `http://bit.ly/ng2be-3-2-services-module-ts`. Also delete all references to the `LocalStorage` service as we plan to add it later in the chapter.
> Finally, import the module into `AppModule` (`app.module.ts`).

# Injecting dependencies

Consuming dependency is easy! More often than not, we use constructor injection to consume a dependency.

## Constructor injection

Add the `import` statement at the top and update the `WorkoutRunnerComponent`'s constructor, as follows:

```
import {WorkoutHistoryTracker} from
'../../services/workout-history-tracker';
...
constructor(private router: Router,
) {
```

As with router, Angular injects `WorkoutHistoryTracker` too when `WorkoutRunnerComponent` is created. Easy!

Before we continue any further with our integration, let's explore some other facts about Angular's DI framework.

### Explicit injection using injector

We can even do explicit injection using Angular's `Injector` service. This is the same injector Angular uses to support DI. Here is how to inject the `WorkoutHistoryTracker` service using `Injector`:

```
constructor(private router: Router, private injector:Injector) {
  this.tracker=injector.get(WorkoutHistoryTracker);
```

We inject the `Injector` first and then ask the `Injector` for the `WorkoutHistoryTracker` instance explicitly.

When would someone want to do this? Well, almost never. Avoid this pattern as it exposes the DI container to your implementation and adds a bit of noise too.

Consuming dependency was easy, but how does the DI framework locate these dependencies?

## Dependency tokens

Remember this expanded version of dependency registration shown earlier?

```
{ provide: WorkoutHistoryTracker, useClass: WorkoutHistoryTracker }
```

The `provide` property value is a token. This token is used to identify the dependency to inject. Whenever Angular sees this statement:

```
constructor(tracker: WorkoutHistoryTracker)
```

It injects the correct dependency based on the class type. This is an example of a class token. The class type is used for dependency searching/mapping. Angular supports some other tokens too.

## String token

Instead of a class, we can use a string literal to identify a dependency. We can register the `WorkoutHistoryTracker` dependency using a string token, as follows:

```
{provide:"MyHistoryTracker", useClass: WorkoutHistoryTracker })
```

If we now do:

```
constructor(private tracker: WorkoutHistoryTracker)
```

Angular does not like it one little bit and fails to inject the dependency. Since the `WorkoutHistoryTracker` seen before was registered with a string token, the token needs to be provided during injection too.

To inject a dependency registered using a string token, we need to use the `@Inject` decorator. This works perfectly fine:

```
constructor(@Inject("MyHistoryTracker")
  private tracker: WorkoutHistoryTracker)
```

When `@Inject()` is not present, the Injector uses the type name of the parameter (class token).

String tokens are useful when registering instances or objects. The app configuration registration examples that we shared earlier can be rewritten using string tokens if there is no such class as `AppConfig`:

```
{ provide: "AppConfiguration", useValue: {name:'Test App', gridSetting:
{...} ...});
```

And then injected using `@Inject`:

```
constructor(@Inject("AppConfiguration") config:any)
```

While any object can act as a token, the most common token types are class and string tokens. Internally, provider turns the token parameter into an instance of the `OpaqueToken` class. Look at the framework documentation to learn more about `OpaqueToken`:
`https://angular.io/docs/ts/latest/api/core/index/OpaqueToken-class.html`.

While the `WorkoutHistoryTracker` injection into `WorkoutRunnerComponent` is done, its integration is still incomplete.

# Integrating with WorkoutRunnerComponent – continued

The history tracker instance (`tracker`) needs to be called when the workout starts, when an exercise is complete, and when the workout finishes.

Add this as the first statement in the `start` function:

```
this.tracker.startTracking();
```

In the `startExerciseTimeTracking` function, add the highlighted code after the `clearInterval` call:

```
clearInterval(this.exerciseTrackingInterval);
if (this.currentExercise !== this.restExercise) {
this.tracker.exerciseComplete(this.workoutPlan.exercises[this.currentExerci
seIndex]);}
```

And the highlighted code inside the workout to complete the `else` condition in the same function:

```
this.tracker.endTracking(true);
this.router.navigate(['/finish']);
```

History tracking is almost complete except for one case. What if the user manually navigates away from the workout page?

When that happens, we can always rely on the component's lifecycle hooks/events to help us. Workout tracking can be stopped when the `NgOnDestroy` event is fired. An appropriate place to perform any clean-up work is before the component is removed from the component tree. Let's do it.

Add this function to `workout-runner.component.ts`:

```
ngOnDestroy() {
  this.tracker.endTracking(false);
}
```

While we now have workout history tracking implemented, we don't have a mechanism to check the history. The pressing need is for a workout history page/component.

# Adding the workout history page

The workout history data that we are collecting during the execution of the workout can now be rendered in a view. Let's add a History component. The component will be available at the `/history` location and can be loaded by clicking on a link in the app header section.

Update the route definition in `app.routes.ts` to include a new route and the related import:

```
import {WorkoutHistoryComponent} from '../workout-history/workout-
history.component';
...
export const routes: Routes = [
  ...,
  { path: 'history', component: WorkoutHistoryComponent }
  { path: '**', redirectTo: '/start' }
])
```

The **History** link needs to be added to the app header section. Let's refactor the header section into its own component. Update the `app.component.ts` template `navbardiv` to:

```
<div class="navbar navbar-default navbar-fixed-top top-navbar">
<div class="container app-container">
<header></header>
</div>
</div>
```

There is a new `HeaderComponent` here. Copy the definition for the header component (`header.component.ts`) from the Git branch `checkpoint3.2`, `app` folder (GitHub location: `http://bit.ly/ng2be-3-2-header-component-ts`). Also add the component to the declarations array in `app.module.ts` as we do for any Angular component:

```
import {HeaderComponent} from './header.component';
...
declarations: [TrainerAppComponent, HeaderComponent],
```

If you look at `HeaderComponent`, the history link is now there. Let's add the workout history component.

The `WorkoutHistoryComponent` implementation is available in the Git branch `checkpoint3.2`; the folder is `workout-history` (GitHub location: `http://bit.ly/ng2be-3-2-workout-history`). Copy all three files from the folder to a corresponding folder locally. Remember to maintain the same folder hierarchy in your local setup too. Make a note that `WorkoutHistoryComponent` has been defined in a separate module (`WorkoutHistoryModule`) and needs to be imported into `AppModule` (`app.module.ts`). Import `WorkoutHistoryModule` into `AppModule` before proceeding further. For now remove all references to `SharedModule` from `WorkoutHistoryModule`.

The `WorkoutHistoryComponent` view code is trivial to say the least: a few Angular constructs, including `ngFor` and `ngIf`. The component implementation too is pretty straightforward. Inject the `WorkoutHistoryTracker` service dependency and set the history data when `WorkoutHistoryComponent` is initialized:

```
ngOnInit() {
  this.history = this.tracker.getHistory();
}
```

And this time, we use the `Location` service instead of `Router` to navigate away from the history component:

```
goBack() {
  this.location.back();
}
```

The Location service is used to interact with the browser URL. Depending upon the URL strategy, either URL paths (such as `/start`, `/workout`) or URL hash segments (such as `#/start`, `#/workout`) are used to track location changes. The router service too uses the location service internally to trigger navigation.

**Router versus Location**
While the `Location` service allows us to perform navigation, using `Router` is a preferred way to perform route navigation. We used the location service here because the need was to navigate to the last route without bothering about how to construct the route.

We are ready to test our workout history implementation. Load the start page and click on the **History** link. The history page is loaded with an empty grid. Start a workout and let an exercise complete. Check the history page again; there should be a workout listed:

7 Minute Workout

History

All Workouts:

| No | Started | Ended | Last Exercise | Exercises Done | Completed |
|---|---|---|---|---|---|
| 1 | 1/10/2016, 9:49 AM | 1/10/2016, 9:57 AM | Side Plank | 12 | Yes |

Looks good, except for one sore point in this listing. It would be better if the historical data were ordered chronologically, with the newest at the top. And it would be great if we had filtering capabilities too.

# Sorting and filtering history data using pipes

In `Chapter 2`, *Building Our First App – 7 Minute Workout*, we explored pipes. We even built our own pipe to format seconds values as hh:mm:ss. Since the primary purpose of pipes is to transform data, this can be used with any input. For arrays, pipes can be used to sort and filter data. We create two pipes, one for each sorting and filtering.

Angular1 has prebuilt filters (filters are pipes in Angular2), `orderBy` and `filter`, for this very purpose. The effort to port these filters in Angular2 has been stalled for now. See this GitHub issue: `http://bit.ly/ng2-issue-2340`.

Let's start with the `orderBy` pipe.

## The orderBy pipe

The `orderBy` pipe we implement is going to order an array of objects based on any of the object's properties. The usage pattern for sorting items in ascending order based on the `fieldName` property is going to be:

```
*ngFor="let item of items| orderBy:fieldName"
```

And for sorting items in descending order, the usage pattern is:

```
*ngFor="let item of items| orderBy:-fieldName"
```

Make note of the extra hyphen before `fieldName`.

Create a folder called `shared` in `src/components` and copy all three files located in the corresponding location in Git branch `checkpoint3.2` (GitHub location: `http://bit.ly/ng2be-3-2-shared`). There are two pipes and a new module definition (`SharedModule`) in this folder. `SharedModule` defines components/directives/pipes that that are shared across the application.

Open `order-by.pipe.ts` and look at the pipe implementation. While we are not going to delve into the pipe's implementation details, some relevant parts need to be highlighted. Look at this pipe outline:

```
@Pipe({ name: 'orderBy' })
export class OrderByPipe {
  transform(value: Array<any>, field:string): any {
   ...
  }
}
```

The preceding `field` variable receives the field on which sorting is required. Check the code below to understand how the `field` argument is passed.

If the field has a – prefix, we truncate the prefix before sorting the array in descending order.

The pipe also uses the spread operator, which may be new to you. Learn more about the spread operator on MDN here: `http://bit.ly/js-spread`.

**To use this pipe in the workout history view, import `SharedModule` into `WorkoutHistoryModule`.**

Update the template HTML:

```
<tr *ngFor="let historyItem of history|orderBy:'-startedOn'; let i =
index">
```

The historical data will now be sorted in descending order on `startedOn`.

Make note of the single quotes around the pipe parameter (`'-startedOn'`). We are passing a literal string to the `orderBy` pipe. Instead, pipe parameters can be bound to component properties too.

That's enough for the `orderBy` pipe. Let's implement filtering.

## The search pipe

The `SearchPipe` that we added earlier does a basic equality-based filtering. Nothing special.

Look at the pipe code; the pipe takes two arguments, the first being the field to search and the second the value to search. We use the array's `filter` function to filter the record, doing a strict equality check.

Let's update the workout history view and incorporate the search pipe too. Open `workout-history.html` and uncomment the div with radio buttons. These radio buttons filter workouts based on whether they were completed or not. This is how the filter selection HTML looks:

```
<label><input type="radio" name="searchFilter" value=""
(change)="completed = null" checked="">All </label>
<label><input type="radio" name="searchFilter" value="true"
(change)="completed = $event.target.value=='true'">
Completed </label>
<label><input type="radio" name="searchFilter" value="false"
(change)="completed = $event.target.value=='true'">
Incomplete </label>
```

We can have three filters: `all`, `completed`, and `incomplete` workouts. The radio selection sets the component's property `completed` using the `change` event expression. `$event.target` is the radio button that was clicked. We do not assign `completed=$event.target.value` as its value is of the string type. The `completed` property (on `WorkoutHistoryComponent`) should be of the `boolean` type for equality comparison with the `WorkoutLogEntry.completed` property to work.

The `search` pipe can now be added to the `ngFor` directive expression. We are going to chain the `search` and `orderBy` pipe. Update the `ngFor` expression to:

```
<tr *ngFor="let historyItem of history
|search:'completed':completed
|orderBy:'-startedOn';
let i = index">
```

The `search` pipe first filters the historical data followed by the `orderBy` pipe reordering it. Pay close attention to the `search` pipe parameters: the first parameter is a string literal denoting the field to search (`'completed'`), whereas the second parameter is derived from the component property `completed`. Having the ability to bind pipe parameters to component properties allows us great flexibility.

Go ahead and verify the search capabilities of the history page. Based on the radio selection, the history records are filtered, and of course they are sorted in reverse chronological order based on the workout start dates.

While pipe usage with arrays looks simple, it can throw up some surprises if we do not understand when pipes are evaluated.

## Pipe gotcha with arrays

To understand the issue with pipes applied to arrays, let's reproduce the problem.

Open `search.pipe.ts` and remove the `@Pipe` decorator attribute `pure`. Also change the following statement:

```
if (searchTerm == null) return [...value];
```

To the following:

```
if (searchTerm == null) return [value];
```

Add a button at the end of the radio list (in `workout-history.html`) that adds a new log entry to the `history` array:

```
<button (click)="addLog()">Add Log</button>
```

Add a function to `WorkoutHistoryComponent`:

```
addLog() {
  this.history.push(Object.assign({},
  this.history[this.history.length-1]));
}
```

The preceding function duplicates the first history item and adds back to the `history` array. If we load the page and click on the button, a new log entry gets added to the history array but it does not show up on the view, unless we change the filter (by clicking on the other radios). Interesting!

Before calling `addLog` make sure at least one history log is already there; otherwise the `addLog` function will fail.

The pipes that we have built thus far are *stateless* (also called pure) in nature. They simply transform input data into output. Stateless pipes are revaluated any time the pipe input changes (the expression on the left side of pipe symbol) or any pipe argument is updated.

For arrays, this happens on an array assignment/reference change and not on the addition or deletion of elements. Switching the filter condition works as it causes the search pipe to evaluate again as the search parameter (the `completed` status) changes. This behavior is something to be aware of.

What's the fix? For starters, we can make the history array immutable, which implies that it cannot be changed once created. To add a new element we need to create a new array with the new value, something like:

```
this.history = [...this.history,
Object.assign({}, this.history[0])];
```

Works perfectly, but we are changing our implementation to make it work with pipes. Instead, we can change the pipe and mark it stateful.

The difference between a stateless and stateful pipe is that stateful pipes are evaluated by Angular every time the framework does a change detection run, which involves checking the complete application for changes. Therefore, with stateful pipes, the check is not limited to the pipe input/argument changes.

To make a `search` pipe stateless, just update the `Pipe` decorator with `pure:false`:

```
@Pipe({
  name: 'search',
  pure:false
})
```

It still does not work! The `search` pipe has one more quirk that needs a fix. The "**All**" radio selection does not work perfectly. Add a new workout log, and it still does not show up, unless we switch filters.

The fix here is to revert this line in the `search` pipe:

```
if (searchTerm == null) return value;
```

To the following:

```
if (searchTerm == null) return [...value];
```

We changed the `if` condition to return a new array every time (using the spread operator), even when `searchTerm` is `null`. If we return the same array reference, Angular does not check for a size change in the array and hence does not update the UI.

That completes our History page implementation. You may now be wondering what the last few fixes on pipes have to do with how change detection works. Or you may be wondering what is change detection? Let's put all of these doubts to rest and introduce everyone to Angular's change detection system.

Angular's change detection will be covered extensively in `Chapter 8`, *Some Practical Scenarios*. The aim of the next section is to introduce the concept of change detection and how Angular performs this process.

# Angular change detection overview

To put it succinctly, change detection is all about tracking changes done to the component model during app execution. This helps Angular's databinding infrastructure to identify what parts of the view need to be updated. Every databinding framework needs to address this issue, and the approach these frameworks take for tracking changes differs. It even differs from Angular1 to Angular2.

To understand how change detection works in Angular, there are a few things that we need to keep in mind.

- Firstly, an Angular app is nothing but a hierarchy of components, from root to the leaf.
- Secondly, there is nothing special about the component properties that we bind to view; therefore Angular needs an efficient mechanism to know when these properties change. It cannot keep polling for changes in these properties.

- And finally to detect changes in a property value Angular does a *strict comparison* (===) between the previous and current value. For reference types it means only the references are compared. No deep comparison is done.

For precisely this reason, we had to mark our search pipe as stateful. Adding elements to an existing array does not change the array reference and hence Angular fails to detect any change to the array. Once the pipe is marked as stateful, the pipe is evaluated irrespective of whether the array has changed or not.

Since Angular cannot know when any bound property is updated automatically, it instead resorts to checking every bound property when a change detection run is triggered. Starting from the root of the component tree, Angular checks each bound property for changes down the component hierarchy. If a change is detected that component is marked for refresh. It's worth reiterating that changes in a bound property do not immediately update the view. Instead, a change detection run works in two phases.

- In the first phase it does the component tree walk and marks components that need to be refreshed due to model updates
- In the second phase, the actual view is synchronized with the underlying model

Model changes and view updates are never interleaved during a change detection run.

We now just need to answer two more questions. When is a change detection run triggered? And how many times does it run?

An Angular change detection run is triggered when any of these events are triggered:

- **User input/ browser events**: We click on a button, enter some text, scroll the content. Each of these actions can update the view (and the underlying model).
- **Remote XHR requests**: This is another common reason for view updates. Getting data from a remote server to show on the grid and getting user data to render a view are examples of this.
- **setTimeout and setInterval**: As it turns out, we can use `setTimeout` and `setInterval` to execute some code asynchronously and at specific intervals. Such code can also update the model. For example, a `setInterval` timer may check for stock quotes at regular interval and updates the stock price on the UI.

And most importantly each component model is checked only once, in a top-down fashion, starting from the root component to the tree leaves.

The last statement is true when Angular is configured to run in production mode. In development mode the component tree is traversed twice for changes. Angular expects that the model be stable after the first tree walk. If that is not the case Angular throws an error in development mode, and ignores the changes in production mode.
We can enable production mode by invoking the `enableProdMode` function before the `bootstrap` function call.

```
import {enableProdMode} from '@angular/core'
enableProdMode();
platformBrowserDynamic().bootstrapModule(AppModule);
```

Let's explore some other façades of the Angular DI framework, starting with hierarchical injectors, a confusing yet very powerful feature of Angular.

# Hierarchical injectors

An injector in Angular is a dependency container that is responsible for storing dependencies and dispensing them when asked for. The provider registration examples shown earlier on modules actually register the dependencies with a global Injector.

## Registering component level dependencies

All of the dependency registration that we have done thus far were done inside a module. Angular goes one step further and allows registration of dependencies at the component level too. There is a similar providers attribute on the `@Component` decorator that allows us to register dependency at the component level.

We could've very well registered the `WorkoutHistoryTracker` dependency on `WorkoutRunnerComponent`. Something on these lines:

```
@Component({
  selector: 'workout-runner',
  providers: [WorkoutRunnerComponent]
})
```

But whether we should do it or not is something that we are going to discuss in this section shortly.

In the context of this discussion on hierarchical injectors, it's important to understand that Angular creates an injector per component (oversimplified). Dependency registration done at the component level is available on the component and its descendants.

We also learned that dependencies are singleton in nature. Once created, the Injector will always return the same dependency every time. This feature is quite evident from the workout history implementation.

`WorkoutHistoryTracker` was registered with the `ServicesModule` and then injected into two components, `WorkoutRunnerComponent` and `WorkoutHistoryComponent`. Both components get the same instance of `WorkoutHistoryTracker`. The next diagram highlights this registration and injection:

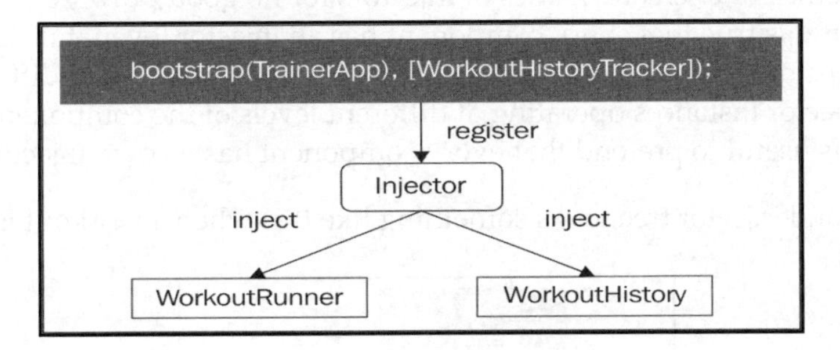


To confirm, just add a `console.log` statement in the `WorkoutHistoryTracker` constructor:

```
console.log("WorkoutHistoryTracker instance created.")
```

Refresh the app and open the history page by clicking on the header link. The message log is generated once, irrespective of how many times we run the workout or open the history page.

We now see a new interaction/data flow pattern! Think carefully; a service is being used to share state between two components. `WorkoutRunnerComponent` is generating data and `WorkoutHistoryComponent` is consuming it. And that too without any interdependence. We are exploiting the fact that dependencies are singleton in nature. This data sharing/interaction/data flow pattern can be used to share state between any number of components. Indeed, this is a very powerful weapon in our arsenal. Next time, there is a need to share state between unrelated components, think of services.

But what does this have to do with hierarchical injectors? Ok, let's not beat around the bush; let's get straight to the point.

While dependencies registered with the injector are singleton, Injector itself is not! At any given point in time, there are multiple injectors active in the application. In fact, injectors are created in the same hierarchy as the component tree. Angular creates an `Injector` instance for every component in the component tree (oversimplification; see the next information box).

Angular does not literally create an injector for each component. As explained in Angular the developer guide:
Every component doesn't need its own injector and it would be horribly inefficient to create masses of injectors for no good purpose.
But it is true that every component has an injector (even if it shares that injector with another component) and there may be many different injector instances operating at different levels of the component tree.
It is useful to pretend that every component has its own injector.

The component and injector tree looks something like this when a workout is running:

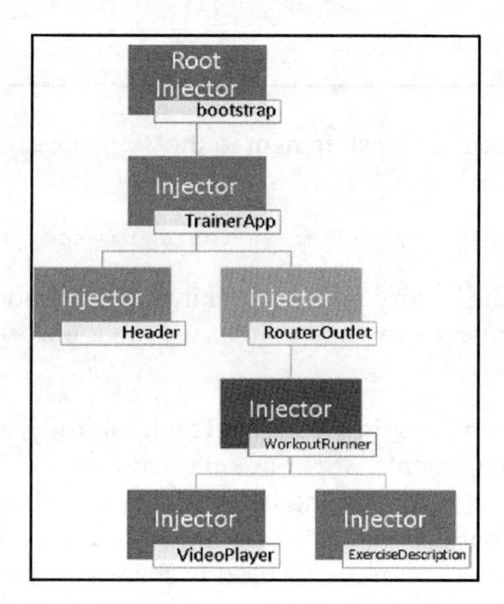

The insert textbox denotes the component name. The **root injector** is the injector created as part of the application bootstrap process.

What is the significance of this injector hierarchy? To understand the implications, we need to understand what happens when a component requests for a dependency.

## Angular DI dependency walk

Whenever requesting for a dependency, Angular first tries to satisfy the dependency from the component's own injector. If it fails to find the requested dependency, it queries the parent component injector for the dependency, and its parent if the probing fails again, and so on and so forth till it finds the dependency or reaches the root injector. The takeaway: any dependency search is hierarchy-based.

Earlier when we registered `WorkoutHistoryTracker`, it was registered with the root injector. The `WorkoutHistoryTracker` dependency request from both `WorkoutRunnerComponent` and `WorkoutHistoryComponent` was satisfied by the root injector, not their own component injectors.

This hierarchical injector structure brings a lot of flexibility. We can configure different providers at different component levels and override the parent provider configuration in child components. This only applies to dependencies registered on components. If the dependency is added on a module it gets registered on the root injector.

Let's try to override the global `WorkoutHistoryTracker` service in components that use it to learn what happens on such overrides. It's going to be fun and we will learn a lot!

Open `workout-runner.component.ts` and to the `@Component` decorator add a `providers` attribute:

```
providers: [WorkoutHistoryTracker]
```

Do this in `workout-history.component.ts` too. Now if we refresh the app, start a workout, and then load the history page, the grid is empty. Irrespective of the times we try to run the workout, the history grid is always empty.

The reason is quite obvious. Post setting the `WorkoutHistoryTracker` provider on each `WorkoutRunnerComponent` and `WorkoutHistoryComponent`, the dependency is being fulfilled by the respective component injectors themselves. Both component injectors create their own instance of `WorkoutHistoryTracker` when requested, and hence history tracking is broken. Look at the following diagram to understand how the request is fulfilled in both scenarios:

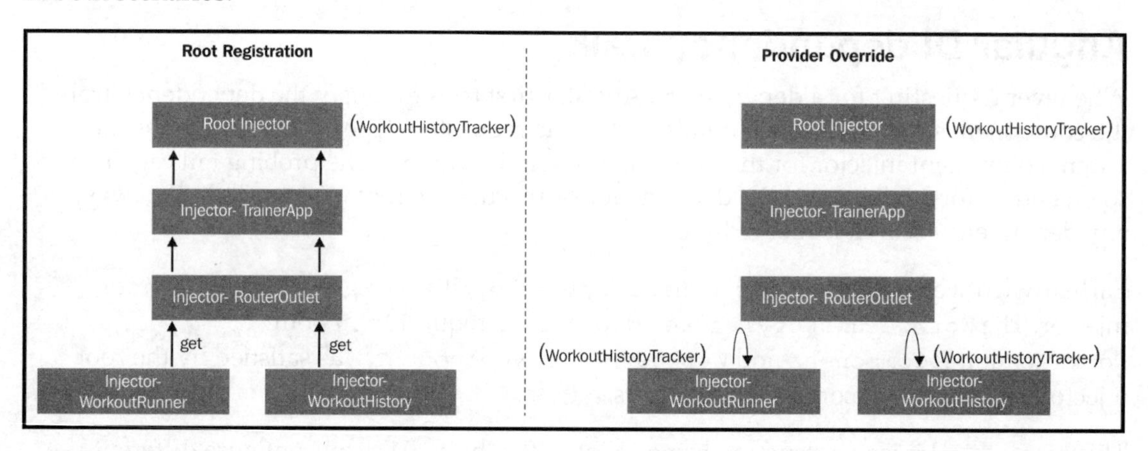


A quick question: What happens if we register the dependency in the root component, `TrainerAppComponent`, instead of doing it during app bootstrapping? Something like:

```
@Component ({
  selector: 'trainer-app',
  providers: [WorkoutHistoryTracker]
}
export class TrainerAppComponent {
```

Interestingly, with this setup too, things work perfectly. That's pretty evident; `TrainerAppComponent` is a parent component for `RouterOutlet` that internally loads `WorkoutRunnerComponent` and `WorkoutHistoryComponent`. Hence in such a setup, the dependency gets fulfilled by the `TrainerAppComponent`'s injector.

Dependency lookup up on the component hierarchy can be manipulated if an intermediate component has declared itself as a host component. We will learn more about it in later chapters.

Hierarchical injectors allow us to register dependencies at a component level, avoiding the need to register all dependencies globally.

This functionality really comes in handy when building an Angular library component. Such components can register their own dependencies without requiring the consumer of the library to register library-specific dependencies.

Remember: if you are having trouble loading the right service/dependency, make sure you check the component hierarchy for overrides done at any level.

We now understand how dependency resolution works in components. But what happens if a service has a dependency? Yet another uncharted territory to explore.

Remove the `provider` registration we did in those two components before continuing further.

## Dependency injection with @Injectable

The `WorkoutHistoryTracker` has a fundamental flaw; the history is not persisted. Refresh the app and the history is lost. We need to add persistence logic to store historical data. To avoid any complex setup, we use the browser local storage to store historical data.

Add a `local-storage.ts` file to the `services` folder. And add the following class definition:

```
export class LocalStorage {
  getItem<T>(key: string): T {
    if (localStorage[key]) {
      return <T>JSON.parse(localStorage[key]);
    }
    return null;
  }
  setItem(key: string, item: any) {
    localStorage[key] = JSON.stringify(item);
  }
}
```

A simple wrapper over the browser's `localStorage` object.

Go ahead and register the `LocalStorage` service in the services module (`services.module.ts`).

And like any other dependency, inject it in the `WorkoutHistoryTracker` constructor (the `workout-history-tracker.ts` file) with the necessary import:

```
import {LocalStorage} from './local-storage';
...
constructor(private storage: LocalStorage) {
```

This is standard DI stuff except that it does not work as expected. If we refresh the app now, Angular throws an error:

```
Cannot resolve all parameters for WorkoutHistoryTracker(?). Make sure they
all have valid type or annotations.
```

Strange! The all so awesome DI failed, and for no good reason! Not really; Angular is not doing any magic. It needs to know the class dependencies, and the only way it can know these is by inspecting the class definition and constructor arguments.

Add a decorator called `@Injectable()` (remember to add parentheses) above `WorkoutHistoryTracker` and add the module import statement:

```
import {Injectable} from '@angular/core';
```

Refresh the page and the DI works perfectly. What made it work?

By putting in the `@Injectable` decorator, we are forcing the TypeScript transpiler to generate metadata for the `WorkoutHistoryTracker` class. This includes details about the constructor arguments. Angular DI consumes this generated metadata to determine the types of dependency the service has, and in future it fulfils these dependencies when the service is created.

What about components using `WorkoutHistoryTracker`? We have not used `@Injectable` there but still the DI works. We don't need to. Any decorator works and there is already an `@Component` decorator applied to all components.

Remember the decorator needs to be added on the calling class (or client class).

The actual integration between the `LocalStorage` service and `WorkoutHistoryTracker` is a mundane process.

Update the constructor for `WorkoutHistoryTracker` as follows:

```
constructor(private storage: LocalStorage) {
  this.workoutHistory =
(storage.getItem<Array<WorkoutLogEntry>>(this.storageKey) || [])
  .map((item: WorkoutLogEntry) => {
    item.startedOn = new Date(item.startedOn.toString());
    item.endedOn = item.endedOn == null ? null
    : new Date(item.endedOn.toString());
    return item;
  });
}
```

And add a declaration for `storageKey`:

```
private storageKey: string = "workouts";
```

The constructor loads the workout logs from the local storage. The `map` function call is necessary as everything stored in `localStorage` is a string. Therefore, while de-serializing we need to convert the string back to the date value.

Add this statement last in the `startTracking`, `exerciseComplete`, and `endTracking` functions:

```
this.storage.setItem(this.storageKey, this.workoutHistory);
```

We save the workout history to local storage every time historical data changes.

That's it! We have built workout history tracking over `localStorage`. Verify it!

Before we move on to our big-ticket item, audio support, there are a few minor fixes that are needed for a better user experience. The first one is related to the **History** link.

# Tracking route changes using the router service

The **History** link in the `Header` component is visible for all routes. It will be better if we hide the link on the workout page. We don't want to lose an in-progress workout by accidentally clicking on the **History** link. Moreover, no one is interested in knowing about the workout history while doing a workout.

The fix is easy. We just need to determine if the current route is the workout route and hide the link. The `Router` service is going to help us with this job.

Open `header.component.ts` and add the necessary imports for router; update the `Header` class definition to:

```
import {Router, Event } from '@angular/router';
...
export class HeaderComponent {
  showHistoryLink: boolean = true;
  private subscription: any;
  constructor(private router: Router) {
    this.router.events.subscribe((data: Event) => {
      this.showHistoryLink=!this.router.url.startsWith('/workout');
    });
  }
```

The `showHistoryLink` property determines if the history link is shown to the user or not. In the constructor, we inject the `Router` service and register a callback on the `events` property using the `subscribe` function.

The `events` property is an observable. We will learn more about observables later in the chapter, but for now it is enough to understand that observables are objects that raise events and can be subscribed to. The `subscribe` function registers a callback function that is invoked every time the route changes.

The callback implementation just toggles the `showHistoryLink` state based on the current route name. The name we derive from the `url` property of the `router` object.

To use the `showHistoryLink` in the view just update the header template line with the anchor tag to:

```
<li *ngIf="showHistoryLink"><a [routerLink]="['History']" ...>...</a></li>
```

And that's it! The **History** link does not show up on the workout page.

Another fix/enhancement is related to the video panel on the workout page.

# Fixing the video playback experience

The current video panel implementation can at best be termed amateurish. The size of the default player is small. When we play the video, the workout does not pause. The video playback is interrupted on exercise transitions. Also, the overall video load experience adds a noticeable lag at the start of every exercise routine. This is a clear indication that this approach to video playback needs some fixing.

This is what we are going to do to fix the video panel:

- Show the image thumbnail for the exercise video instead of loading the video player itself
- When the user clicks on the thumbnail, load a popup/dialog with a bigger video player that can play the selected video
- Pause the workout while the video playback is on

Let's get on with the job!

# Using thumbnails for video

Replace the `ngFor` template html inside `video-player.html` with this snippet:

```
<div *ngFor="let video of videos" class="row video-image">
  <div class="col-sm-12">
    <div id="play-video-overlay">
      <span class="glyphicon glyphicon-play-circle video absolute-center">
      </span>
    </div>
    <img height="220" [src]="'//i.ytimg.com/vi/'+video+'/hqdefault.jpg'" />
  </div>
</div>
```

We have abandoned iframe and instead loaded the thumbnail image of the video (check the `img` tag). All other content shown here is for styling the image.

We have referenced the Stack Overflow post at `http://bit.ly/so-yt-thumbnail` to determine the thumbnail image URL for our videos.

Start a new workout; the images should show up, but the playback functionality is broken. We need to add the video playback dialog.

# Using the angular2-modal dialog library

The Angular framework does not come with any pre-packaged UI library/controls. We need to look outwards and find a community solution for any UI control requirements.

The library we are going to use is angular2-modal, available on GitHub at `http://bit.ly/angular2-modal`. Let's install and configure the library.

From the command line (inside the `trainer` folder), run the following command to install the library:

```
npm i angular2-modal@2.0.0-beta.13 --save
```

To integrate angular2-modal within our app, we need to add the package references for angular2-modal in `systemjs.config.js`. Copy the updated `systemjs.config.js` from Git branch `checkpoint3.2` (GitHub location: `http://bit.ly/ng2be-3-2-system-config-js`) in the `trainer` folder and overwriting the local configuration file. The updated configurations allow SystemJS to know how to load the modal dialog library when it encounters library `import` statements.

The next few steps highlight the configuration ceremony to be performed before angular2-modal can be used:

- In the first step, we configure the root element for angular2-modal. Open `app.component.ts` and add the highlighted code:

  ```
  import {Component, ViewContainerRef} from '@angular/core';
  ...
  import { Overlay } from 'angular2-modal';
  ...
  export class TrainerAppComponent {
    constructor(overlay: Overlay, viewContainer: ViewContainerRef) {
      overlay.defaultViewContainer = viewContainer;
    }
  }
  ```

  This step is essential as the modal dialog needs a container component to host itself. By passing in the `ViewContainerRef` of `TrainerAppComponent`, we allow the dialog to load inside the app root.

- The next step is to add two modules from the library to `AppModule`. Update `app.module.ts` and add this code:

  ```
  import { ModalModule } from 'angular2-modal';
  import { BootstrapModalModule }
    from 'angular2-modal/plugins/bootstrap';
  ...
  imports: [..., ModalModule.forRoot(), BootstrapModalModule]
  ```

  The library is now ready for use.

While angular2-modal has a number of predefined templates for standard dialogs such as alert, prompt, and confirm, these dialogs provide little customization in terms of look and feel. To have better control over the dialog UI, we need to create a custom dialog, which thankfully the library supports.

## Creating custom dialogs with angular2-modal

Custom dialogs in angular2-modal are nothing but Angular components with some special library constructs incorporated.

Copy the `video-dialog.component.ts` file from the `workout-runner/video-player` folder in Git branch `checkpoint3.2` (GitHub location: `http://bit.ly/ng2be-3-2-video-dialog-component-ts`) into your local setup. The file contains the custom dialog implementation.

Next, update `workout-runner.module.ts` and add a new `entryComponents` attribute to the module decorator:

```
import {VideoDialogComponent} from './video-player/video-dialog.component';
...
declarations: [..., VideoDialogComponent],
   entryComponents:[VideoDialogComponent]
```

The `VideoDialogComponent` needs to be added to `entryComponents` as it is not explicitly used in the component tree.

The `VideoDialogComponent` is a standard Angular component, with some modal dialog, specific implementations that we describe later.

The `VideoDialogContext` class has been created to pass the `videoId` of the YouTube video clicked to the dialog instance. The class inherits from `BSModalContext` a configuration class the dialog library uses to alter the behavior and UI of the modal dialog.

To get a better sense of how `VideoDialogContext` is utilized, let's invoke the preceding dialog from workout runner.

Update the `ngFor` div in `video-player.html` and add a `click` event handler:

```
<div *ngFor="let video of videos" (click)="playVideo(video)"
class="row video-image">
```

The preceding handler invokes the `playVideo` method, passing in the video clicked. The `playVideo` function in turn opens the corresponding video dialog. Add the `playVideo` implementation to `video-player.component.ts` as highlighted:

```
import {Modal} from 'angular2-modal';
import { overlayConfigFactory } from 'angular2-modal'
import {VideoDialogComponent, VideoDialogContext}
from './video-dialog.component';
...
export class VideoPlayerComponent {
  @Input() videos: Array<string>;
  constructor(private modal: Modal) { }
    playVideo(videoId: string) {
      this.modal.open(VideoDialogComponent,
      overlayConfigFactory(new VideoDialogContext(videoId)));
  };
}
```

The `playVideo` function calls the `Modal` class `open` function, passing in the dialog component to open and a new instance of the `VideoDialogContext` class with the `videoId` of the YouTube video. Before proceeding, delete the `ngOnChange` function from the file too.

The dialog implementation in `video-dialog.component.ts` implements the `ModalComponent<VideoDialogContext>` interface, required by the modal library. Look at how the context (`VideoDialogContext`) to the dialog is passed to the constructor and how we extract and assign the `videoId` property from the context. Then it's just a matter of binding the `videoId` property to the template view (see the template HTML) and rendering the YouTube player.

And we are good to go. Load the app and start the workout. Then click on any workout video images. The video dialog should load and now we can watch the video!

Before we call the dialog implementation complete, there is one small issue that needs to be fixed. When the dialog opens the workout should pause: that's not happening currently. We will fix it in the next section using Angular eventing support.

If you are having a problem with running the code, look at Git branch `checkpoint3.2` for a working version of what we have done thus far. Or if you are not using Git, download the snapshot of `checkpoint3.2` (a ZIP file) from `http://bit.ly/ng2be-checkpoint3-2`. Refer to the `README.md` file in the `trainer` folder when setting up the snapshot for the first time.

There is one last feature that we plan to add to *7 Minute Workout* before wrapping up the application and building something new with Angular: audio support. It teaches us some new cross-component communication patterns too.

# Cross-component communication using Angular events

We touched upon events in the last chapter when learning about Angular's binding infrastructure. It's time now to look at eventing in more depth. Let's add audio support to *7 Minute Workout*.

## Tracking exercise progress with audio

For the *7 Minute Workout* app, adding sound support is vital. One cannot exercise while constantly staring at the screen. Audio clues helps the user perform the workout effectively as he/she can just follow the audio instructions.

Here is how we are going to support exercise tracking using audio clues:

- A ticking clock sound tracks progress during the exercise
- A half-way indicator sounds, indicating that the exercise is halfway through
- An exercise-completion audio clip plays when the exercise is about to end
- An audio clip plays during the rest phase and informs users about the next exercise

There will be an audio clip for each of these scenarios.

Modern browsers have good support for audio. The HTML5 `<audio>` tag provides a mechanism to embed audio clips into html content. We too will use the `<audio>` tag to playback our clips.

Since the plan is to use the HTML `<audio>` element, we need to create a wrapper directive that allows us to control audio elements from Angular. Remember that directives are HTML extensions without a view.

The Git `checkpoint3.3` folder `trainer/static/audio` contains all the audio files used for playback; copy them first. If you are not using Git, a snapshot of the chapter code is available at `http://bit.ly/ng2be-checkpoint3-3`. Download and unzip the content and copy the audio files.

# Building Angular directives to wrap HTML audio

You may not have realized so far, but we have purposefully shied away from directly accessing the DOM for any of our component implementations. There has not been a need to do it. Angular data binding infrastructure, including property, attribute, and event binding, has helped us manipulate HTML without touching the DOM.

For the audio element too, the access pattern should be Angularish. Let's create a directive that wraps access to audio elements.

Create a folder `workout-audio` inside `workout-runner` and add a new file `my-audio.directive.ts` to it. Next add the implementation for the `MyAudioDirective` directive outlined here to this file:

```
import {Directive, ElementRef} from '@angular/core';

@Directive({
  selector: 'audio',
  exportAs: 'MyAudio'
})
export class MyAudioDirective {
  private audioPlayer: HTMLAudioElement;
  constructor(element: ElementRef) {
    this.audioPlayer = element.nativeElement;
  }
}
```

The `MyAudioDirective` class is decorated with `@Directive`. The `@Directive` decorator is similar to the `@Component` decorator except we cannot have an attached view. Therefore, no `template` or `templateUrl` is allowed!

The preceding `selector` property allows the framework to identify where to apply the directive. Using `audio` as the selector makes our directive load for every `<audio>` tag used in html.

In a standard scenario, directive selectors are attribute-based, which helps us identify where the directive has been applied. We deviate from this norm and use an element selector for the `MyAudioDirective` directive. We want this directive to be loaded for every audio element, and it becomes cumbersome to go to each audio declaration and add a directive-specific attribute. Hence an element selector.

The use of `exportAs` becomes clear when we use this directive in view templates.

The `ElementRef` object injected in constructor is the Angular element for which the directive is loaded. Angular creates the `ElementRef` instance for every component and directive when it compiles and executes the html template. When requested in the constructor, the DI framework locates the corresponding `ElementRef` and injects it. We use `ElementRef` to get hold of the actual audio element in code (the instance of `HTMLAudioElement`). The `audioPlayer` property holds this reference.

The directive now needs to expose an API to manipulate the audio player. Add these functions to the `MyAudioDirective` directive:

```
stop() {
  this.audioPlayer.pause();
}

start() {
  this.audioPlayer.play();
}
get currentTime(): number {
  return this.audioPlayer.currentTime;
}

get duration(): number {
  return this.audioPlayer.duration;
}

get playbackComplete() {
  return this.duration == this.currentTime;
}
```

The `MyAudioDirective` API has two functions (`start` and `stop`) and three getters (`currentTime`, `duration`, and a boolean property called `playbackComplete`). The implementations for these functions and properties just wrap the audio element functions.

Learn about these audio functions from the MDN documentation here: `http://bit.ly/html-media-element`.

To understand how we use the audio directive, let's create a new component that manages audio playback.

Remember to register the `MyAudioDirective` under `WorkoutRunnerModule` before proceeding any further.

# Creating WorkoutAudioComponent for audio support

If we go back and look at the audio cues that are required, there are four distinct audio cues, and hence we are going to create a component with five embedded `<audio>` tags (two audio tags work together for next-up audio).

Open the `workout-audio` folder and create a file called `workout-audio.html` for the component template. Add this HTML snippet:

```
<audio #ticks="MyAudio" loop src="/static/audio/tick10s.mp3"></audio>
<audio #nextUp="MyAudio" src="/static/audio/nextup.mp3"></audio>
<audio #nextUpExercise="MyAudio" [src]= "'/static/audio/'
+ nextupSound">
</audio>
<audio #halfway="MyAudio" src="/static/audio/15seconds.wav"></audio>
<audio #aboutToComplete="MyAudio" src="/static/audio/321.wav"></audio>
```

Five `<audio>` tags one for each:

- **Ticking audio**: This audio produces the ticking sound and is started as soon as the workout starts. Referred using the template variable `ticks`.
- **Next up audio and exercise audio**: There are two audio tags that work together. The first tag with the template variable as nextUp produces the "Next up" sound. And the actual exercise audio (`nextUpExercise`).
- **Halfway audio**: The halfway audio plays halfway through the exercise.
- **About to complete audio**: This audio piece is played to denote the completion of an exercise (`aboutToComplete`).

Did you notice the usage of the `#` symbol in the view? There are some variable assignments prefixed with `#`. In the Angular world, these variables are known as template reference variables or at times template variables.

The platform developer guide describes them thus:

> *A template reference variable is a reference to an DOM element or directive within a template.*

Don't confuse them with the template input variables that we have used with the `ngFor` directive earlier:
`*ngFor="let video of videos"`
Template input variables (`video` in this case) allow us to access the model object from the view. The value assigned to `video` depends upon the context of the `ngFor` directive loop.

Look at the last section where we set the `MyAudioDirective` directive's `exportAs` metadata to `MyAudio`. We repeat that same string while assigning the template reference variable in the preceding view:

```
#ticks="MyAudio"
```

The role of `exportAs` is to define the name that can be used in the view to assign this directive to a variable. Remember, a single element/component can have multiple directives applied to it. `exportAs` allows us to select which directive should be assigned to a template variable.

Template variables, once declared, can be accessed from other parts of the view. We will take up this discussion shortly. But in our case, we will use template variables to refer to the multiple `MyAudioDirective`s from parent component code. Let's understand how that works.

Add the `workout-audio.compnent.ts` file to the `workout-audio` folder with the following outline:

```
import {Component, ViewChild} from '@angular/core';
import {MyAudioDirective} from './my-audio.directive'
import {WorkoutPlan, ExercisePlan, ExerciseProgressEvent,
ExerciseChangedEvent} from '../model';

@Component({
  selector: 'workout-audio',
  templateUrl: '/src/components/workout-runner/workout-audio/
  workout-audio.html'
})
export class WorkoutAudioComponent {
  @ViewChild('ticks') private ticks: MyAudioDirective;
  @ViewChild('nextUp') private nextUp: MyAudioDirective;
  @ViewChild('nextUpExercise')private nextUpExercise: MyAudioDirective;
  @ViewChild('halfway') private halfway: MyAudioDirective;
  @ViewChild('aboutToComplete') private aboutToComplete:
  MyAudioDirective;
  private nextupSound: string;
}
```

The interesting bit here is the `@ViewChild` decorator against the five properties. The `@ViewChild` decorator allows us to inject a child component/directive/element reference into its parent. The parameter passed is the template variable name, which helps DI match the element/directive to inject. When Angular instantiates the `WorkoutAudioComponent`, it injects the corresponding audio component based on the `@ViewChild` decorator. Let's complete the basic class implementation before we look at `@ViewChild` in detail.

Without `exportAs` set on the `MyAudioDirective` directive, the `@ViewChild` injection injects the related `ElementRef` instance instead of the `MyAudioDirective` instance.

The remaining task is to just play the correct audio component at the right time. Add these functions to `WorkoutAudioComponent`:

```
stop() {
  this.ticks.stop();
  this.nextUp.stop();
  this.halfway.stop();
  this.aboutToComplete.stop();
  this.nextUpExercise.stop();
}

resume() {
  this.ticks.start();
  if (this.nextUp.currentTime > 0 && !this.nextUp.playbackComplete)
    this.nextUp.start();
  else if (this.nextUpExercise.currentTime > 0 &&
    !this.nextUpExercise.playbackComplete)
    this.nextUpExercise.start();
  else if (this.halfway.currentTime > 0 &&
    !this.halfway.playbackComplete) this.halfway.start();
  else if (this.aboutToComplete.currentTime > 0 &&
    !this.aboutToComplete.playbackComplete)
    this.aboutToComplete.start();
}

onExerciseProgress(progress: ExerciseProgressEvent) {
  if (progress.runningFor == Math.floor(progress.exercise.duration / 2)
  && progress.exercise.exercise.name != "rest") {
    this.halfway.start();
  }
  else if (progress.timeRemaining == 3) {
    this.aboutToComplete.start();
  }
}
```

```
onExerciseChanged(state: ExerciseChangedEvent) {
  if (state.current.exercise.name == "rest") {
    this.nextupSound = state.next.exercise.nameSound;
    setTimeout(() => this.nextUp.start(), 2000);
    setTimeout(() => this.nextUpExercise.start(), 3000);
  }
}
```

Having trouble writing these functions? They are available in Git branch `checkpoint3.3`.

Next, go ahead and add `WorkoutAudioComponent` to the `declarations` array of `WorkoutRunnerModule`.

There are two new model classes used in the preceding code. Add their declarations to `model.ts`, as follows:

```
export class ExerciseProgressEvent {
  constructor(
  public exercise: ExercisePlan,
  public runningFor: number,
  public timeRemaining: number,
  public workoutTimeRemaining: number) { }
}

export class ExerciseChangedEvent {
  constructor(
  public current: ExercisePlan,
  public next: ExercisePlan) { }
}
```

These are model classes to track progress events. The `WorkoutAudioComponent` implementation consumes this data. How the data is produced becomes clear as we move along.

The `start` and `resume` functions stop and resume audio whenever a workout starts, pauses, or completes. The extra complexity in the resume function it to tackle cases when the workout was paused during next up, when about to complete, or half-way audio playback. We just want to continue from where we left off.

The `onExerciseProgress` function should be called to report the workout progress. It's used to play the halfway audio and about-to-complete audio based on the state of the workout. The parameter passed to it is an object that contains exercise progress data.

The `onExerciseChanged` function should be called when the exercise changes. The input parameter contains the current and next exercise in line, and helps `WorkoutAudioComponent` to decide when to play the next-up exercise audio.

Make note that these two functions are called by the consumer of the component (in this case, `WorkoutRunnerComponent`). We don't invoke them internally.

We touched upon two new concepts in this section: template reference variables and injecting child elements/directives into the parent. It's worth exploring these two concepts in more detail before we continue with the implementation. We'll start with learning more about template reference variables.

## Understanding template reference variables

Template reference variables are created on the view template and are mostly consumed from the view. As you have already learned, these variables can be identified by the `#` prefix used to declare them.

One of the greatest benefits of template variables is that they facilitate cross-component communication at the view template level. Once declared, such variables can be referenced by sibling elements/components and their children. Check out the following snippet:

```
<input #emailId type="email">Email to {{emailId.value}}
<button (click)= "MailUser(emaild.value)">Send</button>
```

This snippet declares a template variable, `emailId`, and then references it in the interpolation and the button `click` expression.

The Angular templating engine assigns the DOM object for `input` (an instance of `HTMLInputElement`) to the `emailId` variable. Since the variable is available across siblings, we use it in a button's `click` expression.

Template variables work with components too. We can easily do this:

```
<trainer-app>
 <workout-runner #runner></workout-runner>
 <button (click)= "runner.start()">Start Workout</button>
</trainer-app>
```

In this case, `runner` has a reference to `WorkoutRunnerComponent` object, and the button is used to start the workout.

The `ref-` prefix is the canonical alternative to `#`. The `#runner` variable can also be declared as `ref-runner`.

## Template variable assignment

There is something interesting about the template variable assignments that we have seen thus far. To recap the examples that we have used:

```
<audio #ticks="MyAudio" loop src="/static/audio/tick10s.mp3"></audio>
<input #emailId type="email">Email to {{emailId.value}}
<workout-runner #runner></workout-runner>
```

What got assigned to the variable depended on where the variable was declared. This is governed by rules in Angular as described here:

- If a directive is present on the element, such as `MyAudioDirective` in the first example shown previously, the directive sets the value. The `MyAudioDirective` directive sets the `ticks` variable to an instance of `MyAudioDirective`.
- If there is no directive present, either the underlying HTML DOM element is assigned or a component object is assigned (as shown in the `email` and `workout-runner` example).

We will be employing this technique to implement workout audio component integration with the workout runner component. This introduction gives us the head start that we need.

The other new concept that we promised to cover is child element/directive injection using the `ViewChild` and `ViewChildren` decorators.

## Using the @ViewChild decorator

The `@ViewChild` decorator informs the Angular DI framework to search for the child component/directive/element in the component tree and inject them into the parent. In the preceding code, the audio element directive (the `MyAudioDirective` class) is injected into the `WorkoutAudioComponent` code.

To establish the context, let's recheck a view fragment from `WorkoutAudioComponent`:

```
<audio #ticks="MyAudio" loop src="/static/audio/tick10s.mp3"></audio>
```

Angular injects the directive (`MyAudioDirective`) into the `WorkoutAudioComponent` property `ticks`. The mapping is done based on the selector passed to the `@ViewChild` decorator.

The selector parameter on `ViewChild` can be a string value, in which case Angular searches for a matching template variable, as before.

Or it can be a type. This is valid:

```
@ViewChild(MyAudioDirective) private ticks: MyAudioDirective;
```

However, it does not work in our case. There are multiple `MyAudioDirective` directives loaded in the `WorkoutAudioComponent` view, one for each of the `<audio>` tags. In such a scenario, the first match is injected. Not very useful. Passing the type selector would have worked if there was only one `<audio>` tag in the view.

Properties decorated with `@ViewChild` are sure to be set before the `ngAfterViewInit` event hook on the component is called. This implies such properties are `null` if accessed inside the constructor.

Like `@ViewChild`, Angular has a decorator to locate multiple child components/directives: `@ViewChildren`.

## The @ViewChildren decorator

`@ViewChildren` works similarly to `@ViewChild` except it's used when the view has multiple child components/directives of one type. Using `@ViewChildren` we can get all the `MyAudioDirective` directive instances in `WorkoutAudioComponent` , as shown here:

```
@ViewChildren(MyAudioDirective) allAudios: QueryList<MyAudioDirective>;
```

Look carefully; `allAudios` is not an array but a custom object, `QueryList<Type>`. `QueryList` is an immutable collection of components/directives that Angular was able to locate. The best thing about this list is that Angular will keep this list in sync with the state of the view. When directives/components get added/removed from the view dynamically this list is updated too. Components/directives generated using `ng-for` are a prime example of this dynamic behavior. Consider the preceding `@ViewChildren` usage and this view template:

```
<audio *ngFor="let clip of clips" src="/static/audio/ "
 +{{clip}}></audio>
```

The number of MyAudioDirective directives created by Angular depends upon the number of clips. When @ViewChildren is used, Angular injects the correct number of MyAudioDirective instances into allAudio property and keeps it in sync when items are added or removed from the clips array.

While the usage of @ViewChildren allows us to get hold of all MyAudioDirective directives, it cannot be used to control the playback. You see, we need to get hold of individual MyAudioDirective instances as the audio playback timing varies. Hence we will stick to the @ViewChild implementation.

Once we get hold of the MyAudioDirective directive attached to each audio element, it is just a matter of playing the audio tracks at the right time.

# Integrating WorkoutAudioComponent

While we have componentized audio playback functionality into WorkoutAudioComponent, it is and always will be tightly coupled to WorkoutRunnerComponent implementation. WorkoutAudioComponent derives its operational intelligence from WorkoutRunnerComponent. Hence the two components need to interact. WorkoutRunnerComponent needs to provide the WorkoutAudioComponent state change data, including when the workout started, exercise progress, workout stopped, paused, and resumed.

One way to achieve this integration would be to use the currently exposed WorkoutAudioComponent API (stop, resume, and other functions) from WorkoutRunnerComponent.

Something can be done by injecting WorkoutAudioComponent into WorkoutRunnerComponent as we did earlier when we injected MyAudioDirective into WorkoutAudioComponent. Check out this code:

```
@ViewChild(WorkoutAudioComponent) workoutAudioPlayer:
WorkoutAudioComponent;
```

The WorkoutAudioComponent functions then can be invoked from WorkoutRunnerComponent from different places in the code. For example, this is how pause would change:

```
pause() {
  clearInterval(this.exerciseTrackingInterval);
  this.workoutPaused = true;
  this.workoutAudioPlay.stop();
```

```
}
```

And to play the next up audio, we would need to change parts of the `startExerciseTimeTracking` function:

```
this.startExercise(next);
this.workoutAudioPlayer.onExerciseChanged(
new ExerciseChangedEvent(next, this.getNextExercise()));
```

This is a perfectly viable option where `WorkoutAudioComponent` becomes a dumb component controlled by `WorkoutRunnerComponent`. The only problem with this solution is that it adds some noise to the `WorkoutRunnerComponent` implementation. `WorkoutRunnerComponent` now needs to manage audio playback too.

There is an alternative, however. `WorkoutRunnerComponent` can expose events that are triggered during different times of workout execution, that is, events such as workout started, exercise started, and workout paused, and so on and so forth. Another advantage of having `WorkoutRunnerComponent` expose events is that it allows us to integrate other components with `WorkoutRunnerComponent` in future, using the same events.

# Exposing WorkoutRunnerComponent events

Angular allows components and directives to expose custom events using the `EventEmitter` class. Add these event declarations to `WorkoutRunnerComponent` at the end of the variable declaration section:

```
workoutPaused: boolean;
@Output() exercisePaused: EventEmitter<number> =
new EventEmitter<number>();
@Output() exerciseResumed: EventEmitter<number> =
new EventEmitter<number>();
@Output() exerciseProgress:EventEmitter<ExerciseProgressEvent> =
new EventEmitter<ExerciseProgressEvent>();
@Output() exerciseChanged: EventEmitter<ExerciseChangedEvent> =
new EventEmitter<ExerciseChangedEvent>();
@Output() workoutStarted: EventEmitter<WorkoutPlan> =
new EventEmitter<WorkoutPlan>();
@Output() workoutComplete: EventEmitter<WorkoutPlan> =
new EventEmitter<WorkoutPlan>();
```

The names of the events are self-explanatory, and within our `WorkoutRunnerComponent` implementation we need to raise these at the appropriate times.

Remember to add `ExerciseProgressEvent` and import to the model module that is already declared on top. And add the `Output` and `EventEmitter` imports to `@angular/core`.

Let's try to understand the role of the `@Output` decorator and the `EventEmitter` class.

## The @Output decorator

We covered a decent amount of Angular eventing capabilities in `Chapter 2`, Building Our First App – 7 Minute Workout. Specifically, we learned how we can consume any event on a component, directive, or DOM element using the `bracketed ()` syntax. How about raising our own events?

In Angular, we can create and raise our own events, events that signify something noteworthy has happened in our component/directive. Using the `@Output` decorator and the `EventEmitter` class we can define and raise custom events.

Now would be a good time to refresh what we learned about events, by revisiting the *Eventing subsection* in the *Angular Event binding infrastructure* section from `Chapter 2`, *Building Our First App – 7 Minute Workout*.

Remember, it is through events that components can communicate with the outside world. When we declare:

```
@Output() exercisePaused: EventEmitter<number> =
new EventEmitter<number>();
```

It signifies that `WorkoutRunnerComponent` exposes an event `exercisePaused` (raised when the workout is paused).

To subscribe to this event, we do:

```
<workout-runner (exercisePaused)="onExercisePaused($event)">
</workout-runner>
```

This looks absolutely similar to how we did DOM event subscription in the workout runner template:

```
<div id="pause-overlay" (click)="pauseResumeToggle()"
(window:keyup)="onKeyPressed($event)">
```

The `@Output` decorator instructs Angular to make this event available for template binding. You can create an event without the `@Output` decorator, but such an event cannot be referenced in html.

The `@Output` decorated can also take a parameter, signifying the name of the event. If not provided, the decorator uses the property name:
`@Output("workoutPaused") exercisePaused: EventEmitter<number> = new EventEmitter<number>();`
This declares an event `workoutPaused` instead of `exercisePaused`.

Like any decorator, the `@Output` decorator too is there just to provide metadata for the Angular framework to work with. The real heavy lifting is done by the `EventEmitter` class.

# Eventing with EventEmitter

Angular embraces Reactive Programming (also dubbed as **Rx**-style programming) to support asynchronous operations and events. If you are hearing this term for the first time or don't have much idea about what reactive programming is, you're not alone.

Reactive programming is all about programming against asynchronous data streams. Such a stream is nothing but a sequence of ongoing events ordered based on the time they occur. We can imagine a stream as a pipe generating data (in some manner) and pushing it to one or more subscribers. Since these events are captured asynchronously by subscribers, they are called asynchronous data streams.

The data can be anything, ranging from browser/DOM element events, to user input to loading remote data loaded using AJAX. With *Rx* style, we consume this data uniformly.

In the Rx world, there are Observers and Observables, a concept derived from the very popular Observer design pattern. Observables are streams that emit data. Observers on the other hand subscribe to these events.

The `EventEmitter` class in Angular is primarily responsible for providing eventing support. It acts both as an observer and observable. We can fire events on it and it can be used to listen to events too.

There are two functions available on `EventEmitter` that are of interest to us:

- `emit`: As the name suggests, use this function to raise events. It takes a single argument that is the event data. `emit` is the observable side.
- `subscribe`: Use this function to subscribe to the events raised by `EventEmitter`. `subscribe` is the observer side.

Let's do some event publishing and subscriptions to understand how the preceding functions work.

## Raising events from WorkoutRunnerComponent

Look at the `EventEmitter` declaration. These have been declared with the `type` parameter. The `type` parameter signifies the type of data emitted.

Let's add the event implementation to `workout-runner.component.ts` in chronological order starting from the top and moving down.

Add this statement to the end of the `start` function:

```
this.workoutStarted.emit(this.workoutPlan);
```

We use the `EventEmitter`'s `emit` function to raise an event `workoutStarted`, with the current workout plan as an argument.

To `pause`, add this line to raise the `exercisePaused` event:

```
this.exercisePaused.emit(this.currentExerciseIndex);
```

To `resume`, add the following line:

```
this.exerciseResumed.emit(this.currentExerciseIndex);
```

Each time, we pass the current exercise index as an argument to `emit` when raising `exercisePaused` and `exerciseResumed` events.

Inside the `startExerciseTimeTracking` function, add the highlighted code after the call to `startExercise`:

```
this.startExercise(next);
this.exerciseChanged.emit(new ExerciseChangedEvent(
                          next, this.getNextExercise()));
```

The argument passed contains the exercise that is going to start (`next`) and the next exercise in line (`this.getNextExercise()`).

To the same function, add the highlighted code:

```
this.tracker.endTracking(true);
this.workoutComplete.emit(this.workoutPlan);
this.router.navigate(['finish']);
```

The event is raised when the workout is completed.

Lastly in the same function, we raise an event that communicates workout progress. Add this statement:

```
--this.workoutTimeRemaining;
this.exerciseProgress.emit(new ExerciseProgressEvent(
this.currentExercise, this.exerciseRunningDuration,
this.currentExercise.duration -this.exerciseRunningDuration,
this.workoutTimeRemaining));
```

That completes our eventing implementation.

As you may have guessed, `WorkoutAudioComponent` now needs to consume these events. The challenge here is how to organize these components so that they can communicate with each other with the minimum dependency on each other.

# Component communication patterns

As the implementation stands now, we have a basic `WorkoutAudioComponent` implementation and have augmented `WorkoutRunnerComponent` by exposing workout lifecycle events. These two just need to talk to each other now.

The first obvious option that comes to mind is to add the `WorkoutAudioComponent` declaration to the `WorkoutRunnerComponent` view. `WorkoutAudioComponent` hence becomes a child component of `WorkoutRunnerComponent`. However, in such a setup, communication between them becomes a bit clunky. Remember, events are a mechanism for a component to communicate with the outside world.

If the parent needs to communicate with its children, it can do this by:

- **Property binding**: The parent component can set up a property binding on the child component to push data to the child component. For example:

  ```
  <workout-audio [stopped]="workoutPaused"></workout-audio>
  ```

  Property binding in this case works fine. When the workout is paused, the audio too is stopped. But not all scenarios can be handled using property bindings. Playing the next exercise audio or halfway audio requires a bit more control.

- **Calling functions on child components**: The parent component can also call functions on the child component if it can get hold of the child component. We have already seen how to achieve this using the `@ViewChild` and `@ViewChildren` decorators in the `WorkoutAudioComponent` implementation.

There is one more not-so-good option where the parent component instance can be injected into the child component. In such a case, the child component can call parent component functions or set up internal event handlers to parent events.

We are going to try this approach and then scrap the implementation for a better one! A lot of learning can be derived from the not-so-optimal solution we plan to implement.

## Injecting a parent component into a child component

Add the `WorkoutAudioComponent` to the `WorkoutRunnerComponent` view just before the last closing div:

```
<workout-audio></workout-audio>
```

Next we need to inject `WorkoutRunnerComponent` into `WorkoutAudioComponent`.

Open `workout-audio.component.ts` and add the following declaration and a constructor:

```
private subscriptions: Array<any>;

constructor( @Inject(forwardRef(() => WorkoutRunnerComponent))
private runner: WorkoutRunnerComponent) {
  this.subscriptions = [
  this.runner.exercisePaused.subscribe((exercise: ExercisePlan) =>
  this.stop()),
  this.runner.workoutComplete.subscribe((exercise: ExercisePlan) =>
  this.stop()),
  this.runner.exerciseResumed.subscribe((exercise: ExercisePlan) =>
  this.resume()),
  this.runner.exerciseProgress.subscribe((
  progress: ExerciseProgressEvent) =>
  this.onExerciseProgress(progress)),
  this.runner.exerciseChanged.subscribe((
  state: ExerciseChangedEvent) =>
  this.onExerciseChanged(state))]
}
```

And remember to add these imports:

```
import {Component, ViewChild, Inject, forwardRef} from '@angular/core';
import {WorkoutRunnerComponent} from '../workout-runner.component'
```

We have used some trickery in the construction injection. If we directly try to inject `WorkoutRunnerComponent` into `WorkoutAudioComponent` it fails with Angular complaining of not being able to find all dependency. Read the code and think carefully; there is a subtle dependency cycle issue lurking. `WorkoutRunnerComponent` already is dependent on `WorkoutAudioComponent` as we have referenced `WorkoutAudioComponent` in the `WorkoutRunnerComponent` view. Now by injecting `WorkoutRunnerComponent` in `WorkoutAudioComponent`, we have created a dependency cycle.

Cyclic dependencies are challenging for any DI framework. When creating a component with a cyclic dependency, the framework has to somehow resolve the cycle. In the preceding example, we resolve the circular dependency issue by using an `@Inject` decorator and passing in the token created using the `forwardRef()` global framework function.

Inside the constructor, we attach a handler to the events, using the `EventEmitter` subscribe function. The arrow function passed to `subscribe` is called whenever the event occurs with a specific event argument. We collect all the subscriptions into an array `subscription`. This array comes in handy when we unsubscribe, which we need to, to avoid memory leaks.

The `EventEmmiter`'s subscription (`subscribe` function) takes three arguments:

```
subscribe(generatorOrNext?: any, error?: any, complete?: any) : any
```

- The first argument is a callback, which is invoked whenever an event is emitted
- The second argument is an error callback function, invoked when the observable (the part that is generating events) errors out
- The final argument takes a callback function that is called when the observable is done publishing events

We have done enough to make audio integration work. Run the app and start the workout. Except for the ticking audio, all other audio clips play at the right time. You may have to wait for some time to hear to the other audio clips. What is the problem?

As it turns out, we never started the ticking audio clip at the start of the workout. We can fix it by either setting the `autoplay` attribute on the `ticks` audio element or using the `component lifecycle events` to trigger the ticking sound. Let's take the second approach.

## Using component lifecycle events

We performed the `MyAudioDirective` injection in `WorkoutAudioComponent`:

```
@ViewChild('ticks') private ticks: MyAudioDirective;
```

This will not be available unit the component's view has been initialized:

We can verify it by accessing the `ticks` variable inside the constructor; it will be null. Angular has still not done its magic and we need to wait for the `WorkoutAudioComponent`'s children to be initialized.

The component's lifecycle hooks can help us here. The `AfterViewInit` event hook is called once the component's view has been initialized and hence is a safe place from which to access the component's child directives/elements. Let's do it quickly.

Update the `WorkoutAudioComponent` by adding the interface implementation, and the necessary imports, as highlighted:

```
import {..., AfterViewInit} from '@angular/core';
...
  export class WorkoutAudioComponent implements AfterViewInit {
    ngAfterViewInit() {
      this.ticks.start();
    }
```

Go ahead and test the app. The app has come to life with full-fledged audio feedback. Nice!

While everything looks fine and dandy on the surface, there is a memory leak in the application now. If in the middle of the workout we navigate away from the workout page (to the start or finish page) and again return to the workout page, multiple audio clips play at random times.

It seems that the `WorkoutRunnerComponent` is not getting destroyed on route navigation, and due to this, none of the child components are destroyed, including `WorkoutAudioComponent`. The net result? A new `WorkoutRunnerComponent` is being created every time we navigate to the workout page, but is never removed from the memory on navigating away.

The primary reason for this memory leak is the event handlers we have added in `WorkoutAudioComponent`. We need to unsubscribe from these events when the audio component unloads, or else the `WorkoutRunnerComponent` reference will not be dereferenced.

Another component lifecycle event comes to our rescue here: `OnDestroy`! Add this implementation to the `WorkoutAudioComponent` class:

```
ngOnDestroy() {
  this.subscriptions.forEach((s) => s.unsubscribe());
}
```

Also remember to add references to the `OnDestroy` event interface as we did for `AfterViewInit`.

Hope the `subscription` array that we created during event subscription makes sense now. One-shot unsubscribe!

This audio integration is now complete. While this approach is not an awfully bad way of integrating the two components, we can do better. Child components referring to the parent component seems to be undesirable.

What if `WorkoutRunnerComponent` and `WorkoutAudioComponent` are organized as sibling components?

Before proceeding, delete the code that we have added to `workout-audio.component.ts` from the *Injecting a Parent Component into a Child Component* section onwards.

## Sibling component interaction using events and template variables

If `WorkoutAudioComponent` and `WorkoutRunnerComponent` become siblings, we can make good use of Angular eventing and template reference variables. Confused? Well, to start with, this is how the components should be laid out:

```
<workout-runner></workout-runner>
<workout-audio></workout-audio>
```

Does it ring any bells? Starting from this template, can you guess how the final template HTML would look? Think about it before you proceed further.

Still struggling? As soon as we make them sibling components, the power of the Angular templating engine comes to the fore. The following template code is enough to integrate `WorkoutRunnerComponent` and `WorkoutAudioComponent`:

```
<workout-runner (exercisePaused)="wa.stop()"
(exerciseResumed)="wa.resume()" (exerciseProgress)=
"wa.onExerciseProgress($event)" (exerciseChanged)=
"wa.onExerciseChanged($event)" (workoutComplete)="wa.stop()"
(workoutStarted)="wa.resume()">
</workout-runner>
<workout-audio #wa></workout-audio>
```

The `WorkoutAudioComponent`'s template variable `wa` is being manipulated on `WorkoutRunnerComponent's` template. Quite elegant! We still need to solve the biggest puzzle in this approach: Where does the preceding code go? Remember, `WorkoutRunnerComponent` is loaded as part of route loading. Nowhere in the code have we had a statement like:

```
<workout-runner></workout-runner>
```

We need to re-organize the component tree and bring in a container component that can host `WorkoutRunnerComponent` and `WorkoutAudioComponent`. The router then loads this container component instead of `WorkoutRunnerComponent`. Let's do it.

Create a folder called `workout-container` inside the `workout-runner` folder and add two new files, `workout-container.component.ts` and `workout-container.html`.

Copy the HTML code with the events described before to the template file, and add the following declaration to `workout-container.component.ts`:

```
import {Component, Input} from '@angular/core';
@Component({
  selector: 'workout-container',
  templateUrl: '/src/components/workout-runner/workout-container.html'
})
export class WorkoutContainerComponent { }
```

The workout container component is ready. Add it to the `declarations` section in `workout-runner.module.ts` and export it instead of `WorkoutRunnerComponent`.

Next, we just need to rewire the routing setup. Open `app.routes.ts`. Change the route for the workout page and add the necessary import:

```
import {WorkoutContainerComponent} from '../workout-runner/
        workout-container/workout-container.component';
..
{ path: '/workout', component: WorkoutContainerComponent },
```

And we have a working audio integration that is clear, concise, and pleasing to the eyes too!

It's time now to wrap up the chapter, but not before addressing the video player dialog glitch introduced in the earlier sections. The workout does not stop/pause when the video player dialog is open.

We are not going to detail the fix here, and urge the readers to give it a try without consulting the `checkpoint3.3` code.

Here is an obvious hint. Use the eventing infrastructure!

And another one: raise events from `VideoPlayerComponent`, one for each playback started and ended.

And the last hint: the `open` function on the dialog service (`Modal`) returns a promise, which is resolved when the dialog is closed.

If you are having a problem with running the code, look at Git branch `checkpoint3.3` for a working version of what we have done thus far. Or if you are not using Git, download the snapshot of `checkpoint3.2` (a ZIP file) from `http://bit.ly/ng2be-checkpoint3-3`. Refer to the `README.md` file in the `trainer` folder when setting up the snapshot for the first time.

# Summary

Bit by bit, piece by piece, we have added a number of enhancements to the *7 Minute Workout* that are imperative for any professional app. There is still scope for new features and improvements, but the core app works just fine.

We started the chapter by exploring the Single Page Application (SPA) capabilities of Angular. Here we learned about basic Angular routing, setting up routes, using route configuration, generating links using the `RouterLink` directive, and using the Angular `Router` and `Location` services to perform navigation.

From the app perspective, we added start, finish, and workout pages to *7 Minute Workout*.

We then built a workout history tracker service that was used to track historical workout executions. During this process, we learned about Angular **Dependency Injection (DI)** in depth. How a dependency is registered, what a dependency token is, and how dependencies are singleton in nature are what we learned in this section. We also learned about injectors and how hierarchical injectors affect dependency probing.

Lastly, we touched upon an important topic: cross component communication, primarily using Angular Eventing. We detailed how to create custom events using the `@Output` decorator and `EventEmitter`.

The `@ViewChild` and `@ViewChildren` decorator that we touched upon in this chapter helped us understand how a parent can get hold of a child component for use. Angular DI also allows injecting a parent component into a child.

We concluded the chapter by building a `WorkoutAudioComponent` and highlighted how sibling component communication can happen using Angular events and template variables.

What next? We are going to build a new app, *Personal Trainer*. This app will allow us to build our own custom workouts. Once we can create our own workout, we are going to morph the *7 Minute Workout* app into a generic *Workout Runner* app that can run workouts that we build using *Personal Trainer*.

For the next chapter, we showcase AngularJS form capabilities while we build a UI that allows us to create, update, and view our own custom workouts/exercises.

# 4
# Personal Trainer

Building Personal Trainer The 7 Minute Workout app has been an excellent opportunity for us to learn about Angular. Working through the app, we have covered a number of Angular constructs. Still, there are areas such as Angular form support and client server communication that remain unexplored. This is partially due to the fact that 7 Minute Workout from a functional standpoint had limited touchpoints with the end user. Interactions were limited to starting, stopping, and pausing the workout. Also, the app neither consumes, nor produces any data (except workout history).

In this chapter, we plan to delve deeper into one of the two aforementioned areas, Angular form support. Keeping up with the health and fitness theme (no pun intended), we plan to build a *Personal Trainer* app. The new app will be an extension to *7 Minute Workout*, allowing us to build our own customized workout plans that are not limited to the *7 Minute Workout* plans that we already have.

This chapter is dedicated to understanding Angular forms and how to put them to use as we build out our *Personal Trainer* app.

The topics that we will cover in this chapter are as follows:

- **Defining Personal Trainer requirements**: Since we are building a new app in this chapter, we start with defining the app requirements.
- **Defining the Personal Trainer model**: Any app design starts with defining its model. We define the model for *Personal Trainer*, which is similar to the *7 Minute Workout* app built earlier.
- **Defining the Personal Trainer layout and navigation**: We define the layout, navigation patterns, and views for the new app. We also set up a navigation system that is integrated with Angular routes and the main view.

- **Adding support pages**: Before we focus on the form capability and build a Workout component, we build some supporting components for workout and exercise listing.
- **Defining the Workout Builder component structure**: We lay out the Workout Builder components that we will use to manage workouts.
- **Building forms**: We make extensive use of HTML forms and input elements to create custom workouts. In the process, we will learn more about Angular Forms. The concepts that we cover include:
- **Form types**: The two types of form that can be built with Angular are template-driven and model-driven. We're working with both template-driven and model-driven forms in this chapter.
- **ngModel**: This provides two-way databinding for template driven forms and allows us to track changes and validate form input.
- **Model Driven Form Controls**: These include the form builder, form control, form group, and form array. These are used to construct forms programmatically.
- **Data formatting**: These are the CSS classes that permit us to style our feedback to the user.
- **Input validation**: We will learn about the validation capabilities of Angular forms.

# The Personal Trainer app – the problem scope

The *7 Minute Workout* app is good, but what if we could create an app that allows us to build more such workout routines customized to our fitness level and intensity requirements? With this flexibility, we can build any type of workout whether it is 7 minutes, 8 minutes, 15 minutes, or any other variations. The opportunities are limitless.

With this premise, let's embark on the journey of building our own *Personal Trainer* app that helps us to create and manage training/workout plans according to our specific needs. Let's start with defining the requirements for the app.

The new *Personal Trainer* app will now encompass the existing *7 Minute Workout* app. The component that supports workout creation will be referred to as `Workout Builder`. The *7 Minute Workout* app itself will also be referred to as `Workout Runner`. In the coming chapters, we will fix *Workout Runner*, allowing it to run any workout created using *Workout Builder*.

# Personal Trainer requirements

Based on the notion of managing workouts and exercises, these are some of the requirements that our *Personal Trainer* app should fulfill:

- The ability to list all available workouts.
- The ability to create and edit a workout. While creating and editing a workout, it should have:
    - The ability to add workout attributes including name, title, description, and rest duration
    - The ability to add/remove multiple exercises for workouts
    - The ability to order exercises in the workout
    - The ability to save workout data
- The ability to list all available exercises.
- The ability to create and edit an exercise. While creating and editing an exercise, it should have:
    - The ability to add exercise attributes such as name, title, description, and procedure
    - The ability to add pictures for the exercise
    - The ability to add related videos for the exercise
    - The ability to add audio clues for the exercise

All the requirements seem to be self-explanatory; hence, let's start with the design of the application. As customary, we first need to think about the model that can support these requirements.

# The Personal Trainer model

No surprises here! The Personal Trainer model itself was defined when we created the *7 Minute Workout* app. The two central concepts of workout and exercise hold good for *Personal Trainer* too.

The only problem with the existing workout model is that it is in the directory for workout-runner. This means that in order to use it, we will have to import it from that directory. It makes more sense to move the model into the `service` folder so that it is clear that it can be used across features.

Let's understand how we can share the model across the application.

# Sharing the workout model

We are going to share the workout model as a **service**. As mentioned in the previous chapter, a **service** has no particular definition. Basically, it is a class that holds functionality that might be useful to have in multiple locations throughout our application. Since it will be used in both Workout Runner and Workout Builder, our workout model fits that definition. There is not much ceremony involved in making our model into a **service** – so let's get started by doing that.

First, download the base version of the new *Personal Trainer* app from `checkpoint4.1` in the GitHub repository for the book.

The code is available on GitHub `https://github.com/chandermani/angular2byexample` for everyone to download. Checkpoints are implemented as branches in GitHub. The branch to download is as follows: **GitHub Branch: checkpoint4.1**. If you are not using Git, download the snapshot of Checkpoint 4.1 (a ZIP file) from the following GitHub location: `https://github.com/chandermani/angular2byexample/archive/checkpoint4.1.zip`. Refer to the `README.md` file in the `trainer` folder when setting up the snapshot for the first time.

This code has the complete *7 Minute Workout (Workout Runner)* app. We have added some more content to support the new *Personal Trainer* app. Some of the relevant updates are:

- Adding the new `WorkoutBuilder` feature. This feature contains implementations pertaining to *Personal Trainer*.
- Updating the layout and styles of the app.
- Adding some components and HTML templates with placeholder content for *Personal Trainer* in the `workout-builder` folder under `trainer/src/components`.
- Defining a new route to the `WorkoutBuilder` feature. We will cover setting up this route within the app in the coming section.

Let's get back to defining the model.

# The model as a service

In the last chapter, we dedicated a complete section to learning about Angular services, and one thing we found out was that services are useful for sharing data across controllers and other Angular constructs. We essentially do not have any data but a blueprint that describes the shape of the data. The plan, hence, is to use services to expose the model structure. Open the `model.ts` file present in the `services` folder under `app`.

The `model.ts` file has been moved into the `services` folder as the service is shared across the *Workout Builder* and *Workout Runner* apps. Note: in `checkpoint4.1` we have updated the import statements in `workout-runner.component.ts`, `workout-audio.component.ts` and `workout-audio0.component.ts` in the `workout-runner` folder under `trainer/src/components` to reflect this change.

In `Chapter 2`, *Building Our First App – 7 Minute Workout*, we reviewed the class definitions in the model file: `Exercise`, `ExercisePlan`, and `WorkoutPlan`. As we then mentioned, these three classes constitute our base model. We will now start using this base model in our new app.

That's all on the model design front. The next thing we are going to do is define the structure for the new app.

# The Personal Trainer layout

The skeleton structure of *Personal Trainer* looks like this:

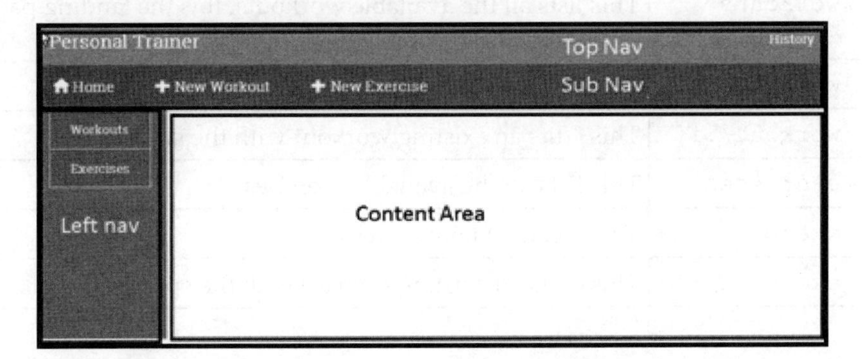

This has the following components:

- **Top Nav**: This contains the app branding title and history link.
- **Sub Nav**: This has navigation elements that change based on the active component.
- **Left Nav**: This contains elements that are dependent upon the active component.
- **Content Area**: This is where the main view for our component will display. This is where most of the action happens. We will create/edit exercises and workouts and show a list of exercises and workouts here.

Look at the source code files; there is a new folder `workout-builder` under `trainer/src/components`. It has files for each component that we described previously, with some placeholder content. We will be building these components as we go along in this chapter.

However, we first need to link up these components within the app. This requires us to define the navigation patterns for the Workout Builder app and accordingly define the app routes.

# Personal Trainer navigation with routes

The navigation pattern that we plan to use for the app is the list-detail pattern. We will create list pages for the exercises and workouts available in the app. Clicking on any list item takes us to the detailed view for the item where we can perform all CRUD operations (create/read/update/delete). The following routes adhere to this pattern:

| **Route** | **Description** |
|---|---|
| `/builder` | This just redirects to **builder/workouts**. |
| `/builder/workouts` | This lists all the available workouts. It is the landing page for *Workout Builder*. |
| `/builder/workout/new` | This creates a new workout. |
| `/builder/workout/:id` | This edits an existing workout with the specific ID. |
| `/builder/exercises` | This lists all the available exercises. |
| `/builder/exercise/new` | This creates a new exercise. |
| `/builder/exercise/:id` | This edits an existing exercise with the specific ID. |

# Getting started

At this point, if you look at the route configuration in `app.routes.ts` in the `src/components/app` folder, you will find one new route definition – `builder`:

```
export const routes: Routes = [
  ...
{ path: 'builder', component: WorkoutBuilderComponent },
  ...
];
```

And if you run the application, you will see that the start screen shows another link: **Create a Workout:**

Behind the scenes, we have added another router link for this link into `start.html`:

```
<a [routerLink]="['/builder']">
  <span>Create a Workout</span>
  <span class="glyphicon glyphicon-plus"></span>
</a>
```

And if you click on this link, you will be taken to the following view:

Again behind the scenes, we have added a `WorkoutBuilderComponent` in the `trainer/src/components/workout-builder` folder with the following related template in `workout-builder.component.html`:

```
<div class="row">
    <div class="col-sm-3"></div>
    <div class="col-sm-6">
        <h1 class="text-center">Workout Builder</h1>
    </div>
    <div class="col-sm-3"></div>
</div>
```

And this view is displayed on the screen under the header using the router outlet in our `app.component.ts` view template:

```
<div class="container body-content app-container">
    <router-outlet></router-outlet>
</div>`
```

We have wrapped this component in a new module named `workout-builder.module.ts`:

```
import { NgModule }      from '@angular/core';
import { CommonModule } from '@angular/common';

import { WorkoutBuilderComponent } from "./workout-builder.component";

@NgModule({
    imports: [CommonModule],
    declarations: [
        WorkoutBuilderComponent,
    ],
    exports: [WorkoutBuilderComponent],
})
export class WorkoutBuilderModule { }
```

The only thing that might look different here from the other modules that we have created is that we are importing `CommonModule` instead of `BrowserModule`. This avoids our importing the whole of `BrowserModule` a second time, which would generate an error when we get to implementing lazy loading for this module.

Finally, we have added an import for this module to `app.module.ts`:

```
...
@NgModule({
  imports: [
  ...
    WorkoutBuilderModule],
  ...
```

So nothing surprising here. These are the basic component building and routing patterns that we introduced in the previous chapters. Following these patterns, we should now begin to think about adding the additional navigation outlined previously for our new feature. However, before we jump into doing that, there are a couple of things we need to consider.

First, if we start adding our routes to the `app.routes.ts` file then the number of routes stored there will grow. These new routes for *Workout Builder* will also be intermixed with the routes for *Workout Runner*. While the number of routes we are now adding might seem insignificant, over time this could get to be a maintenance problem.

Second, we need to take into consideration that our application now consists of two features – *Workout Runner* and *Workout Builder*. We should be thinking about ways to separate these features within our application so that they can be developed independently of each other.

Put differently, we want **loose-coupling** between the features that we build. Using this pattern allows us to swap out a feature within our application without affecting the other features. For example, somewhere down the line we may want to convert the *Workout Runner* into a mobile app but leave the *Workout Builder* intact as a web-based application.

Going back to the first chapter, we emphasized that this ability to separate our components from each other is one of the key advantages of using the **component design pattern** that Angular implements. Fortunately, Angular's router gives us the ability to separate out our routing into logically organized **routing configurations** that closely match the features in our application.

In order to accomplish this separation, Angular allows us to use **child routing** where we can isolate the routing for each of our features. In this chapter, we will use **child routing** to separate out the routing for *Workout Builder*.

# Introducing child routes to Workout Builder

Angular supports our goal of isolating the routing for our new *Workout Builder* by providing us with the ability to create a hierarchy of router components within our application. We currently have just one router component, which is in the root component of our application. But Angular allows us to add what are called **child router components** under our root component. This means that one feature can be ignorant of the routes the other is using and each is free to adapt its routes in response to changes within that feature.

Getting back to our application, we can use **child routing** in Angular to match the routing for the two features of our application with the code that will be using them. So in our application we can structure the routing into the following routing hierarchy for our *Workout Builder* (at this point, we are leaving the *Workout Runner* as-is to show the before and after comparison):

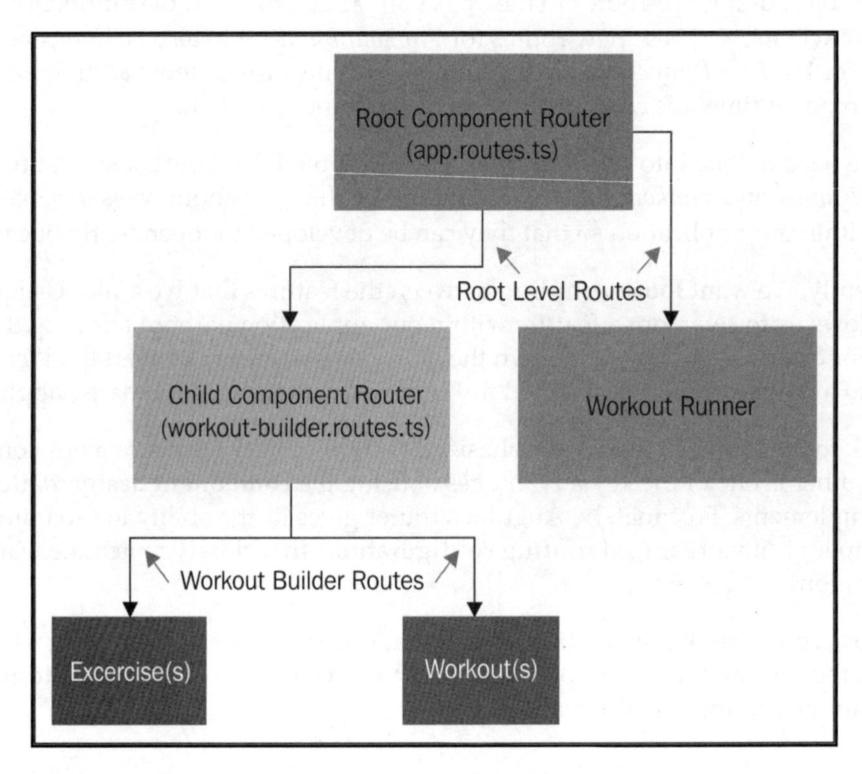


With this approach, we can create a logical separation of our routes by feature and make them easier to manage and maintain.

So let's get started by adding child routing to our application.

From this point on in this section, we'll be adding to the code that we downloaded earlier for this chapter. If you want to see the complete code for this next section, you can download it from checkpoint 4.2 in the GitHub repository. If you want to work along with us as we build out the code for this section, still be sure to add the changes in `app.css` in the `trainer/static/css` folder that are part of this checkpoint since we won't be discussing them here. Also be sure and add the files for exercise(s) and workout(s) from the `trainer/src/components/workout-builder` folder in the repository. At this stage these are just stub files, which we will implement later in this chapter. However, you will need these stub files here in order to implement navigation for the *Workout Builder* module.
The code is available for everyone to download on GitHub at `https://github.com/chandermani/angular2byexample`. Checkpoints are implemented as branches in GitHub. The branch to download is as follows: **GitHub Branch: checkpoint4.2.** . If you are not using Git, download the snapshot of Checkpoint 4.2 (a ZIP file) from the following GitHub location: `https://github.com/chandermani/angular2byexample/archive/checkpoint4.2.zip`. Refer to the `README.md` file in the `trainer` folder when setting up the snapshot for the first time.

## Adding the child routing component

In the `workout-builder` directory, add a new TypeScript file named `workout-builder.routes.ts` with the following imports:

```
import { ModuleWithProviders } from '@angular/core';
import { Routes, RouterModule } from '@angular/router';

import { WorkoutBuilderComponent}  from "./workout-builder.component";
import { ExerciseComponent} from './exercise/exercise.component';
import { ExercisesComponent} from './exercises/exercises.component';
import { WorkoutComponent} from './workout/workout.component';
import { WorkoutsComponent} from './workouts/workouts.component';
```

As you can see, we are importing the components we just mentioned; they will be part of our *Workout Builder* (exercise, exercises, workout, and workouts). Along with those imports, we are also importing `ModuleWithProviders` from the Angular core module and `Routes` and `RouterModule` from the Angular router module. These imports will give us the ability to add and export child routes.

Then add the following route configuration to the file:

```
export const workoutBuilderRoutes: Routes = [
    {
        path: 'builder',
        component: WorkoutBuilderComponent,
        children: [
            {path:'', pathMatch: 'full', redirectTo: 'workouts'},
            {path:'workouts', component: WorkoutsComponent },
            {path:'workout/new',  component: WorkoutComponent },
            {path:'workout/:id', component: WorkoutComponent },
            {path:'exercises', component: ExercisesComponent},
            {path:'exercise/new', component: ExerciseComponent },
            {path:'exercise/:id', component: ExerciseComponent }
        ]
    }
];
```

The first configuration, path: 'builder', sets the base URL for the child routes so that each of the child routes prepends it. The next configuration identifies the `WorkoutBuilder` component as the feature area root component for the child components in this file. This means it will be the component in which each of the child components is displayed using `router-outlet`. The final configuration is a list of one or more children that defines the routing for the child components.

One thing to note here is that we have set up `Workouts` as the default for the child routes with the following configuration:

```
{path:'', pathMatch: 'full', redirectTo: 'workouts'},
```

This configuration indicates that if someone navigates to `builder`, they will be redirected to the `builder/workouts` route. The `pathMatch: 'full'` setting means that the match will only be made if the path after workout/builder is an empty string. This prevents the redirection from happening if the routes are something else such as `workout/builder/exercises` or any of the other routes we have configured within this file.

Finally, add the following export statement:

```
export const workoutBuilderRouting: ModuleWithProviders =
RouterModule.forChild(workoutBuilderRoutes);
```

This export registers our child routes with the router and is very similar to the one in app.routes.ts, with one difference: instead of `RouterModule.forRoot` we are using `RouterModule.forChild`. The reason for the difference may seem self-explanatory: we are creating child routes, not the routes in the root of the application, and this is how we signify that. Under the hood, however, there is a significant difference. This is because we cannot have more than one router service active in our application. `forRoot` creates the router service but `forChild` does not.

# Updating the WorkoutBuilder component

We next need to update the `WorkoutBuilder` component to support our new child routes. To do so, change the `@Component` decorator for Workout Builder to:

- Remove the `selector`
- Replace the reference to a `templateUrl` with a `template` reference
- Add a `<sub-nav>` custom element to the template
- Add a `<router-outlet>` tag to the template
- The decorator should now look like the following:

```
@Component({
    template: `<div class="navbar navbar-default
               navbar-fixed-top second-top-nav">
           <sub-nav></sub-nav>
        </div>
        <div class="container body-content app-container">
           <router-outlet></router-outlet>
        </div>`
})
```

We are removing the selector because the `WorkoutBuilderComponent` will not be embedded in the application root: `app.component.ts`. Instead, it will be reached from `app.routes.ts` through routing. And while it will handle incoming routing requests from `app.routes.ts`, it will in turn be routing them to the other components contained in the Workout Builder feature.

And those components will display their views using the `<router-outlet>` tag that we have just added to the `WorkoutBuilder` template. Given that the template for `WorkoutBuilderComponent` will be simple, we are also swapping out the `templateUrl` for an inline template.

Typically, for a component's view we recommend using a `templateUrl` that points to a separate HTML template file. This is especially true when you anticipate that the view will involve more than a few lines of HTML. In that situation, it is much easier to work with a view inside its own HTML file. The separate HTML file allows you to use an HTML editor with features such as color coding and tag completion. In contrast, an inline template is simply a string within a TypeScript file and the editor does not give you these benefits.

We are also adding a `<sub-nav>` element that will be used to create a secondary top-level menu for navigating within the *Workout Builder* feature. We'll discuss that a little later in this chapter.

## Updating the Workout Builder module

Now let's update `WorkoutBuilderModule`. This is going to involve some significant changes since we will be turning this module into a feature module. So this module will import all the components that we will be using for building a workout. We won't cover all those imports here, but be sure to add them from `workout-builder.ts` in the `trainer/src/components/workout-builder` folder in `checkpoint 4.2` of the GitHub repository.

One import that is worth highlighting is the following:

```
import { workoutBuilderRouting } from './workout-builder.routes';
```

It imports the child routing that we just set up.

Now let's update the `@NgModule` decorator to the following:

```
@NgModule({
    imports: [
        CommonModule,
        workoutBuilderRouting
    ],
    declarations: [
       WorkoutBuilderComponent,
       WorkoutComponent,
```

```
        WorkoutsComponent,
        ExerciseComponent,
        ExercisesComponent,
        SubNavComponent,
        LeftNavExercisesComponent,
        LeftNavMainComponent
        ],
    exports: [WorkoutBuilderComponent],
})
```

## Updating app.routes

One last step: return to `app.routes.ts` and remove the import of the `WorkoutBuilderComponent` and its route from that file.

## Putting it all together

From the previous chapter, we already know how to set up root routing for our application. But now what we have instead of root routing is area or feature routing that contains child routes. We have been able to achieve the separation of concerns we discussed earlier, so that all the routes related to the *Workout Builder* are now separately contained in their own routing configuration. This means that we can manage all the routing for *Workout Builder* in the `WorkoutBuilderRoutes` component without affecting other parts of the application.

We can see how the router combines the routes in `app.routes.ts` with the default route in `workout-builder.routes.ts`, if we now navigate from the start page to the Workout Builder.

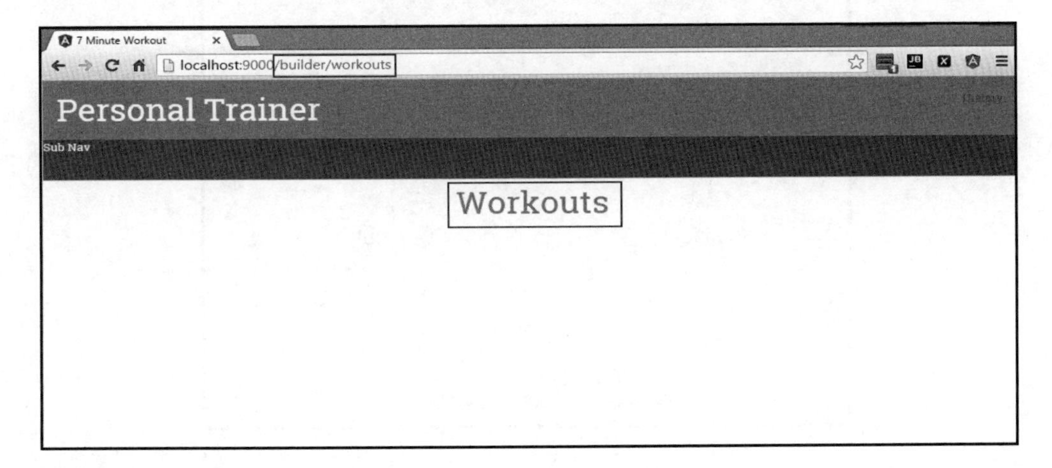

If we look at the URL in the browser, it is `/builder/workouts`. You'll recall that the router link on the start page is `['/builder']`. So how did the router takes us to this location?

It does it this way: when the link is clicked, the Angular router first looks to `app.routes.ts` for the `builder` path because that file contains the configuration for the root routes in our application. The router does not find that path because we have removed it from the routes in that file.

However, the `WorkoutBuilderComponent` has been imported into our `AppModule` and that component in turn imports `workoutBuilderRouting` from `workout-builder-routes.ts`.The latter file contains the child routes that we just configured. The router finds that `builder` is the parent route in that file and so it uses that route. It also finds the default setting that redirects to the child path `workouts` in the event that the `builder` path ends with an empty string, which it does in this case.

If you look at the screen, you will see it is displaying the view for `Workouts` (and not as previously *Workout Builder*). This means that the router has successfully routed the request to `WorkoutsComponent`, which is the component for the default route in the child route configuration that we set up in `workout-builder.routes.ts`.

This process of route resolution is illustrated here:

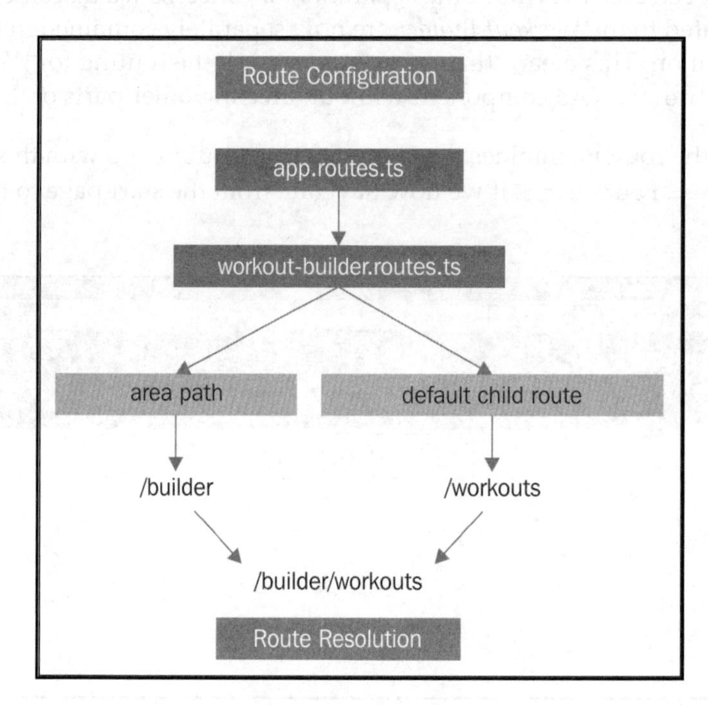

One final thought on child routing. When you look at our child routing component, `workout-builder.component.ts`, you will see that it has no references to its `parent` component, `app.component.ts` (as we mentioned earlier, the `<selector>` tag has been removed, so the `Workout Builder` component is not being embedded in the root component). This means that we have successfully encapsulated `Workout Builder` (and all of the components that it imports) in a way that will allow us to move all of it elsewhere in the application or even into a new application.

Now it's time for us to move on to converting our routing for the Workout Builder to use lazy loading and building out its navigation menus. If you want to see the completed code for this next section, you can download it from the companion codebase in `checkpoint 4.3`. Again, if you are working along with us as we build the application, be sure and update the `app.css` file, which we are not discussing here.

The code is also available for everyone to download on GitHub at `https://github.com/chandermani/angular2byexample`. Checkpoints are implemented as branches in GitHub. The branch to download is as follows: **GitHub Branch: checkpoint4.3** (folder – `trainer`). If you are not using Git, download the snapshot of Checkpoint 4.3 (a ZIP file) from the following GitHub location: `https://github.com/chandermani/angular2byexample/archive/checkpoint4.3.zip`. Refer to the `README.md` file in the `trainer` folder when setting up the snapshot for the first time.

# Lazy loading of routes

When we roll out our application, we expect that our users will be accessing the **Workout Runner** every day (and we know that this will be the case for you!). But we anticipate that they will only occasionally be using the *Workout Builder* to construct their exercises and workout plans. It would therefore be nice if we could avoid the overhead of loading the **Workout Builder** when our users are just doing their exercises in the **Workout Runner**. Instead, we would prefer to load *Workout Builder* only on demand when a user wants to add to or update their exercises and workout plans. This approach is called **lazy loading**.

Under the hood, Angular uses SystemJS to accomplish this lazy loading. It allows us to employ an asynchronous approach when loading our modules. This means that we can load just what is required to get the application started and then load other modules as we need them.

So in our *Personal Trainer*, we want to change the application so that it only loads the **Workout Builder** on demand. And the Angular router allows us to do just that using lazy loading.

But before we get started implementing lazy loading, let's take a look at our current application and how it is loading our modules. With the developer tools open in the **Sources** tab, start up the application; when the start page appears in your browser, you see that all the files in the application have loaded, including both the *Workout Runner* and *Workout Builder* files:

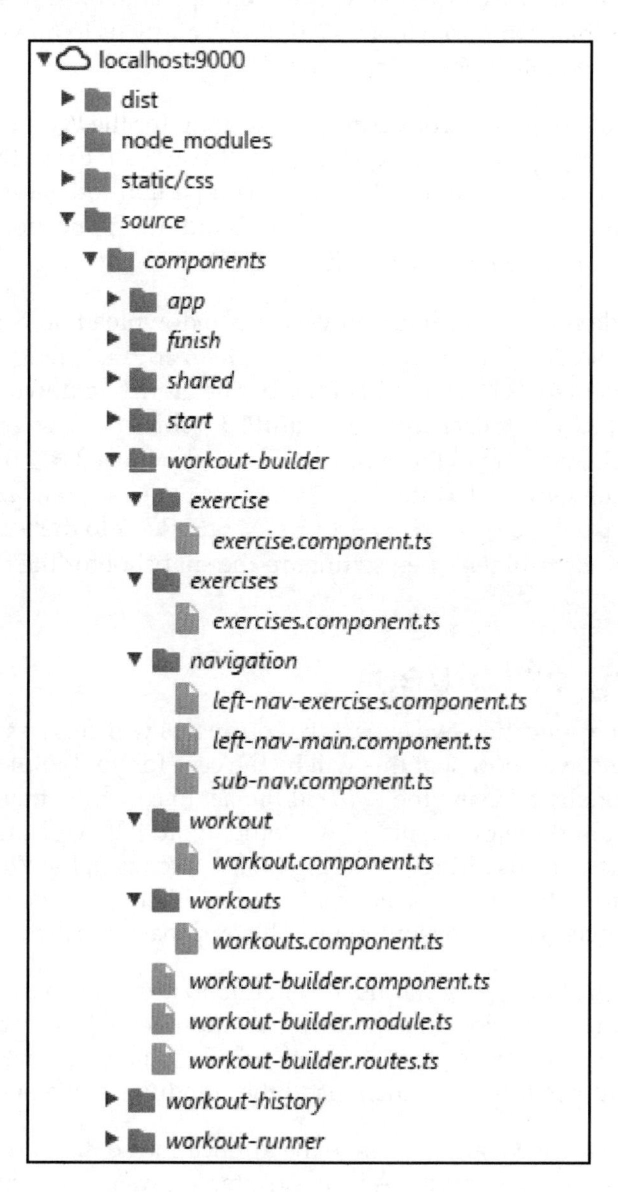

So even though we may just want to use the *Workout Runner*, we have to load the *Workout Builder* as well. In a way, this makes sense if you think of our application as a **Single Page Application** (**SPA**). In order to avoid round trips to the server, an SPA will typically load all the resources that will be needed to use the application when it is first started up by a user. But in our case, the important point is that we do not **need** the *Workout Builder* when the application is first loaded. Instead, we would like to load those resources only when the user decides that they want to add or change a workout or exercise.

So let's get started with making that happen.

First, modify the `app.routes.ts` to add the following separate route configuration for our `workoutBuilderRoutes`:

```
const workoutBuilderRoutes: Routes = [
  {
    path: 'builder',
    loadChildren: 'dist/components/workout-builder/workout-
builder.module#WorkoutBuilderModule'
  }
];
```

Notice that the `loadChildren` property is:

```
component:  file path + # + component name
```

This configuration provides the information that will be needed to load and instantiate our component. Pay particular attention to the file path; it points to the location of our code in the `dist` folder when it is deployed as a JavaScript file and not to the folder where the TypeScript version of that file is located.

Next, update the Routes configuration to add the following:

```
export const routes: Routes = [
    { path: 'start', component: StartComponent },
    { path: 'workout', component: WorkoutContainerCompnent },
    { path: 'finish', component: FinishComponent },
    { path: 'history', component: WorkoutHistoryComponent },
    ...workoutBuilderRoutes,
    { path: '**', redirectTo: '/start' }
];
```

You will notice that we have added a reference to WorkoutBuilderRoutes, which we just configured and prepended with three dots. With the three dots we are using the ES2015 spread operator to insert an array of routes – specifically the routes for the WorkoutBuilder feature. These routes will be contained in WorkoutBuilderRoutes and will be maintained separately and apart from the routes in the root of our application. Finally , remove the import of the WorkoutBuilderComponent from this file.

Next go back to workout-builder.routes.ts and change the path property to an empty string:

```
export const workoutBuilderRoutes: Routes = [
    {
        path: '',
. . .
    }
];
```

We are making this change because we are now setting the path ('builder') to the WorkoutBuilderRoutes in the new configuration for them that we added in app.routes.ts.

Finally go back to the app-module.ts and remove the WorkoutBuilderModule import in the @NgModule configuration in that file. What this means is that instead of loading the **Workout Builder** feature when the application first starts, we only load it when a user accesses the route to *Workout Builder*.

Let's go back and run the application again keeping the **Sources** tab open in the Chrome developer tools. When the application begins and the start page loads, only the files related to the *Workout Runner* appear and not those related to the *Workout Builder*, as shown here:

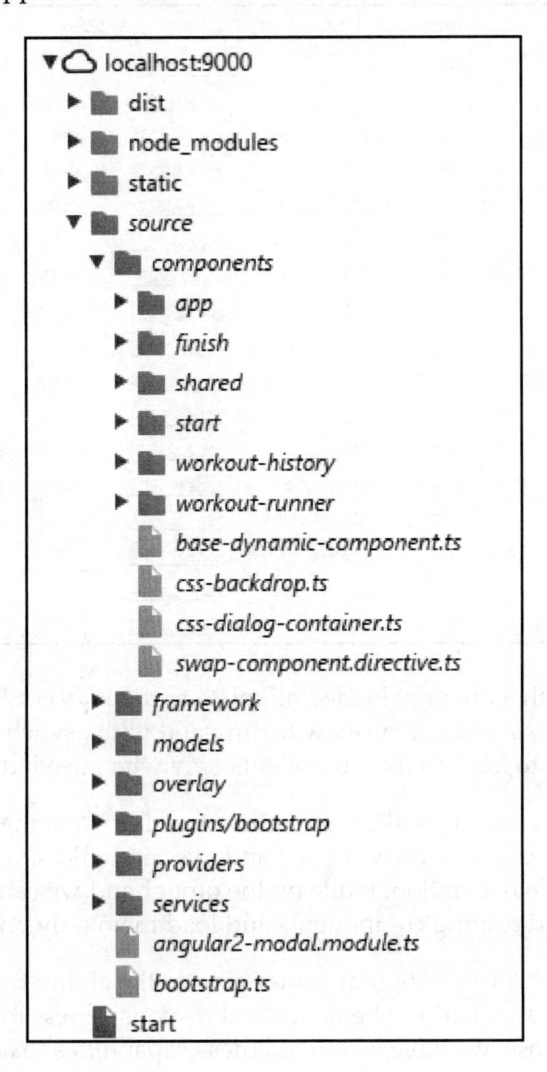

Then, if we clear the **Network** tab and click on the **Create a Workout link**, we'll see only the files related to the *Workout Builder* load:

| Name | Method▲ | Status | Type | Initiator | Size | Time | Timeline |
|---|---|---|---|---|---|---|---|
| workout-builder.module.js | GET | 200 | xhr | zone.js:1274 | 7.0 KB | 3 ms | |
| exercise.component.js | GET | 200 | xhr | zone.js:1274 | 2.8 KB | 5 ms | |
| exercises.component.js | GET | 200 | xhr | zone.js:1274 | 2.8 KB | 5 ms | |
| left-nav-exercises.component.js | GET | 200 | xhr | zone.js:1274 | 2.9 KB | 6 ms | |
| left-nav-main.component.js | GET | 200 | xhr | zone.js:1274 | 2.9 KB | 6 ms | |
| sub-nav.component.js | GET | 200 | xhr | zone.js:1274 | 2.8 KB | 6 ms | |
| workout-builder.component.js | GET | 200 | xhr | zone.js:1274 | 3.2 KB | 6 ms | |
| workout.component.js | GET | 200 | xhr | zone.js:1274 | 2.8 KB | 4 ms | |
| workouts.component.js | GET | 200 | xhr | zone.js:1274 | 2.8 KB | 4 ms | |
| workout-builder.routes.js | GET | 200 | xhr | zone.js:1274 | 5.3 KB | 4 ms | |
| workout.component.html | GET | 200 | xhr | zone.js:1274 | 461 B | 6 ms | |
| workouts.component.html | GET | 200 | xhr | zone.js:1274 | 452 B | 5 ms | |
| exercise.component.html | GET | 200 | xhr | zone.js:1274 | 449 B | 5 ms | |
| exercises.component.html | GET | 200 | xhr | zone.js:1274 | 453 B | 3 ms | |
| sub-nav.component.html | GET | 200 | xhr | zone.js:1274 | 728 B | 3 ms | |
| left-nav-exercises.component.html | GET | 200 | xhr | zone.js:1274 | 335 B | 3 ms | |
| left-nav-main.component.html | GET | 200 | xhr | zone.js:1274 | 571 B | 3 ms | |

As we can see, the files that are now loaded all relate to the *Workout Builder*. This means that we have achieved encapsulation of our new feature and with asynchronous routing we are able to use lazy loading to load all its components only when needed.

Child and asynchronous routing make it straightforward to implement applications that allow us to *have our cake and eat it too*. On one hand, we can build single-page applications with powerful client-side navigation, while on the other hand we can also encapsulate features in separate child routing components and load them only on demand.

This power and flexibility of the Angular router give us the ability to meet user expectations by closely mapping our application's behavior and responsiveness to the ways they will use the application. In this case, we have leveraged these capabilities to achieve what we set out to do: immediately load *Workout Runner* so that our users can get to work on their exercises right away, but avoid the overhead of loading *Workout Builder* and instead only serve it when a user wants to build a workout.

Now that we have the routing configuration in place in the *Workout Builder*, we will turn our attention to creating the sub-level and left navigation; this will enable us to use this routing. The next sections cover implementing this navigation.

# Integrating sub- and side-level navigation

The basic idea around integrating sub- and side-level navigation into the app is to provide context-aware sub-views that change based on the active view. For example, when we are on a list page as opposed to editing an item, we may want to show different elements in the navigation. An e-commerce site is a great example of this. Imagine Amazon's search result page and product detail page. As the context changes from a list of products to a specific product, the navigation elements that are loaded also change.

## Sub-level navigation

We'll start by adding sub-level navigation to the *Workout Builder*. We have already imported our `SubNavComponent` into the *Workout Builder*. But currently it is just displaying placeholder content:

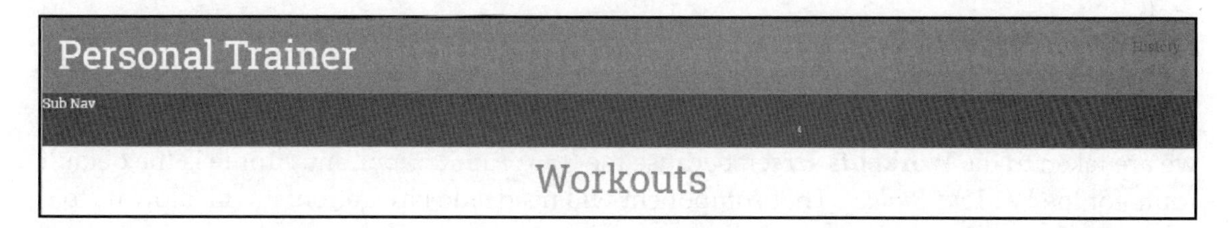


We'll now replace that content with three router links: **Home**, **New Workout**, and **New Exercise**.

Open the `sub-nav.component.html` file and change the HTML in it to the following:

```
<div>
    <a [routerLink]="['/builder/workouts']" class="btn btn-primary">
        <span class="glyphicon glyphicon-home"></span> Home
    </a>
    <a [routerLink]="['/builder/workout/new']" class="btn btn-primary">
        <span class="glyphicon glyphicon-plus"></span> New Workout
    </a>
    <a [routerLink]="['/builder/exercise/new']" class="btn btn-primary">
        <span class="glyphicon glyphicon-plus"></span> New Exercise
    </a>
</div>
```

Now rerun the application and you will see the three navigation links. If we click on the **New Exercise** link button, we will be routed to `ExerciseComponent` and its view will appear in the **Router Outlet** in the *Workout Builder* view:

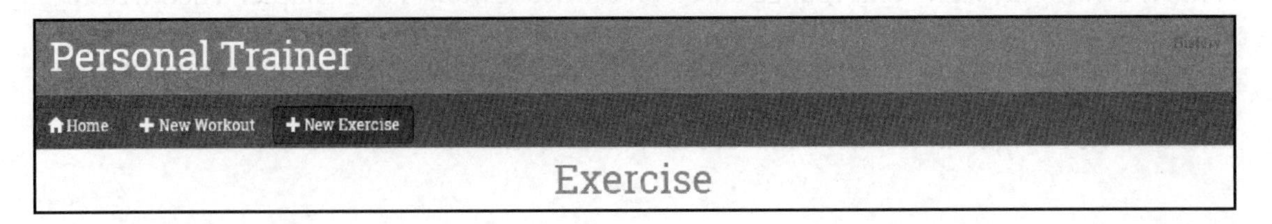


The **New Workout** link button will work in a similar fashion; when clicked on, it will take the user to the `WorkoutComponent` and display its view in the router outlet. Clicking on the **Home** link button will return the user to the `WorkoutsComponent` and view.

## Side navigation

Side level navigation within the *Workout Builder* will vary depending on the child component that we navigate to. For instance, when we first navigate to the *Workout Builder*, we are taken to the **Workouts** screen because the `WorkoutsComponent`'s route is the default route for the *Workout Builder*. That component will need side navigation; it will allow us to select to view a list of workouts or a list of exercises.

The component-based nature of Angular gives us an easy way to implement these context-sensitive menus. We can define new components for each of the menus and then import them into the components that need them. In this case, we have three components that will need side menus: **Workouts**, **Exercises**, and **Workout**. The first two of these components can actually use the same menu so we really only need two side menu components: `LeftNavMainComponent`, which will be like the preceding menu and will be used by the `Exercises` and `Workouts` components, and `LeftNavExercisesComponent`, which will contain a list of existing exercises and will be used by the `Workouts` component.

We already have files for the two menu components, including template files, and have imported them into `WorkoutBuilderModule`. We will now integrate these into the components that need them.

First, modify the `workouts.component.html` template to add the selector for the menu:

```
div class="container-fluid">
    <div id="content-container" class="row">
        <left-nav-main></left-nav-main>
        <h1 class="text-center">Workouts</h1>
    </div>
</div>
```

Then replace the placeholder text in the `left-nav-main.component.html` with the navigation links to `WorkoutsComponent` and `ExercisesComponent`:

```
<div class="col-sm-2 left-nav-bar">
    <div class="list-group">
        <a [routerLink]="['/builder/workouts']" class="list-group-item
list-group-item-info">Workouts</a>
        <a [routerLink]="['/builder/exercises']" class="list-group-item
list-group-item-info">Exercises</a>
    </div>
</div>
```

Run the application and you should see the following:

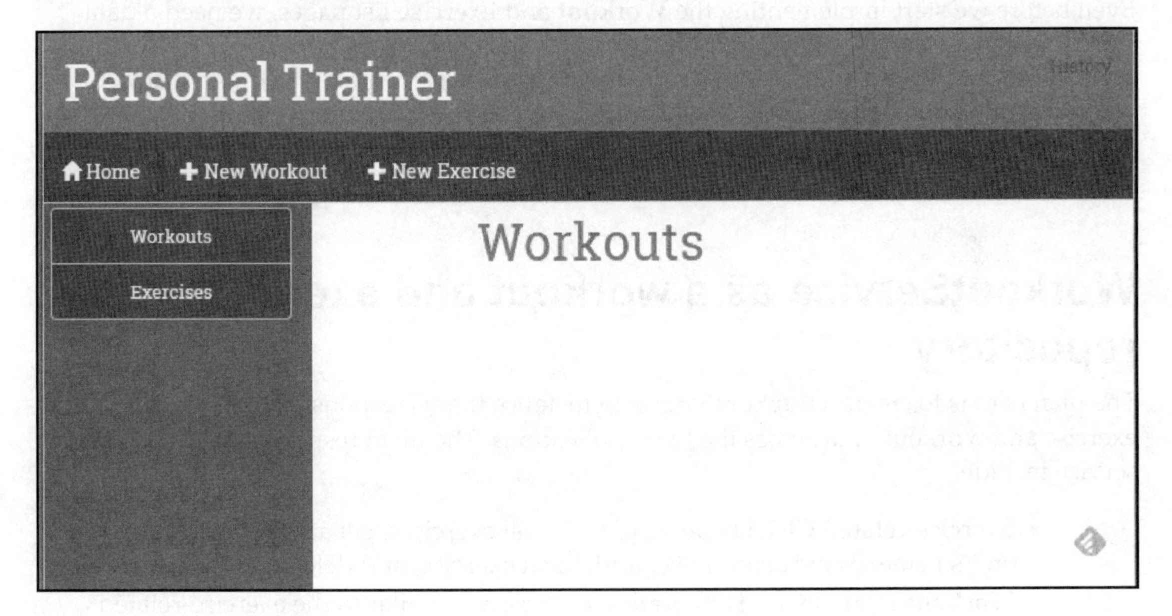

Follow the exact same steps to complete the side menu for the `Exercises` component.

We won't show the code for these two menus here but you can find it in the `workout-builder/exercises` folder under `trainer/src/components` in `checkpoint 4.3` of the GitHub repository.

For the menu for the **Workouts** screen, the steps are the same except that you should change `left-nav-exercises.component.html` to the following:

```
<div class="col-sm-2 left-nav-bar">
    <h3>Exercises</h3>
</div>
```

We will use this template as the starting point for building out a list of exercises that will appear on the left-hand side of the screen and can be selected for inclusion in a workout.

# Implementing workout and exercise lists

Even before we start implementing the **Workout** and **Exercise** list pages, we need a data store for exercise and workout data. The current plan is to have an in-memory data store and expose it using an Angular service. In `Chapter 5`, *Supporting Server Data Persistence*, where we talk about server interaction, we will move this data to a server store for long-term persistence. For now, the in-memory store will suffice. Let's add the store implementation.

## WorkoutService as a workout and exercise repository

The plan here is to create a `WorkoutService` instance that is responsible for exposing the exercise and workout data across the two applications. The main responsibilities of the service include:

- **Exercise-related CRUD operations**: Get all exercises, get a specific exercise based on its name, create an exercise, update an exercise, and delete it
- **Workout-related CRUD operations**: These are similar to the exercise-related operations, but targeted toward the workout entity

The code is available to download on GitHub at `https://github.com/chandermani/angular2byexample`. The branch to download is as follows: **GitHub Branch: checkpoint4.4** (folder – `trainer`). If you are not using Git, download the snapshot of Checkpoint 4.4 (a ZIP file) from the following GitHub location: `https://github.com/chandermani/angular2byexample/archive/checkpoint4.4.zip`. Refer to the `README.md` file in the `trainer` folder when setting up the snapshot for the first time. Again, if you are working along with us as we build the application, be sure to update the `app.css` file, which we are not discussing here. Because some of the files in this section are rather long, rather than showing the code here, we are also going to suggest at times that you simply copy the files into your solution.

Locate `workout-service.ts` in the `trainer/src/services` folder. The code in that file should look like the following except for the implementation of the two methods `setupInitialExercises` and `setupInitialWorkouts`, which we have left out because of their length:

```
import {Injectable} from '@angular/core';
import {ExercisePlan} from './model';
import {WorkoutPlan} from './model';
import {Exercise} from "./model";

@Injectable()
export class WorkoutService {
    workouts: Array<WorkoutPlan> = [];
    exercises: Array<Exercise> = [];

    constructor() {
        this.setupInitialExercises();
        this.setupInitialWorkouts();
    }

    getExercises(){
        return this.exercises;
    }

    getWorkouts(){
        return this.workouts;
    }
    setupInitialExercises(){
     // implementation of in-memory store.
    }

    setupInitialWorkouts(){
```

```
        // implementation of in-memory store.
    }
}}
```

As we have mentioned before, the implementation of an Angular service is straightforward. Here we are declaring a class with the name `WorkoutService` and decorating it with `@Injectable` to support injecting it throughout our application. In the class definition, we first create two arrays: one for `Workouts` and one for `Exercises`. These arrays are of types `WorkoutPlan` and `Exercise` respectively, and we therefore need to import `WorkoutPlan` and `Exericse` from `model.ts` to get the type definitions for them.

The constructor calls two methods to set up the **Workouts** and **Services List**. At the moment, we are just using an in-memory store that populates these lists with data.

The two methods, `getExercises` and `getWorkouts`, as the names suggest, return a list of exercises and workouts respectively. Since we plan to use the in-memory store to store workout and exercise data, the `Workouts` and `Exercises` arrays store this data. As we go along, we will be adding more functions to the service.

There is one more thing we need to do to make the service available to be injected throughout our application.

Open `services.module.ts` in the same folder, and then import `WorkoutService` and add it as a provider:

```
---- other imports ----
import { WorkoutService } from "./workout-service";

@NgModule({
    imports: [],
    declarations: [],
    providers: [
        LocalStorage,
        WorkoutHistoryTracker,
        WorkoutService],
})
```

This registers `WorkoutService` as a provider with Angular's **Dependency Injection** framework.

Time to add the components for the workout and exercise lists!

# Workout and exercise list components

First, open the `workouts.component.ts` file in the `trainer/src/components/workout-builder/workouts` folder and update the imports as follows:

```
import { Component, OnInit} from '@angular/core';
import { Router } from '@angular/router';

import { WorkoutPlan } from "../../../services/model";
import { WorkoutService } from "../../../services/workout-service";
```

This new code imports `OnInit` from the Angular core as well as `Router`, `WorkoutService` and the `WorkoutPlan` type.

Next replace the class definition with the following code:

```
export class WorkoutsComponent implements OnInit {
    workoutList:Array<WorkoutPlan> = [];

    constructor(
        public router:Router,
        public workoutService:WorkoutService) {}

    ngOnInit() {
        this.workoutList = this.workoutService.getWorkouts();
    }

    onSelect(workout: WorkoutPlan) {
        this.router.navigate( ['./builder/workout', workout.name] );
    }
}
```

This code adds a constructor into which we are injecting the `WorkoutService` and the `Router`. The `ngOnInit` method then calls the `getWorkouts` method on the `WorkoutService` and populates a `workoutList` array with a list of `WorkoutPlans` returned from that method call. We'll use that `workoutList` array to populate the list of Workout plans that will display in the `Workouts` component's view.

You'll notice that we are putting the code for calling `WorkoutService` into an `ngOnInit` method. We want to avoid placing this code in the constructor. Eventually, we will be replacing the in-memory store that this service uses with a call to an external data store and we do not want the instantiation of our component to be affected by this call. Adding these method calls to the constructor would also complicate testing the component.

To avoid such unintended side effects, we instead place the code in the `ngOnInit` method. This method implements one of Angular's lifecycle hooks, `OnInit`, which Angular calls after creating an instance of the service. This way we rely on Angular to call this method in a predictable way that does not affect the instantiation of the component.

Next we'll make almost identical changes to the `Exercises` component. As with the `Workouts` component, this code injects the workout service into our component. This time, we then use the workout service to retrieve the exercises.

Because it so similar to what we just showed you for the `Workouts` component, we won't show that code here. Just add it from the `workout-builder/exercises` folder `checkpoint 4.4`.

# Workout and exercise list views

Now we need to implement the list views that have so far been empty!

In this section, we will be updating the code from `checkpoint 4.3` with what is found in `checkpoint 4.4`. So if you are coding along with us, simply follow the steps laid out in this section. If you want to see the finished code, then just copy the files from `checkpoint 4.4` into your solution.

## Workouts list views

To get the view working, open `workouts.component.html` and add the following markup:

```
<div class="container-fluid">
  <div id="content-container" class="row">
    <left-nav-main></left-nav-main>
    <h1 class="text-center">Workouts</h1>
    <div class="workouts-container">
      <div *ngFor="let workout of workoutList|orderBy:'title'"
       class="workout tile" (click)="onSelect(workout)">
```

```
            <div class="title">{{workout.title}}</div>
            <div class="stats">
              <span class="duration" >
              <span class="glyphicon glyphicon-time"></span> -
              {{workout.totalWorkoutDuration()|secondsToTime}}</span>
              <span class="length pull-right" >
              <span class="glyphicon glyphicon-th-list">
              </span> - {{workout.exercises.length}}</span>
                  </div>
             </div>
          </div>
      </div>
  </div>
```

We are using one of the Angular core directives, `ngFor`, to loop through the list of workouts and display them in a list on the page. We add the `*` sign in front of `ngFor` to identify it as an Angular directive. Using a `let` statement we assign `workout` as a local variable that we use to iterate through the worklist and identify the values to be displayed for each workout (for example, `workout.title`). We then use one of our custom pipes, `orderBy`, to display a list of workouts in alphabetical order by title. We are also using another custom pipe, `secondsToTime`, to format the time displayed for the total workout duration.

If you are coding along with us, you will need to move the `secondsToTime` pipe into the shared folder and include it in the `SharedModule`. Then add `SharedModule` to `WorkoutBuilderModule` as an additional import. That change has already been made in `checkpoint 4.4` in the GitHub repository.

Finally, we bind the click event to the following `onSelect` method that we add to our component:

```
onSelect(workout: WorkoutPlan) {
    this.router.navigate( ['./builder/workout', workout.name] );
}
```

This sets up navigation to the workout detail page. This navigation happens when we double-click on an item in the workout list. The selected workout name is passed as part of the route/URL to the workout detail page.

Go ahead and refresh the builder page (`/builder/workouts`); one workout is listed, the 7 Minute Workout. Click on the tile for that workout. You'll be taken to the **Workout** screen and the workout name `7MinWorkout` will appear at the end of the URL:

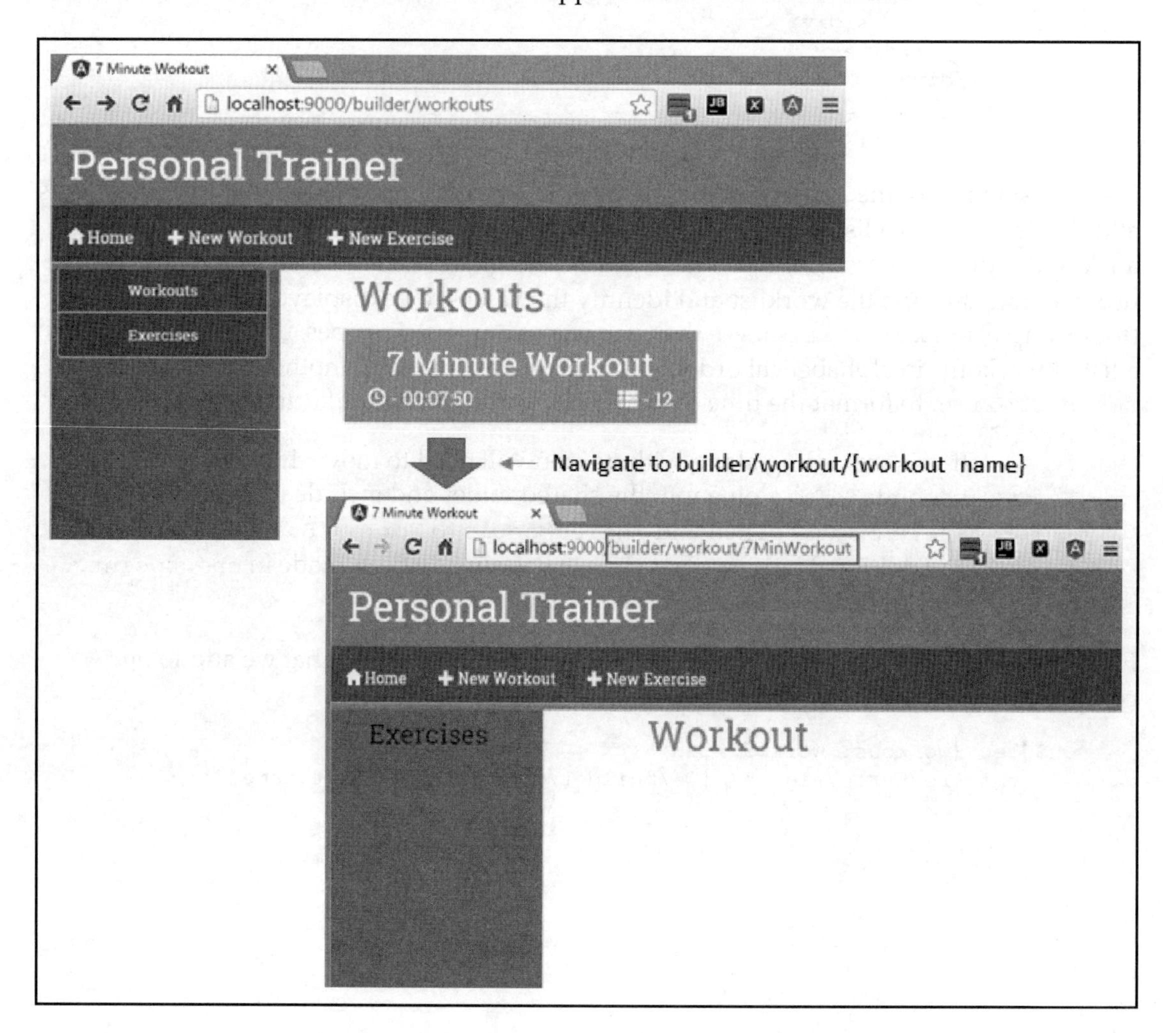

## Exercises list views

We are going to follow the same approach for the `Exercises` list view as we did for the `Workouts` list view. Except that in this case, we will actually be implementing two views: one for the `Exercises` component (which will display in the main content area when a user navigates to that component) and one for the `LeftNavExercisesComponent` exercises context menu (that will display when user navigates to the `Workouts` component to create/edit a workout).

For the `Exercises` component, we will follow an approach that is almost identical to what we did to display a list of workouts in the `Workouts` component. So we won't show that code here. Just add the files for `exercise.conponent.ts` and `exercise.component.html` from `checkpoint 4.4`.

When you are done copying the files, click on the **Exercises** link in the left navigation to load the 12 exercises that you have already configured in `WorkoutService`.

As with the `Workouts` list, this sets up the navigation to the exercise detail page. Double-clicking on an item in the exercises list takes us to the exercise detail page. The selected exercise name is passed as part of the route/URL to the exercise detail page.

In the final list view, we will add a list of exercises that will display in the left context menu for the *Workout Builder* screen. This view is loaded in the left navigation when we create or edit a workout. Using Angular's component-based approach, we will update the `leftNavExercisesComponent` and its related view to provide this functionality. Again we won't show that code here. Just add the files for `left-nav-exercises.component.ts` and `left-nav-exercises.component.html` from the `trainer/src/components/navigation` folder in `checkpoint 4.4`.

Once you are done copying those files, click on the **New Workout** button on the sub-navigation menu in the *Workout Builder* and you will now see a list of exercises, displayed in the left navigation menu exercises that we have already configured in `WorkoutService`.

Time to add the ability to load, save, and update exercise/workout data!

# Building a workout

The core functionality *Personal Trainer* provides centers around workout and exercise building. Everything is there to support these two functions. In this section, we focus on building and editing workouts using Angular.

The `WorkoutPlan` model has already been defined, so we are aware of the elements that constitute a workout. The *Workout Builder* page facilitates user input and lets us build/persist workout data.

Once complete, the *Workout Builder* page will look like this:

The page has a left navigation that lists all the exercises that can be added to the workout. Clicking on the arrow icon on the right adds the exercise to the end of the workout.

The center area is designated for workout building. It consists of exercise tiles laid out in order from top to bottom and a form that allows the user to provide other details about the workout such as name, title, description, and rest duration.

This page operates in two modes:

- **Create/New**: This mode is used for creating a new workout. The URL is `#/builder/workout/new`.
- **Edit**: This mode is used for editing the existing workout. The URL is `#/builder/workout/:id`, where `:id` maps to the name of the workout.

With this understanding of the page elements and layout, it's time to build each of these elements. We will start with left nav (navigation).

# Finishing left nav

At the end of the previous section, we updated the left navigation view for the `Workout` component to show a list of exercises. Our intention was to let the user click on an arrownext to an exercise to add it to the workout. At the time, we deferred implementing the `addExercise` method in the `LeftNavExercisesComponent` that was bound to that click event. Now we will go ahead and do that.

We have a couple of options here. The `LeftNavExercisesComponent` is a child component of the `WorkoutComponent`, so we can implement child/parent inter-component communication to accomplish that. We covered this technique in the previous chapter while working on *7 Minute Workout*.

However, adding an exercise to the workout is part of a larger process of building the workout and using child/parent inter-component communication would make the implementation of the `AddExercise` method differ from the other functionality that we will be adding going forward.

For this reason, it makes more sense to follow another approach for sharing data, one that we can use consistently throughout the process of building a workout. That approach involves using a service. As we get into adding the other functionality for creating an actual workout, such as save/update logic, and implementing the other relevant components, the benefits of going down the service route will become increasingly clear.

So we introduce a new service into the picture: the WorkoutBuilderService. The ultimate aim of the WorkoutBuilderService service is to co-ordinate between the WorkoutService (which retrieves and persists the workout) and the components (such as LeftNavExercisesComponent and others we will add later), while the workout is being built, hence reducing the amount of code in WorkoutComponent to the bare minimum.

# Adding WorkoutBuilderService

WorkoutBuilderService tracks the state of the workout being built. It:

- Tracks the current workout
- Creates a new workout
- Loads the existing workout
- Saves the workout

Copy workout-builder-service.ts from the workout-builder/builder-services folder under trainer/src/components in checkpoint 4.5.

The code is also available for everyone to download on GitHub at https://github.com/chandermani/angular2byexample. Checkpoints are implemented as branches in GitHub. The branch to download is as follows: **GitHub Branch: checkpoint4.5** (folder – trainer). If you are not using Git, download the snapshot of Checkpoint 4.5 (a ZIP file) from the following GitHub location: https://github.com/chandermani/angular2byexample/archive/checkpoint4.5.zip. Refer to the README.md file in the trainer folder when setting up the snapshot for the first time. Again, if you are working along with us as we build the application, be sure to update the app.css file, which we are not discussing here.

While we normally make services available application-wide, WorkoutBuilderService will only be used in the *Workout Builder* feature. Therefore, instead of registering it with the providers in AppModule, we will register it in the provider array of WorkoutBuilderModule as follows (after adding it as an import at the top of the file):

```
providers: [
    WorkoutBuilderService,
    . . .
]
```

Adding it as a provider here means that it will only be loaded when the *Workout Builder* feature is being accessed and it cannot be reached outside this module. This means that it can be evolved independently of other modules in the application and can be modified without affecting other parts of the application.

Let's look at some of the relevant parts of the service.

`WorkoutBuilderService` needs the type definitions for `WorkoutPlan`, `Exercise`, and `WorkoutService`, so we import these into the component:

```
import { WorkoutPlan, Exercise } from '../../../services/model';
import { WorkoutService } from "../../../services/workout-service";
```

`WorkoutBuilderService` has a dependency on `WorkoutService` to provide persistence and querying capabilities. We resolve this dependency by injecting `WorkoutService` into the constructor for `WorkoutBuilderService`:

```
constructor(public workoutService:WorkoutService){}
```

`WorkoutBuilderService` also needs to track the workout being built. We use the `buildingWorkout` property for this. The tracking starts when we call the `startBuilding` method on the service:

```
startBuilding(name: string){
    if(name){
        this.buildingWorkout = this.workoutService.getWorkout(name)
        this.newWorkout = false;
    }else{
        this.buildingWorkout = new WorkoutPlan("", "", 30, []);
        this.newWorkout = true;
    }
    return this.buildingWorkout;
}
```

The basic idea behind this tracking function is to set up a `WorkoutPlan` object (`buildingWorkout`) that will be made available to components to manipulate the workout details. The `startBuilding` method takes the workout name as a parameter. If the name is not provided, it implies we are creating a new workout, and hence a new `WorkoutPlan` object is created and assigned; if not, we load the workout details by calling `WorkoutService.getWorkout(name)`. In any case, the `buildingWorkout` object has the workout being worked on.

The `newWorkout` object signifies whether the workout is new or an existing one. It is used to differentiate between save and update situations when the `save` method on this service is called.

The rest of the methods, `removeExercise`, `addExercise`, and `moveExerciseTo`, are self-explanatory and affect the exercise list that is part of the workout (`buildingWorkout`).

`WorkoutBuilderService` is calling a new method, `getWorkout`, on `WorkoutService`, which we have not added yet. Go ahead and copy the `getWorkout` implementation from the `workout-service.ts` file in the `services` folder under `trainer/src` in `checkpoint 4.5`. We will not dwell on the new service code as the implementation is quite simple.

Let's get back to left nav and implement the remaining functionality.

# Adding exercises using ExerciseNav

To add exercises to the workout we are building, we just need to import and inject `WorkoutBuilderService` into the `LeftNavExercisesComponent` and call its `addExercise` method, passing the selected exercise as a parameter:

```
constructor(
    public workoutService:WorkoutService,
    public workoutBuilderService:WorkoutBuilderService) {}
. . .
addExercise(exercise:Exercise) {
    this.workoutBuilderService.addExercise(new ExercisePlan(exercise, 30));
}
```

Internally, `WorkoutBuilderService.addExercise` updates the `buildingWorkout` model data with the new exercise.

The preceding implementation is a classic case of sharing data between independent components. The shared service exposes the data in a controlled manner to any component that requests it. While sharing data, it is always a good practice to expose the state/data using methods instead of directly exposing the data object. We can see that in our component and service implementations too. `LeftNavExercisesComponent` does not update the workout data directly; in fact, it does not have direct access to the workout being built. Instead, it relies upon the service method, `addExercise`, to change the current workout's exercise list.

Since the service is shared, there are pitfalls to be aware of. As services are injectable through the system, we cannot stop any component from taking dependency on any service and calling its functions in an inconsistent manner, leading to undesired results or bugs. For example, `WorkoutBuilderService` needs to be initialized by calling `startBuilding` before `addExercise` is called. What happens if a component calls `addExercise` before the initialization takes place?

# Implementing the Workout component

The `Workout` component is responsible for managing a workout. This includes creating, editing, and viewing the workout. Due to the introduction of `WorkoutBuilderService`, the overall complexity of this component will be reduced. Other than the primary responsibility of integrating with, exposing, and interacting with its template view, we will delegate most of the other work to `WorkoutBuilderService`.

The `Workout` component is associated with two routes/views, namely `/builder/workout/new` and `/builder/workout/:id`. These routes handle both creating and editing workout scenarios. The first job of the component is to load or create the workout that it needs to manipulate.

# Route parameters

But before we get to building out the `WorkoutComponent`and its associated view, we need to touch briefly on the navigation that brings a user to the screen for that component. This component handles both creating and editing workout scenarios. The first job of the component is to load or create the workout that it needs to manipulate. We plan to use Angular's routing framework to pass the necessary data to the component, so that it will know whether it is editing an existing workout or creating a new one, and in the case of an existing workout, which component it should be editing.

How is this done? `WorkoutComponent` is associated with two routes, namely `/builder/workout/new` and `/builder/workout/:id`. The difference in these two routes lies in what is at the end of these routes; in one case, it is `/new`, and in the other `/:id`. These are called **route parameters.** The `:id` in the second route is a token for a route parameter. The router will convert the token to the ID for the workout component. As we saw earlier, this means that the URL that will be passed to the component in the case of *7 Minute Workout* will be `/builder/workout/7MinuteWorkout`.

How do we know that this workout name is the right parameter for the ID? As you recall, when we set up the event for handling a click on the **Workout** tiles on the **Workouts** screen that takes us to the **Workout** screen, we designated the workout name as the parameter for the ID, like so:

```
onSelect(workout: WorkoutPlan) {
    this.router.navigate( ['./builder/workout', workout.name] );
}
```

Here, we are constructing the route using the programmatic interface for the router (we covered routing in detail in the previous chapter, so we won't go over that again here). The `router.navigate` method accepts an array. This is called the **link parameters array**. The first item in the array is the path of the route, and the second is a route parameter that specifies the ID of the workout. In this case, we set the `id` parameter to the workout name. From our discussion of routing in the previous chapter, we know that we can also construct the same type of URL as part of a router link or simply enter it in the browser to get to the **Workouts** screen and edit a particular workout.

The other of the two routes ends with `/new`. Since this route does not have a `token` parameter, the router will simply pass the URL unmodified to the `WorkoutComponent`. The `WorkoutComponent` will then need to parse the incoming URL to identify that it should be creating a new component.

# Route guards

But before the link takes the user to the `WorkoutComponent`, there is another step along the way that we need to consider. The possibility always exists that the ID that is passed in the URL for editing a workout could be incorrect or missing. In those cases, we do not want the component to load, but instead we want to have the user redirected to another page or back to where they came from.

Angular offers a way to accomplish this result with **route guards**. As the name implies, route guards **provide a way to prevent navigation to a route**. A route guard can be used to inject custom logic that can do things such as check authorization, load data, and make other verifications to determine if the navigation to the component needs to be cancelled or not. And all of this is done before the component loads so it is never seen if the routing is canceled.

Angular offers several route guards, including `CanActivate`, `CanActivateChild`, `CanDeActivate`, `Resolve`, and `CanLoad`. At this point we are interested in the first of these these hooks: `CanActivate`.

## Implementing the CanActivate route guard

The `CanActivate` guard permits navigation to proceed or stops it based on the conditions that we set up in the implementation that we provide. In our case, what we want to do is use `CanActivate` to check the validity of any id that is passed for an existing workout. Specifically, we will run a check on that ID by making a call to the `WorkoutService` to retrieve the Workout Plan and see if it exists. If it exists, we will let the navigation proceed; if not we will stop it.

Copy `workout.guard.ts` from the workout-builder/workout folder under `trainer/src/components` in `checkpoint 4.5` and you will see the following code:

```
import { Injectable } from '@angular/core';
import { ActivatedRouteSnapshot, CanActivate, Router, RouterStateSnapshot }
from '@angular/router';

import { WorkoutPlan } from "../../../services/model";
import { WorkoutService } from "../../../services/workout-service";

@Injectable()
export class WorkoutGuard implements CanActivate {
    publicworkout: WorkoutPlan;

    constructor(
        public workoutService: WorkoutService,
        public router: Router) {}

    canActivate(
        route: ActivatedRouteSnapshot,
        state: RouterStateSnapshot
    ) {
        this.workout = this.workoutService.getWorkout(route.params['id']);
        if(this.workout){ return true; }
        this.router.navigate(['/builder/workouts']);
        return false;
    }
}
```

As you can see the guard is an injectable class that implements the `CanActivate` interface. We implement the interface with the `CanActivate` method. The `CanActivate` method accepts two parameters; the `ActivatedRouteSnapshot` and the `RouterStateSnapshot`. In this case, we are only interested in the first of these two parameters. This parameter contains a `params` array from which we extract the id parameter for the route.

The `CanActivate` method can return either a simple `boolean` value or an `Observable<boolean>`. The latter is useful if we have to make an asynchronous call as part of the method. If we return an `Observable`, the router will wait until the asynchronous call is resolved before proceeding with navigation. In this case, however, we are not making such an asynchronous call, as we are using a local in-memory data store. So we are just returning a simple true/false `boolean`.

In the next chapter, when we start using the HTTP module to make asynchronous calls to an external data store, we will refactor this code to return an `Observable<boolean>`.

This code injects the `WorkoutService` into the guard. The `CanActivate` method then calls the `GetWorkout` method of the `WorkoutService` using the parameter supplied in the route. If the workout exists, then `canActivate` returns true and the navigation proceeds; if not, it re-routes the user to the workouts page and returns false.

The final step in implementing `WorkoutGuard` is to add it to the route configuration for `WorkoutComponent`. So update `workout-builder.routes.ts` as follows:

```
export const workoutBuilderRoutes: Routes = [
    {
        path: '',
        component: WorkoutBuilderComponent,
        children: [
            {path:'', pathMatch: 'full', redirectTo: 'workouts'},
            {path:'workouts', component: WorkoutsComponent },
            {path:'workout/new',  component: WorkoutComponent },
            {path:'workout/:id', component: WorkoutComponent,
             canActivate: [WorkoutGuard] },
            {path:'exercises', component: ExercisesComponent},
            {path:'exercise/new', component: ExerciseComponent },
            {path:'exercise/:id', component: ExerciseComponent }
        ]
    }
];
```

With this configuration, we are assigning `WorkoutGuard` to the `canActivate` property of the route for `WorkoutComponent`. This means that `WorkoutGuard` will be called prior to the router navigating to `WorkoutComponent`.

# Implementing the Workout component continued...

Now that we have established the routing that takes us to the `Workout` component, let's turn to completing its implementation. So copy the `workout.component.ts` file from the `workout-builder/workout` folder under trainer/src/components in `checkpoint 4.5`. (Also copy `workout-builder.module.ts` from the `workout-builder` folder. We'll discuss the changes in this file a little later when we get to Angular forms.)

Open `workout.component.ts` and you'll see that we have added a constructor that injects `ActivatedRoute` and `WorkoutBuilderService`:

```
constructor(
public route: ActivatedRoute,
public workoutBuilderService:WorkoutBuilderService){ }
```

In addition, we have added the following `ngOnInit` method:

```
ngOnInit() {
    this.sub = this.route.params.subscribe(params => {
        let workoutName = params['id'];
        if (!workoutName) {
            workoutName = "";
        }
        this.workout = this.workoutBuilderService.startBuilding(
        workoutName);
    });
}
```

The method subscribes to the route parameters and extracts the `id` parameter for the workout. If no ID is found, then we treat it as a new workout since `workout/new` is the only path that is configured in `WorkoutBuilderRoutes` that we allow to reach this screen without an ID. In that case, we provide an empty string as a parameter in the call to the `StartBuilding` method of the `WorkoutBuilderService` , which will cause it to return a new Workout.

We are subscribing to the route parameters because they are `Observables`, which can change during the lifetime of the component. This gives us the ability to reuse the same component instance with different parameters even though the `OnInit` life cycle event for that component is called only once. We'll cover `Observables` in detail in the next chapter.

In addition to this code, we have also added a series of methods to the `Workout Component` for adding, removing, and moving a workout. These methods all call corresponding methods on the `WorkoutBuilderService` and we will not review them in detail here. We've also added an array of `durations` for populating the duration drop-down list.

For now, this is enough for the **component** class implementation. Let's update the associated `Workout` template.

# Implementing the Workout template

Now copy the `workout.component.html` files from the `workout-builder/workout` folder under `trainer/src/components` in `checkpoint 4.5`. Run the app, navigate to `/builder/workouts`, and double-click on the *7 Minute Workout* tile. This should load the *7 Minute Workout* details with a view similar to the one shown at the start of the *Building a workout* section.

In the event of any problem, you can refer to the `checkpoint4.5` code in the **GitHub repository: Branch: checkpoint4.5** (folder – `trainer`).

We will be dedicating a lot of time to this view, so let's understand some specifics here.

The exercise list div (`id="exercise-list"`) lists the exercises that are part of the workout in order. We display them as top-to-bottom tiles in the left part of the content area. Functionally, this template has:

- The Delete button to delete the exercise
- Reorder buttons to move the exercise up and down the list as well as to the top and bottom

We use `ngFor` to iterate over the list of exercises and display them:

```
<div *ngFor="let exercisePlan of workout.exercises; let i=index"
class="exercise-item">
```

You will notice that we are using the `*` asterisk in front of `ngFor`, which is shorthand for the `<template>` tag. We are also using `let` to set two local variables: `exerisePlan` to identify an item in the list of exercises and `i` to set up an index value that we will use to show a number for the exercises as they are displayed on the screen. We will also use the index value to manage reordering and deleting exercises from the list.

The second div element for workout data (`id="workout-data"`) contains the HTML input element for details such as name, title, and rest duration and a button to save workout changes.

The complete list has been wrapped inside the HTML form element so that we can make use of the form-related capabilities that Angular provides. So what are these capabilities?

# Angular forms

Forms are such an integral part of HTML development that any framework that targets client-side development just cannot ignore them. Angular provides a small but well-defined set of constructs that make standard form-based operations easier.

If we think carefully, any form of interaction boils down to:

- Allowing user inputs
- Validating those inputs against business rules
- Submitting the data to the backend server

Angular has something to offer for all the preceding use cases.

For user input, it allows us to create two-way bindings between the form input elements and the underlying model, hence avoiding any boilerplate code that we may have to write for model input synchronization.

It also provides constructs to validate the input before it can be submitted.

Lastly, Angular provides `http` services for client-server interaction and persisting data to the server. We'll cover those services in `Chapter 5`, *Supporting Server Data Persistence*.

Since the first two use cases are our main focus in this chapter, let's learn more about Angular user input and data validation support.

# Template-driven and model-driven forms

Angular offers two types of forms: **template-driven** and **model-driven**. We'll be discussing both types of form in this chapter. Because the Angular team is indicating that many of us will primarily use **template-driven forms**, that is what we will start with in this chapter.

# Template-driven forms

As the name suggests, **template-driven forms** place the emphasis on developing a form within an HTML template and handling most of the logic for the form-inputs, data validation, saving, and updating-in form directives placed within that template. The result is that very little form-related code is required in the component class that is associated with the form's template.

**Template-driven forms** make heavy use of the `ngModel` form directive. We will be discussing it in the next sections. It provides two-way databinding for form controls, which is a nice feature indeed. It allows us to write much less boilerplate code to implement a form. It also helps us to manage the state of the form (such as whether the form controls have changed and whether these changes have been saved). And it also gives us the ability to easily construct messages that display if the validation requirements for a form control have not been met (for example, a required field not provided, e-mail not in the right format, and so on).

# Getting started

In order to use Angular forms in our `Workout` component, we must first add some additional configurations. First, open the `systemjs.config.js` file in the `trainer` folder in `checkpoint 4.5` and add forms to the `ngPackageNames` array:

```
var ngPackageNames = [
  'common',
  'compiler',
  'core',
  'forms',
  'http',
  'platform-browser',
  'platform-browser-dynamic',
```

```
    'router',
    'testing'
];
```

With this in place, SystemJS will download this module for use in our application.

Next, open the copy of `workout-buider.module.ts` from the `workout-builder` folder under `trainer/src/components` in `checkpoint 4.5`. You will see that it adds the following highlighted code:

```
@NgModule({
    imports: [
        CommonModule,
        FormsModule,
        SharedModule,
        workoutBuilderRouting
    ],
```

This indicates that we will be using the forms module. Once we make this change, we will not have to do any further imports related to forms in the `Workout` component.

This brings in all the directives that we will need to implement our form including:

- `NgForm`
- `ngModel`

Let's start using these to build our form.

## Using NgForm

In our template, we have added the following `form` tag:

```
<form #f="ngForm" class="row" name="formWorkout" (ngSubmit)="save(f.form)"
novalidate>. . .
</form>
```

Let's take a look at what we have here. One interesting thing is that we are still using a standard `<form>` tag and not a special Angular tag. We've also used `#` to define a local variable `#f` to which we have assigned `ngForm`. Creating this local variable provides us with the convenience of being able to use it for form-related activity in other places within the form. For example, you can see that we are using it at the end of the opening `form` tag in a parameter, `f.form`, which is being passed to the `onSubmit` event bound to `(ngSubmit)`.

That last binding to `(ngSubmit)` should tell us that something different is going on here. Even though we did not explicitly add the `NgForm` directive, our `<form>` now has additional events such as `ngSubmit` to which we can bind actions. How did this happen? Well, this was **not** triggered by our assigning `ngForm` to a local variable. Instead, it happened *auto-magically* because we imported the forms module into `workout-builder.module.ts`.

With that import in place, Angular scanned our template for a `<form>` tag and wrapped that `<form>` tag within an `NgForm` directive. The Angular documentation indicates that `<form>` elements in the component will be upgraded to use the Angular form system. This is important because it means that various capabilities of the `NgForm` directive are now available to use with the form. These include the `ngSubmit` event, which signals when a user has triggered a form submission and provides the ability to validate the entire form before submitting it.

## ngModel

One of the fundamental building blocks for template-driven forms is `ngModel`, and you will find it being used throughout our form. One of the primary roles of `ngModel` is to support two-way binding between user input and an underlying model. With such a setup, changes in the model are reflected in the view, and updates to the view too are reflected back on the model. Most of the other directives that we have covered so far only support one-way binding from models to views. This is also due to the fact that `ngModel` is only applied to elements that allow user input.

As you know, we already have a model that we are using for the **Workout** page-`WorkoutPlan`. Here is the `WorkoutPlan` model from `model.ts`:

```
@Injectable()
export class WorkoutPlan {
  constructor(
    public name: string,
    public title: string,
    public restBetweenExercise: number,
    public exercises: ExercisePlan[],
    public description?: string) {
  }
totalWorkoutDuration(): number{
 . . . . . .
}
```

Note the use of the ? after description. This means that it is an optional property in our model and is not required to create a WorkoutPlan. In our form, this will mean that we will not require that a description be entered and everything will work fine without it.

Within the WorkoutPlan model we also have a reference to an array made up of instances of another type of model: ExercisePlan. ExercisePlan in turn is made up of a number (duration) and another model (Exercise), which looks like this:

```
@Injectable()
export class ExercisePlan {
  constructor(public exercise: Exercise, public duration: any) {
  }
}
```

Notice that we have decorated both model classes with @Injectable. This is used so that TypeScript will generate the necessary metadata for the entire object hierarchy, in this case the nested classes ExercisePlan within WorkoutPlan and Exercise within ExercisePlan. What this means is that we can create complex hierarchies of models that can all be data-bound within our form using NgModel.

So throughout the form, whenever we need to update one of the values in a WorkoutPlan or an ExercisePlan, we can use NgModel to do that (the WorkoutPlan model will be represented by a local variable named workout in the following examples).

## Using ngModel with input and textarea

Open workout-component.html and look for ngModel. Here too, it has only been applied to HTML elements that allow user data input. These include input, textarea, and select. The workout name input setup looks like this:

```
<input type="text" name="workoutName" class="form-control" id="workout-
name" placeholder="Enter workout name. Must be unique."
[(ngModel)]="workout.name">
```

The preceding `[(ngModel)]` directive sets up a two-way binding between the input control and the `workout.name` model property. The brackets and parentheses should each look familiar. Previously we used them separately from each other: the `[]` brackets for property binding and the `()` parentheses for event binding. In the latter case, we usually bound the event to a call to a method in the component associated with the template. You can see an example of this in the form with the button that a user clicks on to remove an exercise:

```
<div class="pull-right" (click)="removeExercise(exercisePlan)"><span
class="glyphicon glyphicon-trash"></span></div>
```

Here the click event is explicitly bound to a method called `removeExercise` in our `Workout` component class. But for the `workout.name` input, we do not have an explicit binding to a method on the component. So what's going on here and how does the update happen without us calling a method on the component? The answer to that question is that the combination `[( )]` is shorthand for both binding a model property to the input element and wiring up an event that updates the model.

Put differently, if we reference a model element in our form, `ngModel` is smart enough to know that what we want to do is update that element (`workout.name` here) when a user enters or changes the data in the input field to which it is bound. Under the hood, Angular creates an update method similar to what we would otherwise have to write ourselves. Nice! This approach keeps us from having to write repetitive code to update our model.

Angular supports most of the HTML5 input types, including text, number, select, radio, and checkbox. This means binding between a model and any of these input types just works out of the box.

The `textarea` element works the same as the input:

```
<textarea name="description" . . .
[(ngModel)]="workout.description"></textarea>
```

Here we bind `textarea` to `workout.description`. Under the hood, `ngModel` updates the workout description in our model with every change we type into the text area.

To test out how this works, why don't we verify this binding? Add a model interpolation expression against any of the linked inputs such as this one:

```
<input type="text". . . [(ngModel)]="workout.name">{{workout.name}}
```

Open the **Workout** page, type something in the input, and see how the interpolation is updated instantaneously. The magic of two-way binding!

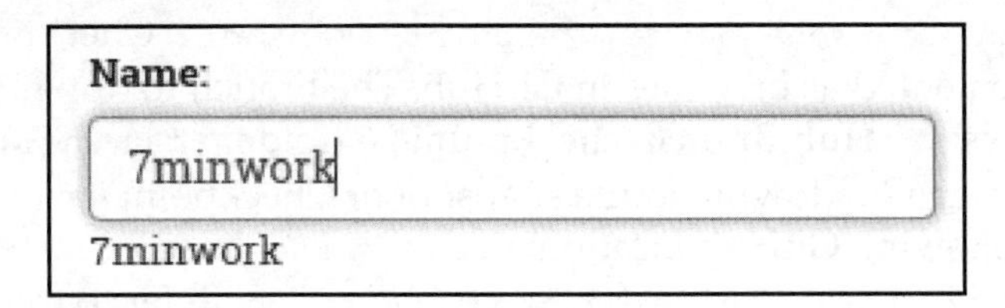


## Using ngModel with select

Let's look at how select has been set up:

```
<select . . . name="duration" [(ngModel)]="exercisePlan.duration">
    <option *ngFor="let duration of durations"
[value]="duration.value">{{duration.title}}</option>
</select>
```

We are using `ngFor` here to bind to an array, `durations`, which is in the `Workout` component class. The array looks like this:

```
[{ title: "15 seconds", value: 15 },
 { title: "30 seconds", value: 30 }, ...]
```

The `ngFor` component will loop over the array and populate the dropdown values with the corresponding values in the array with the title for each item being displayed using interpolation -- `{{duration.title}}`. And `[(ngModel)]` then binds the drop-down selection to the `exercisePlan.duration` in the model. Notice here that we are binding to the nested model: `ExercisePlan`. This is powerful stuff that enables us to create complicated forms with nested models, all of which can use `ngModel` for databinding.

Like input, select too supports two-way binding. We saw how changing select updates a model, but the model-to-template binding may not be apparent. To verify that a model to a template binding works, open the *7 Minute Workout* app and verify the duration dropdowns. Each one has a value that is consistent with the model value (30 seconds).

Angular does an awesome job of keeping the model and view in sync using `ngModel`. Change the model and see the view updated; change the view and watch the model updated instantaneously.

Now let's add validation to our form.

The code is also available for everyone to download on GitHub at `https://github.com/chandermani/angular2byexample`. Checkpoints are implemented as branches in GitHub. The branch to download is as follows: **GitHub Branch: checkpoint4.6** (folder - `trainer`). Or if you are not using Git, download the snapshot of Checkpoint 4.6 (a ZIP file) from the following GitHub location: `https://github.com/chandermani/angular2byexample/archive/checkpoint4.5.zip`. Refer to the `README.md` file in the `trainer` folder when setting up the snapshot for the first time. Again, if you are working along with us as we build the application, be sure and update the `app.css` file, which we are not discussing here.

# Angular validation

As the saying goes, *Never trust user input*. Angular has support for validation, including the standard required, min, max, and pattern as well as custom validators.

# ngModel

`ngModel` is the building block that we will use to implement validation. It does two things for us: it maintains the model state and provides a mechanism for identifying validation errors and displaying validation messages.

To get started, we need to assign `ngModel` to a local variable in all of our form controls that we will be validating. In each case, we need to use a unique name for this local variable. For example, for workout name we add `#name="ngModel"` within the `input` tag for that control. The workout name `input` tag should now look like this:

```
<input type="text" name="workoutName" #name="ngModel" class="form-control"
id="workout-name" placeholder="Enter workout name. Must be unique."
[(ngModel)]="workout.name" required>
```

Continue through the form, assigning `ngModel` to local variables for each of the inputs. Also add the required attribute for all the required fields.

# The Angular model state

Whenever we use `NgForm`, every element within our form, including input, text area, and select, has some states defined on the associated model. `ngModel` tracks these states for us. The states tracked are:

- `pristine`: The value of this is `true` as long as the user does not interact with the input. Any update to the `input` field and `ng-pristine` is set to `false`.
- `dirty`: This is the reverse of `ng-pristine`. This is `true` when the input data has been updated.
- `touched`: This is `true` if the control ever had focus.
- `untouched`: This is `true` if the control has never lost focus. This is just the reverse of `ng-touched`.
- `valid`: This is `true` if there are validations defined on the `input` element and none of them are failing.
- `invalid`: This is `true` if any of the validations defined on the element are failing.

`pristine\dirty` or `touched\untouched` are useful properties that can help us decide when error labels are shown.

# Angular CSS classes

Based on the model state, Angular adds some CSS classes to an input element. These include the following:

- `ng-valid`: This is used if the model is valid
- `ng-invalid`: This is used if the model is invalid
- `ng-pristine`: This is used if the model is pristine
- `ng-dirty`: This is used if the model is dirty
- `ng-untouched`: This is used when the input is never visited
- `ng-touched`: This is used when the input has focus

To verify it, go back to the `workoutName` input tag and add a template reference variable named `spy` inside the `input` tag:

```
<input type="text" name="workoutName" #name="ngModel" class="form-control"
id="workout-name" placeholder="Enter workout name. Must be unique."
[(ngModel)]="workout.name" required #spy>
```

Then, below the tag, add the following label:

```
<label>{{spy.className}}</label>
```

Reload the application and click on the **New Workout** link in the *Workout Builder*. Before touching anything on the screen, you will see the following displayed:

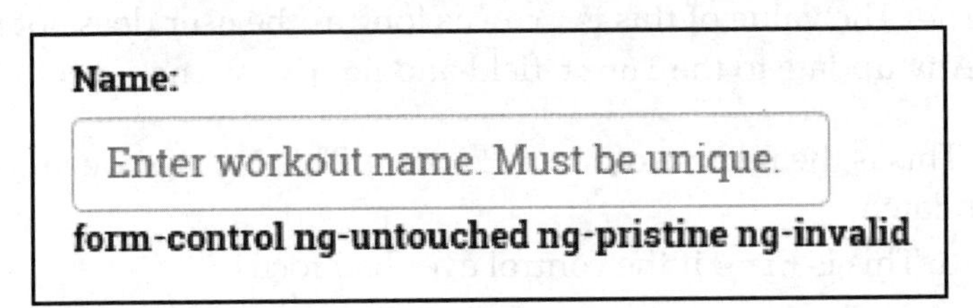


Add some content into the **Name** input box and tab away from it. The label changes to this:

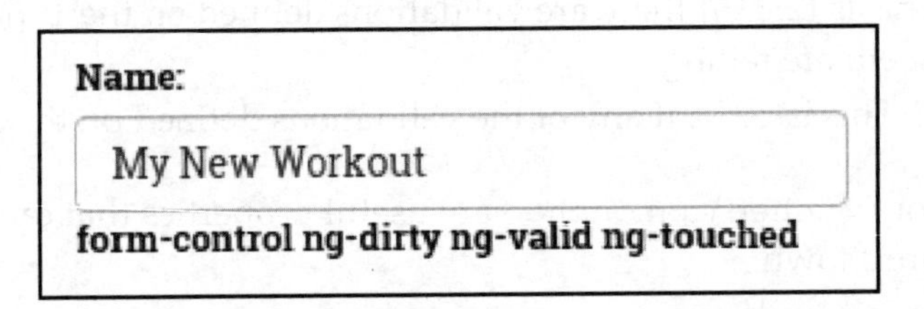


What we are seeing here is Angular changing the CSS classes that apply to this control as the user interacts with it. You can also see these changes by inspecting the `input` element in the developer console.

These CSS class transitions are tremendously useful if we want to apply visual clues to the element depending on its state. For example, look at this snippet:

```
input.ng-invalid {  border:2px solid red; }
```

This draws a red border around any input control that has invalid data.

As you add more validations to the Workout page, you can observe (in the developer console) how these classes are added and removed as the user interacts with the `input` element.

Now that we have an understanding of model states and how to use them, let's get back to our discussion of validations (before moving on, remove the variable name and label that you just added).

## Workout validation

The workout data needs to be validated for a number of conditions.

After taking the step of adding the local variable references for `ngModel` and the required attribute to our `input` fields, we have been able to see how `ngModel`tracks changes in the state of these controls and how it toggles the CSS styles.

## Displaying appropriate validation messages

Now the input needs to have a value; otherwise, the validation fails. But how can we know if the validation has failed? `ngModel` comes to our rescue here. It can provide the validation state of the particular input. And that gives us what we need to display an appropriate validation message.

Let's go back to the input control for the Workout name. In order to get a validation message to display, we have to first modify the input tag to the following:

```
<input type="text" name="workoutName" #name="ngModel" class="form-control"
id="workout-name" placeholder="Enter workout name. Must be unique."
[(ngModel)]="workout.name" required>
```

We have added a local variable called `#name` and assigned `ngModel` to it. This is called a template reference variable and we can use it with the following label to display a validation message for the input:

```
<label *ngIf="name.control.hasError('required') && (name.touched)"
class="alert alert-danger validation-message">Name is required</label>
```

We are showing the validation message in the event that the name is not provided **and** the control has not been touched. To check the first condition, we retrieve the `hasError` property of the control and see if the error type is `required`. We check to see if the name input has been `touched` because we do not want the message to display when the form first loads for a new workout.

You will notice that we are using a somewhat more verbose style to identify validation errors than is required in this situation. Instead of `name.control.hasError('required')`, we could have used `!name.valid` and it would have worked perfectly fine. However, using the more verbose approach allows us to identify validation errors with greater specificity, which will be essential when we start adding multiple validators to our form controls. We'll look at using multiple validators a little later in this chapter. For consistency, we'll stick with the more verbose approach.

Load the new Workout page (`/builder/workouts/new`) now. Enter a value in the name input box and then delete it. The error label appears as shown in the following screenshot:

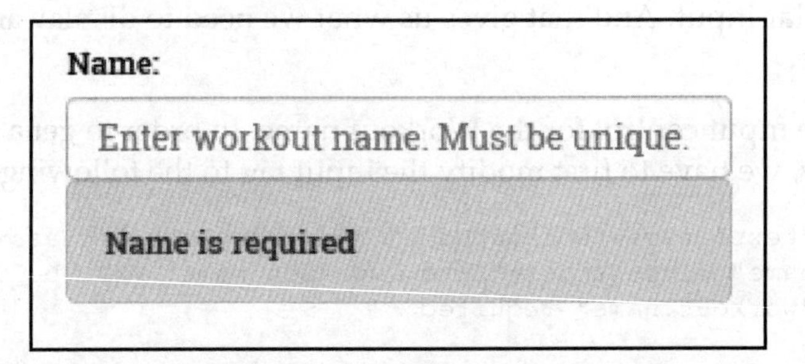


## Adding more validation

Angular provides four out-of-the-box validators:

- `required`
- `minLength`
- `maxLength`
- `pattern`

We've seen how the `required` validator works. Now let's look at two of the other out-of-the-box validators: `minLength` and `maxLength`. In addition to making it required, we want the title of the workout to be between 5 and 20 characters (we'll look at the `pattern` validator a little later in this chapter).

So in addition to the `required` attribute we added previously to the title input box, we will add the `minLength` attribute and set it to `5`, and add the `maxLength` attribute and set it to `20`, like so:

```
<input type="text" . . . minlength="5" maxlength="20" required>
```

Then we add another label with a message that will display when this validation is not met:

```
<label *ngIf="(title.control.hasError('minlength') ||
title.control.hasError('maxlength')) && workout.title.length > 0"
class="alert alert-danger validation-message">Title should be between 5 and
20 characters long.</label>
```

## Managing multiple validation messages

You'll see that the conditions for displaying the message now test for the length not being zero. This prevents the message from displaying in the event that the control is touched but left empty. In that case, the title required message should display. This message only displays if nothing is entered in the field and we accomplish this by checking explicitly to see if the control's `hasError` type is `required`:

```
<label *ngIf="title.control.hasError('required')" class="alert alert-danger
validation-message">Title is required.</label>
```

Since we are attaching two validators to this input field, we can consolidate the check for the input being touched by wrapping both validators in a div tag that checks for that condition being met:

```
<div *ngIf="title.touched">
  . . . [the two validators] . . .
</div>
```

What we just did shows how we can attach multiple validations to a single input control and also display the appropriate message in the event that one of the validation conditions is not met. However, it's pretty clear that this approach will not `scale` for more complicated scenarios. Some inputs contain a lot of validations and controlling when a validation message shows up can become complex. As the expressions for handling the various displays get more complicated, we may want to refactor and move them into a custom directive. Creating a custom directive will be covered in detail in `Chapter 6`, *Angular 2 Directives in Depth*.

### Custom validation messages for an exercise

A workout without any exercise is of no use. There should at least be one exercise in the workout and we should validate this restriction.

The problem with exercise count validation is that it is not something that the user inputs directly and the framework validates. Nonetheless, we still want a mechanism to validate the exercise count in a manner similar to other validations on this form.

What we will do is add a hidden input box to the form that contains the count of the exercises. We will then bind this to `ngModel` and add a pattern validator that will check to make sure that there is more than one exercise. We will set the value of the input box to the count of the exercises:

```
<input type="hidden" name="exerciseCount" #exerciseCount="ngModel"
ngControl="exerciseCount" class="form-control" id="exercise-count"
[(ngModel)]="workout.exercises.length" pattern="[1-9][0-9]*">
```

Then we will attach a validation message to it similar to what we just did with our other validators:

```
<label *ngIf="exerciseCount.control.hasError('pattern')" class="alert
alert-danger extended-validation-message">The workout should have at least
one exercise!</label>
```

We are not using `ngModel` in its true sense here. There is no two-way binding involved. We are only interested in using it to do custom validation.

Open the new Workout page, add an exercise, and remove it; we should see the error:

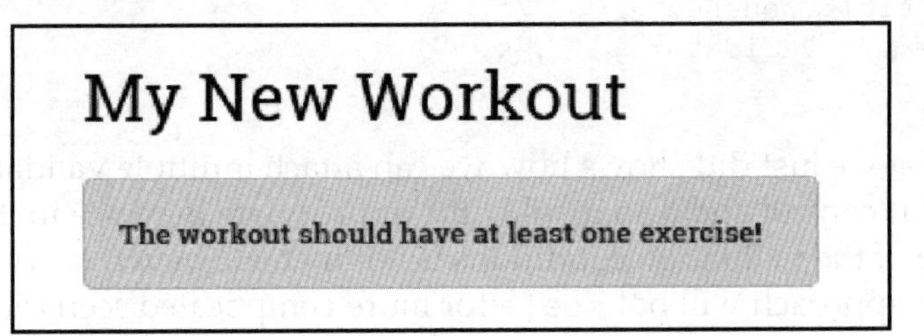


What we did here could have been easily done without involving any model validation infrastructure. But by hooking our validation into that infrastructure, we do derive some benefits. We can now determine errors with a specific model and errors with the overall form in a consistent and familiar manner. Most importantly, if our validation fails here, the entire form will be invalidated.

Implementing custom validation the way we just did is not what you would want to do very often. Instead, it will usually make more sense to implement this kind of complicated logic inside a custom directive. We'll cover creating custom directives in detail in `Chapter 6`, *Angular 2 Directives in Depth.*

One nuisance with our newly implemented `Exercise Count` validation is that it shows when the screen for a new `Workout` first appears. With this message, we are not able to use `ng-touched` to hide the display. This is because the exercises are being added programmatically and the hidden input we are using to track their count never changes from untouched as exercises are added or removed.

To fix this problem, we need an additional value to check when the state of the exercise list has been reduced to zero except when the form is first loaded. The only way that situation can happen is if the user adds and then removes exercises from a workout to the point that there are no more exercises. So we'll add another property to our component that we can use to track whether the remove method has been called. We call that value `removeTouched` and set its initial value to `false`:

```
removeTouched: boolean = false;
```

Then in the remove method we will set that value to `true`:

```
removeExercise(exercisePlan: ExercisePlan) {
    this.removeTouched = true;
    this.workoutBuilderService.removeExercise(exercisePlan);
}
```

Next will add `removeTouched` to our validation message conditions, like so:

```
<label *ngIf="exerciseCount.control.hasError('pattern') && (removeTouched)"
```

Now, when we open a new workout screen, the validation message will not display. But if the user adds and then removes all the exercises, then it will display.

To understand how model validation rolls up into form validation, we need to understand what form-level validation has to offer. However, even before that, we need to implement saving the workout and calling it from the workout form.

# Saving the workout

The workout that we are building needs to be persisted (in-memory only). The first thing that we need to do is extend `WorkoutService` and `WorkoutBuilderService`.

`WorkoutService` needs two new methods: `addWorkout` and `updateWorkout`:

```
addWorkout(workout: WorkoutPlan){
    if (workout.name){
        this.workouts.push(workout);
        return workout;
    }
}

updateWorkout(workout: WorkoutPlan){
    for (var i = 0; i < this.workouts.length; i++) {
        if (this.workouts[i].name === workout.name) {
            this.workouts[i] = workout;
            break;
        }
    }
}
```

The `addWorkout` method does a basic check on the workout name and then pushes the workout into the workout array. Since there is no backing store involved, if we refresh the page, the data is lost. We will fix this in the next chapter where we persist the data to a server.

The `updateWorkout` method looks for a workout with the same name in the existing workouts array and if found, updates and replaces it.

We only add one save method to `WorkoutBuilderService` as we are already tracking the context in which workout construction is going on:

```
save(){
    let workout = this.newWorkout ?
        this._workoutService.addWorkout(this.buildingWorkout) :
        this._workoutService.updateWorkout(this.buildingWorkout);
    this.newWorkout = false;
    return workout;
}
```

The `save` method calls either `addWorkout`, or `updateWorkout` in the `Workout` service based on whether a new workout is being created or an existing one is being edited.

From a service perspective, that should be enough. Time to integrate the ability to save workouts into the `Workout` component and learn more about the form directive!

Before we look at - `NgForm` in more detail, let's add the save method to `Workout` to save the workout when the `Save` button is clicked on. Add this code to the `Workout` component:

```
save(formWorkout:any){
    if (!formWorkout.valid) return;
    this.workoutBuilderService.save();
    this.router.navigate(['/builder/workouts']);
}
```

We check the validation state of the form using its invalid property and then call the `WorkoutBuilderService.save` method if the form state is valid.

## More on NgForm

Forms in Angular have a different role to play as compared to traditional forms that post data to the server. If we go back and look again at the form tag, we will see that it is missing the standard action attribute. The `novalidate` attribute on the form directive tells the browser not to do inbuilt input validations (this is not Angular-specific but is an HTML 5 attribute).

The standard form behavior of posting data to the server using full-page post-back does not make sense with an SPA framework such as Angular. In Angular, all server requests are made through asynchronous invocations originating from directives or services.

The form here plays a different role. When the form encapsulates a set of input elements (such as input, textarea, and select) it provides an API for:

- Determining the state of the form, such as whether the form is dirty or pristine based on the input controls on it
- Checking validation errors at the form or control level

If you still want the standard form behavior, you can add an `ngNoForm` attribute, but this will definitely cause a full-page refresh. We'll explore the specifics of the `NgForm` API a little later in this chapter when we look at saving the form and implementing validation.

The state of the `FormControl` objects within the form is being monitored by `NgForm`. If any of them are invalid, then `NgForm` sets the entire form to invalid. In this case, we have been able to use `NgForm` to determine that one or more of the `FormControl` objects is invalid and therefore the state of the form as a whole is invalid too.

Let's look at one more issue before we finish this chapter.

# Fixing the saving of forms and validation messages

Open a new Workout page and directly click on the **Save** button. Nothing is saved as the form is invalid, but validations on individual form input do not show up at all. It now becomes difficult to know what elements have caused validation failure. The reason behind this behavior is pretty obvious. If we look at the error message binding for the name input element, it looks like this:

```
*ngIf="name.control?.hasError('required') && name.touched"
```

Remember that earlier in the chapter, we explicitly disabled showing validation messages until the user has touched the input control. The same issue has come back to bite us and we need to fix it now.

We do not have a way to explicitly change the touched state of our controls to untouched. Instead, we will resort to a little trickery to get the job done. We'll introduce a new property called `submitted`. Add it at the top of `Workout` class definition and set its initial value to `false`, like so:

```
submitted: boolean = false;
```

The variable will be set to true on the `Save` button click. Update the save implementation by adding the highlighted code:

```
save(formWorkout){
    this.submitted = true;
    if (!formWorkout.valid) return;
    this._workoutBuilderService.save();
    this.router.navigate(['/builder/workouts']);
}
```

Nonetheless, how does this help? Well, there is another part to this fix that requires us to change the error message for each of the controls we are validating. The expression now changes to:

```
*ngIf="name.control.hasError('required') && (name.touched || submitted)"
```

With this fix, the error message is shown when the control is touched or the form submit button is pressed (`submitted` is `true`). This expression fix now has to be applied to every validation message where a check appears.

If we now open the new **Workout** page and click on the **Save** button, we should see all validation messages on the input controls:

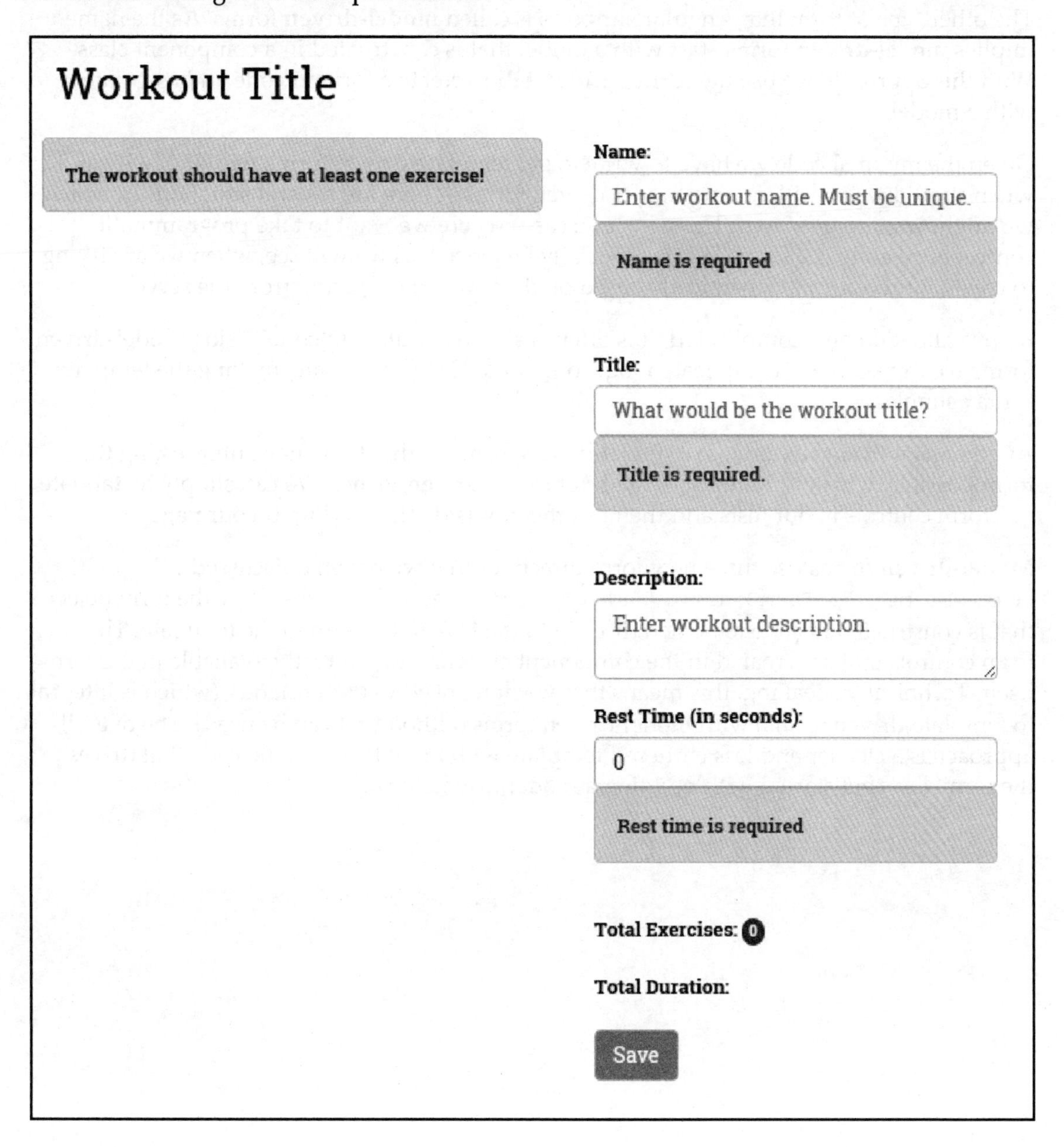

# Model-driven forms

The other type of form that Angular supports is called **model-driven** forms. As the name implies **model-driven forms** start with a model that is constructed in a component class. With this approach, we use the **form builder API** to create a form in code and associate it with a model.

Given the minimal code we have to write to get template-driven forms working, why and when should we consider using model-driven forms? There are several situations in which we might want to use them. These include cases where we want to take programmatic control of creating the form. This is especially beneficial, as we will see, when we are trying to create form controls dynamically based on data we are retrieving from the server.

If our validation gets complicated, it is often easier to handle it in code. Using model-driven forms we can keep this complicated logic out of the HTML template, making the template syntax simpler.

Another significant advantage of model-driven forms, is that they make unit-testing the form possible, which is not the case with **template-driven forms.** We can simply instantiate our form controls in our tests and then test them outside the markup on our page.

**Model-driven forms** use three new form directives that we haven't discussed before: `FormGroup`, `FormControl`, and `FormArray`. These directives allow the form object that is constructed in code to be tied directly to the HTML markup in the template. The form controls that are created in the component class are then directly available in the form itself. Technically speaking, this means that we don't need to use `ngModel` (which is integral to template-driven forms) with model-driven forms (although it can be used). The overall approach is a cleaner and less cluttered template with more focus on the code that drives the form. Let's get started with building a model-driven form.

## Getting started with model-driven forms

We'll make use of model-driven forms to build the form to add and edit **Exercises**. Among other things this form will allow the user to add links to exercise videos on YouTube. And since they can add any number of video links, we will need to be able to add controls for these video links dynamically. This challenge will present a good test of how effective model-driven forms can be in developing more complex forms. Here is how the form will look:

To get started, open workout-builder.module.ts and add the following import:

```
import { FormsModule, ReactiveFormsModule }   from '@angular/forms';
 ...
@NgModule({
    imports: [
        CommonModule,
        FormsModule,
        ReactiveFormsModule,
        SharedModule,
        workoutBuilderRouting
    ],
```

ReactiveFormsModule contains what we will need to build model-driven forms.

Next copy exercise-builder-service.ts from the workout-builder/builder-services folder under trainer/src/components in checkpoint 4.6 and import it into workout-builder.module.ts:

```
import { ExerciseBuilderService } from "./builder-services/exercise-builder-service";
```

Then add it as an additional provider to the providers array in that same file:

```
@NgModule({
   . . .
    providers: [
        ExerciseBuilderService,
        ExerciseGuard,
        WorkoutBuilderService,
        WorkoutGuard
    ]
})
```

You will notice here that we also have added ExerciseGuard as a provider. We won't be covering that here, but you should copy it from the exercise folder as well and also copy the updated workout-builder.routes.ts, which adds it as a route guard for the navigation to ExerciseComponent.

Now open `exercise.component.ts` and add the following import statement at the top of the file:

```
import { Validators, FormArray, FormGroup, FormControl, FormBuilder } from
'@angular/forms';
```

This brings in the following, which we will use to construct our form:

- `FormBuilder`
- `FormGroup`
- `FormControl`
- `FormArray`

Finally, we inject `FormBuilder` (as well as `Router`, `ActivatedRoute`, and `ExerciseBuilderService`) into the constructor of our class:

```
constructor(
    public route: ActivatedRoute,
    public router: Router,
    public exerciseBuilderService:ExerciseBuilderService,
    public formBuilder: FormBuilder
){}
```

With these preliminary steps out of the way, we can now go ahead and start building out our form.

## Using the FormBuilder API

The `FormBuilder` API is the foundation for model-driven forms. You can think of it as a factory for turning out the forms we are constructing in our code. Go ahead and add the `ngOnInit` lifecycle hook to your class, as follows:

```
ngOnInit():any{
    this.sub = this.route.params.subscribe(params => {
        let exerciseName = params['id'];
        if (exerciseName === 'new') {
            exerciseName = "";
        }
        this.exercise =
this.exerciseBuilderService.startBuilding(exerciseName);
    });
    this.buildExerciseForm();
}
```

When `ngOnInit` fires it will call a method for building our form (in addition to setting up the exercise we are building). So, at this stage of the component lifecycle, we are starting the process of constructing our form in code.

Now let's implement the `buildExerciseForm` method by adding the following code:

```
buildExerciseForm(){
    this.exerciseForm = this.formBuilder.group({
        'name': [this.exercise.name, [Validators.required,
AlphaNumericValidator.invalidAlphaNumeric]],
        'title': [this.exercise.title, Validators.required],
        'description': [this.exercise.description, Validators.required],
        'image': [this.exercise.image, Validators.required],
        'nameSound': [this.exercise.nameSound],
        'procedure': [this.exercise.procedure],
        'videos': this.addVideoArray()
    })
}
```

Let's examine this code. To start with, we are using the injected instance of `FormBuilder` to construct the form and assign it to a local variable `exerciseForm`. Using `formBuilder.group`, we add several form controls to our form. We add each of them by a simple key/value mapping:

```
'name': [this.exercise.name, Validators.required],
```

The left side of the mapping is the name of the `FormControl`, and the right is an array containing as its first element the value of the control (in our case, the corresponding element on our exercise model) and the second a validator (in this case, the out-of-the-box required validator). Nice and neat! It's definitely easier to see and reason about our form controls with setting them up outside the template.

We can not only build up `FormControls` in our form this way, but also add `FormControlGroups` and `FormControlArray` that contain `FormControls` within them. This means we can create complex forms that contain nested input controls. In our case, as we have mentioned, we are going to need to accommodate the possibility of our users adding multiple videos to an exercise. We can do this by adding the following code:

```
'videos': this.addVideoArray()
```

What we are doing here is assigning a `FormArray` to videos, which means we can assign multiple controls in this mapping. To construct this new `FormArray`, we add the following `addVideoArray` method to our class:

```
addVideoArray(){
    if(this.exercise.videos){
        this.exercise.videos.forEach((video : any) => {
            this.videoArray.push(new FormControl(video,
Validators.required));
        });
    }
    return this.videoArray;
}
```

This method constructs a `FormControl` for each video; each is then added each to a `FormArray` that is assigned to the videos control in our form.

# Adding the form model to our HTML view

So far, we have been working behind the scenes in our class to construct our form. The next step is to wire up our form to the view. To do this, we use the same controls that we used to build the form in our code: `formGroup`, `formControl`, and `formArray`.

Open `exercise.component.html` and add a `form` tag as follows:

```
<form [formGroup]="exerciseForm" (ngSubmit)="onSubmit(exerciseForm)"
novalidate>
```

Within the tag we are first assigning the `exerciseForm` that we just built in code to `formGroup`. This establishes the connection between our coded model and the form in the view. We also wire up the `ngSubmit` event to an `onSubmit` method in our code (we'll discuss this method a little later). Finally we turn off the browser's form validation using `novalidate`.

# Adding form controls to our form inputs

Next we start constructing the inputs for our form. We'll start with the input for the name of our exercise:

```
<input name="name" formControlName="name" class="form-control" id="name"
placeholder="Enter exercise name. Must be unique.">
```

We assign the name of our coded form control to `formControlName`. This establishes the link between the control in our code and the `input` field in the markup. Another item of interest here is that we do not use the `required` attribute.

## Adding validation

The next thing that we do is add a validation message to the control that will display in the event of a validation error:

```
<label *ngIf="exerciseForm.controls.name.hasError('required') &&
(exerciseForm.controls.name.touched || submitted)" class="alert alert-
danger validation-message">Name is required</label>
```

Notice that this markup is very similar to what we used in template-driven forms for validation, except that the syntax for identifying the control is somewhat more verbose Again, it checks the state of the `hasError` property of the control to make sure it is valid.

But wait a minute! How can we validate this input? Haven't we have removed the required attribute from our tag? This is where the control mappings that we added in our code come into play. If you look back at the code for the form model, you can see the following mapping for the `name` control:

```
'name': [this.exercise.name, Validators.required],
```

The second element in the mapping array assigns the required validator to the name form control. This means that we don't have to add anything to our template; instead, the form control itself is attached to the template with a required validator. The ability to add a validator in our code enables us to conveniently add validators outside our template. This is especially useful when it comes to writing custom validators with complex logic behind them.

## Adding dynamic form controls

As we mentioned earlier, the **Exercise** form that we are building requires that we allow the user to add one or more videos to the exercise. Since we don't know how many videos the user may want to add, we will have to build the `input` fields for these videos dynamically as the user clicks on the **Add Video** button. Here's how it will look:

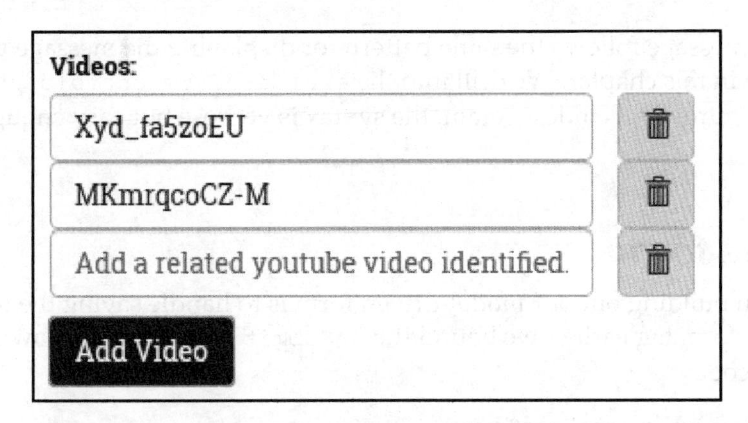


We have already seen the code in our component class that we use to do this. Now let's take a look at how it is implemented in our template.

We first use `ngFor` to loop through our list of videos. Then we assign the index in our videos to a local variable, `i`. No surprises so far:

```
<div *ngFor="let video of videoArray.controls; let i=index" class="form-
group">
```

Inside the loop we do three things. First, we add a button to allow the user to delete a video:

```
<button type="button" (click)="deleteVideo(i)"  class="btn alert-danger
pull-right">
    <span class="glyphicon glyphicon-trash text-danger"></span>
</button>
```

We bind a `deleteVideo` method in our component class to the button's `click` event and pass to it the index of the video being deleted.

Next, we dynamically add a video `input` field for each of the videos currently in our exercise:

```
<input type="text" class="form-control" [formControlName]="i"
placeholder="Add a related youtube video identified."/>
```

We then add a validation message for each of the video `input` fields.

```
<label
*ngIf="exerciseForm.controls['videos'].controls[i].hasError('required') &&
(exerciseForm.controls['videos'].controls[i].touched || submitted)"
class="alert alert-danger validation-message">Video identifier is
required</label>
```

The validation message follows the same pattern for displaying the message that we have used elsewhere in this chapter. We drill into the `exerciseFormControls` group to find the particular control by its index. Again, the syntax is verbose but easy enough to understand.

# Saving the form

The final step in building out our model-driven form is to handle saving the form. When we constructed the form tag earlier, we bound the `ngSubmit` event to the following `onSubmit` method in our code:

```
onSubmit(formExercise:FormGroup){
    this.submitted = true;
    if (!formExercise.valid) return;
    this.mapFormValues(formExercise);
    this.exerciseBuilderService.save();
    this.router.navigate(['/builder/exercises']);
}
```

This method sets `submitted` to `true`, which will trigger the display of any validation messages that might have been previously hidden because the form had not been touched. It also returns without saving in the event that there are any validation errors on the form. If there are none, then it calls the following `mapFormValues` method, which assigns the values from our form to the `exercise` that will be saved:

```
mapFormValues(form: FormGroup){
    this.exercise.name = form.controls['name'].value;
    this.exercise.title = form.controls['title'].value;
    this.exercise.description = form.controls['description'].value;
    this.exercise.image = form.controls['image'].value;
    this.exercise.nameSound = form.controls['nameSound'].value;
    this.exercise.procedure = form.controls['procedure'].value;
    this.exercise.videos = form.controls['videos'].value;
}
```

It then calls the save method in `ExerciseBuilderService` and routes the user back to the exercise list screen (remember that any new exercise will not display in that list because we have not yet implemented data persistence in our application).

We hope this makes it clear; model-driven forms offer many advantages when we are trying to build more complicated forms. They allow programming logic to be removed from the template. They permit validators to be added to the form programmatically. And they support building forms dynamically at runtime.

# Custom validators

Now we'll take a look at one more thing before we conclude this chapter. As anyone who has worked on building web forms (either in Angular or any other web technology) knows, we are often called on to create validations that are unique to the application we are building. Angular provides us with the flexibility to enhance our model-driven form validation by building custom validators.

In building our exercise form, we need to be sure about what is entered, as a name contains only alphanumeric characters and no spaces. This is because when we get to storing the exercises in a remote data store, we are going to use the name of the exercise as its key. So, in addition to the standard required field validator, let's build another validator that checks to make sure that the name entered is in alphanumeric form only.

Creating a custom control is quite straightforward. In its simplest form, an Angular custom validator is a function that takes a control as an input parameter, runs the validation check, and returns true or false. So let's start by adding a TypeScript file with the name `alphanumeric-validator.ts`. In that file, first import `FormControl` from `@angular/forms`; then add the following class to that file:

```
export class AlphaNumericValidator {
    static invalidAlphaNumeric(control: FormControl):{ [key:string]:boolean
} {
        if ( control.value.length && !control.value.match(/^[a-z0-9]+$/i)
){
            return {invalidAlphaNumeric: true };
        }
        return null;
    }
}
```

The code follows the pattern for creating a validator that we just mentioned. The only thing that may be a little surprising is that it returns true when the validation fails! As long as you are clear on this one quirk, you should have no problem writing your own custom validator.

## Integrating a custom validator into our forms

So how do we plug our custom validator into our form? If we are using model-driven forms, the answer is pretty simple. We add it just like a built-in validator when we build our form in code. Let's do that. Open `exercise.component.ts` and first add an import for our custom validator:

```
import {AlphaNumericValidator} from "./alphanumeric-validator";
```

Then, modify the form builder code to add the validator to the `name` control:

```
buildExerciseForm(){
    this.exerciseForm = this._formBuilder.group({
'name': [this.exercise.name, [Validators.required,
AlphaNumericValidator.invalidAlphaNumeric]],
  . . . [other form controls] . . .
    });
}
```

Since the name control already has a required validator, we add `AlphaNumericValidator` as a second validator using an array that contains both validators. The array can be used to add any number of validators to a control.

The final step is to incorporate the appropriate validation message for the control into our template. Open `workout.component.html` and add the following label just below the label that displays the message for the required validator:

```
<label *ngIf="exerciseForm.controls.name.hasError('invalidAlphaNumeric') &&
(exerciseForm.controls.name.touched || submitted)" class="alert alert-
danger validation-message">Name must be alphanumeric</label>
```

The exercise screen will now display a validation message if a non-alphanumeric value is entered in the name input box:

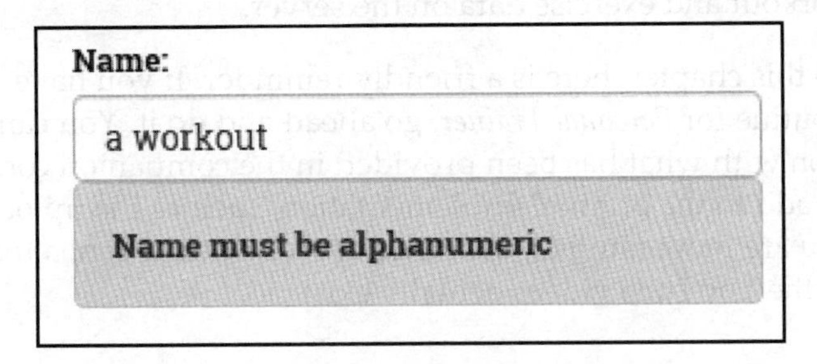


As we hope you can see, model-driven forms give us the ability to add custom validators to our forms in a straightforward manner that allows us to maintain the validation logic in our code and easily integrate it into our templates.

You may have noticed that in this chapter, we have not covered how to use custom validators in template-driven forms. That is because implementing them requires the additional step of building a custom directive. We'll cover that in `Chapter 6`, *Angular 2 Directives in Depth.*

# Summary

We now have a *Personal Trainer* app. The process of converting a specific *7 Minute Workout* app to a generic *Personal Trainer* app helped us learn a number of new concepts.

We started the chapter by defining the new app requirements. Then, we designed the model as a shared service.

We defined some new views and corresponding routes for the *Personal Trainer* app. We also used both child and asynchronous routing to separate out *Workout Builder* from the rest of the app.

We then turned our focus to workout building. One of the primary technological focuses in this chapter was on Angular forms. The *Workout Builder* employed a number of form input elements and we implemented a number of common form scenarios using both template-driven and model-driven forms. We also explored Angular validation in depth, and implemented a custom validator.

The next chapter is all about client-server interaction. The workouts and exercises that we create need to be persisted. In the next chapter, we build a persistence layer, which will allow us to save workout and exercise data on the server.

Before we conclude this chapter, here is a friendly reminder. If you have not completed the exercise building routine for *Personal Trainer*, go ahead and do it. You can always compare your implementation with what has been provided in the companion code base. There are also things you can add to the original implementation, such as file uploads for the exercise image, and once you are more familiar with client-server interaction, a remote check to determine whether the YouTube videos actually exist.

# 5
# Supporting Server Data Persistence

It's now time to talk to the server! There is no fun in creating a workout, adding exercises, and saving it to later realize that all our efforts are lost because the data did not persist anywhere. We need to fix this.

Seldom are applications self-contained. Any consumer app, irrespective of its size, has parts that interact with elements outside its boundary. With web-based applications, the interaction is mostly with a server. Apps interact with the server to authenticate, authorize, store/retrieve data, validate data, and perform other such operations.

This chapter explores the constructs that Angular provides for client-server interaction. In the process, we add a persistence layer to *Personal Trainer* that loads and saves data to a backend server.

The topics we cover in this chapter include the following:

- **Provisioning a backend to persist workout data**: We set up a MongoLab account and use its Data API to access and store workout data.
- **Understanding the Angular HTTP client library**: The HTTP client library allows us to interact with a server over HTTP. You'll learn how to make all types of `GET`, `POST`, `PUT`, and `DELETE` requests with the HTTP client library's `XMLHttpRequest` class.
- **Implementing the loading and saving of workout data**: We use the HTTP module to load and store workout data in the MongoLab databases.
- **Two ways in which we can use the HTTP module's XMLHttpRequest**: Either Observables or with promises.

- **Using RxJS and Observables**: To subscribe to and query streams of data.
- **Using promises**: In this chapter, we will see how to use promises as part of HTTP invocation and response.
- **Working with cross-domain access**: As we are interacting with a MongoLab server in a different domain, you will learn about browser restrictions on cross-domain access. You will also learn how JSONP and CORS help us make cross-domain access easy and about Angular JSONP support.

Let's set the ball rolling.

# Angular and server interactions

Any client-server interaction typically boils down to sending HTTP requests to a server and receiving responses from a server. For heavy JavaScript apps, we depend on the AJAX request/response mechanism to communicate with the server. To support AJAX-based communication, Angular provides the Angular HTTP module. Before we delve into the HTTP module, we need to set up our server platform that stores the data and allows us to manage it.

# Setting up the persistence store

For data persistence, we use a document database called MongoDB (`https://www.mongodb.org/`), hosted over MongoLab (`https://mongolab.com/`), as our data store. The reason we zeroed in on MongoLab is that it provides an interface to interact with the database directly. This saves us the effort of setting up server middleware to support MongoDB interaction.

It is never a good idea to expose the data store/database directly to the client. But in this case, since our primary aim is to learn about Angular and client-server interaction, we take this liberty and directly access the MongoDB instance hosted in MongoLab.
There is also a new breed of apps that are built over **noBackend** solutions. In such a setup, frontend developers build apps without the knowledge of the exact backend involved. Server interaction is limited to making API calls to the backend. If you are interested in knowing more about these noBackend solutions, do check out `http://nobackend.org/`.

Our first task is to provision an account on MongoLab and create a database:

1. Go to `https://mongolab.com` and sign up for a MongoLab account by following the instructions on the website.
2. Once the account is provisioned, log in and create a new Mongo database by clicking on the **Create New** button on the home page.
3. On the database creation screen, you need to make some selections to provision the database. See the following screenshot to select the free database tier and other options:

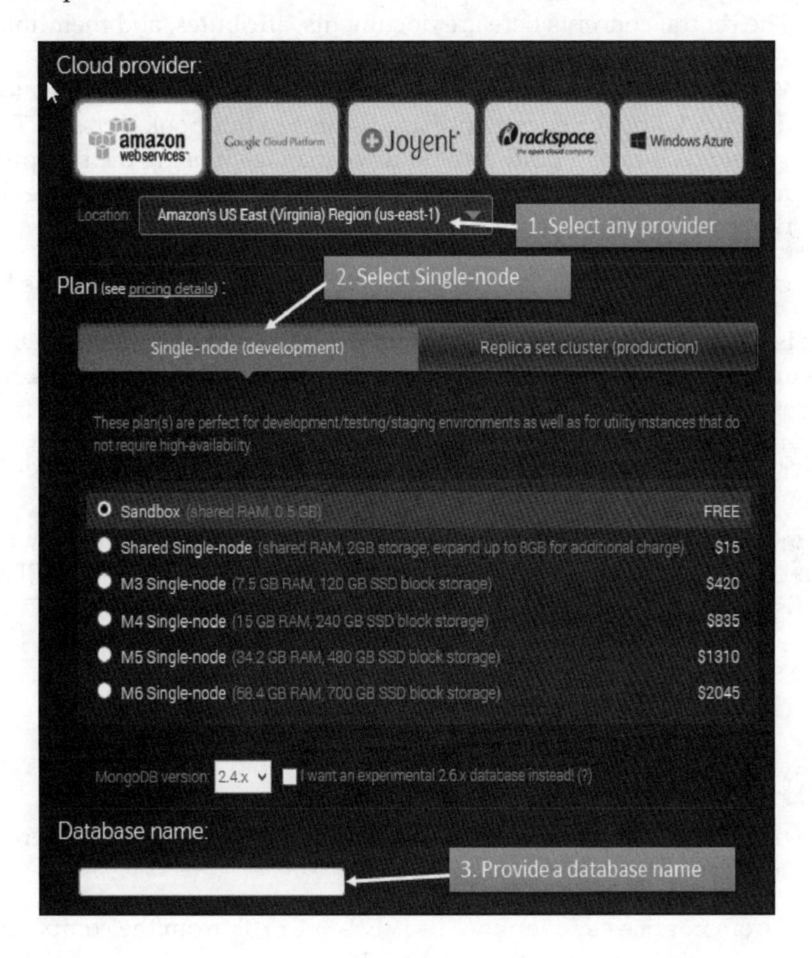

4. Create the database and make a note of the database name that you create.
5. Once the database is provisioned, open the database and add two collections to it from the **Collection** tab:
    - `exercises`: This stores all *Personal Trainer* exercises
    - `workouts`: This stores all *Personal Trainer* workouts

Collections in the MongoDB world equate to a database table.

MongoDB belongs to a breed of databases called **document databases**. The central concepts here are documents, attributes, and their linkages. And unlike traditional databases, the schema is not rigid.
We will not be covering what document databases are and how to perform data modeling for document-based stores in this book. *Personal Trainer* has a limited storage requirement and we manage it using the two previously mentioned document collections. We may not even be using the document database in its true sense.

Once the collections are added, add yourself as a user to the database from the **Users** tab.

The next step is to determine the API key for the MongoLab account. The provisioned API key has to be appended to every request made to MongoLab. To get the API key, perform the following steps:

1. Click on the username (not the account name) in the top-right corner to open the user profile.
2. In the section titled **API Key**, the current API key is displayed; copy it. At the same time, click on the button below the API key to enable Data API access. This is disabled by default.

The data store schema is complete. We now need to seed these collections.

# Seeding the database

The *Personal Trainer* app already has a predefined workout and a list of 12 exercises. We need to seed the collections with this data.

Open `seed.js` from `chapter5/checkpoint1/trainer/db` from the companion code base. It contains the seed JSON script and detailed instructions on how to seed data into the MongoLab database instance.

Once seeded, the database will have one workout in the workouts collection and 12 exercises in the exercises collection. Verify this on the MongoLab site; the collections should show this:

| NAME | DOCUMENTS | CAPPED? | SIZE |
|---|---|---|---|
| exercises | 12 | false | 19.30 KB |
| workouts | 1 | false | 8.97 KB |

Everything has been set up now, so let's start our discussion of the HTTP module and implement workout/exercise persistence for the *Personal Trainer* app.

# The basics of the HTTP module

At the core of the HTTP module is the HTTP client. It performs HTTP requests using `XMLHttpRequest` as the default backend (JSONP is also available, as we will see later in this chapter). It supports requests such as `GET`, `POST`, `PUT`, and `DELETE`. In this chapter, we will use the HTTP client to make all of these types of requests. As we will see, the HTTP client makes it easy to make these calls with a minimal amount of setup and complexity. None of this terminology will come as a surprise to anyone who has previously worked with Angular or built JavaScript applications that communicate with a backend data store.

However, there is a substantial change in the way Angular handles HTTP requests. Calling a request now returns an Observable of HTTP responses. It does so by using the **RxJS** library, which is a well-known open source implementation of the asynchronous Observable pattern.

You can find the RxJS project on GitHub at `https://github.com/Reactive-Extensions/RxJS`. The site indicates that the project is being actively developed by Microsoft in collaboration with a community of open source developers. We will not be covering the asynchronous Observable pattern in great detail here, and we encourage you to visit that site to learn more about the pattern and how RxJS implements it. The version of RxJS that Angular is using is beta 5.

In the simplest of terms, using Observables allows a developer to think about the data that flows through an application as streams of information that the application can dip into and use whenever it wants. These streams change over time, which allows the application to react to these changes. This quality of Observables provides a foundation for **functional reactive programming** (**FRP**), which fundamentally shifts the paradigm for building web applications from imperative to reactive.

The `RxJS` library provides operators that allow you to subscribe to and query these data streams. Moreover, you can easily mix and combine them, as we will see in this chapter. Another advantage of Observables is that it is easy to cancel or unsubscribe from them, making it possible to seamlessly handle errors inline.

While it is still possible to use promises, the default method in Angular uses Observables. We will also cover promises in this chapter.

# Personal Trainer and server integration

As described in the previous section, client-server interaction is all about asynchronicity. As we alter our *Personal Trainer* app to load data from the server, this pattern becomes self-evident.

In the previous chapter, the initial set of workouts and exercises was hardcoded in the `WorkoutService` implementation. Let's see how to load this data from the server first.

## Loading exercise and workout data

Earlier in this chapter, we seeded our database with a data form, the `seed.js` file. We now need to render this data in our views. The MongoLab Data API is going to help us here.

The MongoLab Data API uses an API key to authenticate access requests. Every request made to the MongoLab endpoints needs to have a query string parameter, `apikey=<key>`, where `key` is the API key that we provisioned earlier in the chapter. Remember that the key is always provided to a user and associated with his/her account. Avoid sharing your API keys with others.

The API follows a predictable pattern to query and update data. For any MongoDB collection, the typical endpoint access pattern is one of the following (given here is the base URL: `https://api.mongolab.com/api/1/databases`):

- `/<dbname>/collections/<name>?apiKey=<key>`: This has the following requests:
    - `GET`: This action gets all objects in the given collection name.
    - `POST`: This action adds a new object to the collection name. MongoLab has an `_id` property that uniquely identifies the document (object). If not provided in the posted data, it is autogenerated.
- `/<dbname>/collections/<name>/<id>?apiKey=<key>`: This has the following requests:
    - `GET`: This gets a specific document/collection item with a specific ID (a match done on the `_id` property) from the collection name
    - `PUT`: This updates the specific item (`id`) in the collection name
    - `DELETE`: This deletes the item with a specific ID from the collection name

For more details on the Data API interface, visit the MongoLab Data API documentation at `http://docs.mongolab.com/data-api`.

Now we are in a position to start implementing exercise/workout list pages.

The code we are starting with in this chapter is `checkpoint 4.6` (folder: `trainer`) in the GitHub repository for this book. It is available on GitHub (`https://github.com/chandermani/angular2byexample`). Checkpoints are implemented as branches in GitHub. If you are not using Git, download the snapshot of Checkpoint 4.6 (a ZIP file) from the following GitHub location: `https://github.com/chandermani/angular2byexample/tree/checkpoint4.6`. Refer to the `README.md` file in the `trainer` folder when setting up the snapshot for the first time.

# Loading exercise and workout lists from a server

To pull exercise and workout lists from the MongoLab database, we have to rewrite our `WorkoutService` service methods: `getExercises` and `getWorkouts`. But before we can do that, we have to set up our service to work with Angular's HTTP module.

# Adding the HTTP module and RxJS to our project

The Angular HTTP module is included in the Angular bundles that you have already installed. To use it, we need to import it into `app.module.ts`, like so:

```
import { HttpModule } from '@angular/http';
. . .
@NgModule({
  imports: [
. . .
    HttpModule,
. . .
})
```

We also need an external third-party library: **Reactive Extensions for JavaScript** (**RxJS**). RxJS implements the Observable pattern and is used by Angular with the HTTP module. It is included in the Angular bundles that are already part of our project.

# Updating workout-service to use the HTTP module and RxJS

Now, open `workout-service.ts` from `trainer/src/services`. In order to use the HTTP module and RxJS within `WorkoutService`, we need to add the following imports to that file:

```
import { Http, Response } from '@angular/http';
import { Observable } from 'rxjs/Observable';
import 'rxjs/add/operator/map';
import 'rxjs/add/operator/catch';
```

We are importing HTTP and Response from the HTTP module along with `Observable` from RxJS and two additional RxJS operators: `map` and `catch`. We'll see how these operators are used as we work through the code in this section.

In the class definition, add the following properties, which include a workout property and ones that set the URL for the collections in our Mongo database and the key to that database as well as another property: `params`, which sets up the API key as a query string for API access:

```
workout: WorkoutPlan;
collectionsUrl = "https://api.mongolab.com/api/1/
databases/<dbname>/collections";
apiKey = <key>
params = '?apiKey=' + this._apiKey;
```

We replace the <dbname> and <key> tokens with the database name and API key of the database that we provisioned earlier in the chapter.

Next, inject the HTTP module into the `WorkoutServiceconstructor` using the following line of code:

```
constructor(public http: Http) {
}
```

Then change the `getExercises()` method to this:

```
getExercises(){
    return this.http.get(this.collectionsUrl + '/exercises' + this.params)
        .map((res: Response) => <Exercise[]>res.json())
        .catch(WorkoutService.handleError);
}
```

If you are used to working with promises for asynchronous data operations, what you see here will look different. Instead of a promise that has a call to `then()` chained to it, where we expect to receive the data being returned asynchronously, we have a `map()` method.

According to the Angular documentation (`https://angular.io/docs/ts/latest/guide/server-communication.html`), what happens here is that the `http.get` method returns an Observable of HTTP responses (`Observable<Response>`) from the RxJS library.

Returning an Observable is the default response when using the HTTP module's `get` method. The Observable can, however, be converted to a promise. And, as we will see later in this chapter, the option to return JSONP also exists.

The `map` method that we can see in the code is an RxJS operator included in the `RxJS` library we imported earlier. It is needed here because what is retrieved from the remote server is not in a form that is required in our application. As you can see, we are using a `json()` method to convert the response body, `response`, into a JavaScript object.

The Angular documentation also makes it clear that we should not be returning the `Observable<Response>` response object to the components that are calling this method. Instead, we should be hiding the particulars of our data access operations from the rest of the application. This way, we can change those operations if need be without having to make modifications in the other code within our application.

Before we move on, there is one more thing to touch upon in this code. The Observable also has a `catch` operator that accepts a method, `handleError`, for handling a failed response. The `handleError` method takes the failed response as a parameter. At the moment, we'll follow the same approach that is laid out in the documentation for this `handleError` method. We log the error to the console and use `Observable.throw` to convert the JSON error into a user-friendly message, which we then return:

```
static handleError (error: Response) {
    console.error(error);
    return Observable.throw(error.json().error || 'Server error');
}
```

To be clear, this is not production code but it will give us the opportunity to show how to write code upstream to handle errors that are generated as part of data access.

It is important to understand that at this stage, our Observable is what is called *cold*. This means that no data is flowing through it until there is a subscription to it. This can bring about a gotcha moment for things such as adds and updates if you are not careful to add subscriptions to your Observables.

## Modifying getWorkouts() to use the HTTP module

The change in the code for retrieving workouts is almost identical to that for the exercises:

```
getWorkouts(){
    return this.http.get(this.collectionsUrl + '/workouts' + this.params)
        .map((res:Response) => <WorkoutPlan[]>res.json())
        .catch(WorkoutService.handleError);
}
```

Now that the `getExercises` and `getWorkouts` methods are updated, we need to make sure that they work with the upstream callers.

# Updating the workout/exercise list pages

The exercises and workouts list pages (as well as `LeftNavExercises`) call either the `getExercises` or `getWorkouts` method in `model.ts`. In order to get these working with the remote calls that are now being made using the HTTP module, we need to modify those calls to subscribe to the Observable that is being returned by the HTTP module. So, update the code in the `ngOnInit` method in `exercises.component.ts` to the following:

```
ngOnInit() {
    this.workoutService.getExercises()
        .subscribe(
            exerciseList=> {
                this.exerciseList = exerciseList;
            },
            (err: any) => console.error(err)
        );
}
```

Our method now subscribes to the Observable that is being returned by the `getExercises` method; at the point when the response arrives, it assigns the results to `exerciseList`. If there is an error, it assigns it to a `console.error` call that displays the error in the console. All of this is now being handled asynchronously using the HTTP module with RxJS.

Go ahead and make similar changes to the `ngOnInit` methods in `workouts.component,ts` and `left-nav-exercises.component.ts`.

Refresh the workout/exercise list page and the workout and exercise data will be loaded from the database server.

Look at the complete implementation in checkpoint 5.1 in the GitHub repository if you are having difficulty in retrieving/showing data. Note that in this checkpoint, we have disabled navigation links to the workout and exercise screens because we still have to add the Observable implementation to them. We'll do that in the next section.
Also remember to replace the database name and API key before you run the code from Checkpoint 5.1. If you are not using Git, download the snapshot of Checkpoint 5.1 (a ZIP file) from the following GitHub location: `https://github.com/chandermani/angular2byexample/tree/checkpoint5.1`. Refer to the `README.md` file in the `trainer` folder when setting up the snapshot for the first time.

This looks good and the lists are loading fine. Well, almost! There is a small glitch in the workout list page. We can easily spot it if we look carefully at any list item (in fact, there is only one item):

The workout duration calculations are not working anymore! What could be the reason? We need to look back at how these calculations were implemented. The `WorkoutPlan` service (in `model.ts`) defines a `totalWorkoutDuration` method that does the math for this.

The difference is in the workout array that is bound to the view. In the previous chapter, we created the array with model objects that were created using the `WorkoutPlan` service. But now, since we are retrieving data from the server, we bind a simple array of JavaScript objects to the view, which for obvious reasons has no calculation logic.

We can fix this problem by mapping a server response to our model class objects and returning them to any upstream caller.

# Mapping server data to application models

Mapping server data to our model and vice versa may be unnecessary if the model and server storage definition match. If we look at the `Exercise` model class and the seed data that we have added for the exercise in MongoLab, we will see that they do match and hence mapping becomes unnecessary.

Mapping a server response to the model data becomes imperative if:

- Our model defines any methods
- A stored model is different from its representation in code
- The same model class is used to represent data from different sources (this can happen for mashups, where we pull data from disparate sources)

The `WorkoutPlan` service is a prime example of an impedance mismatch between a model representation and its storage. Look at the following screenshot to understand these differences:

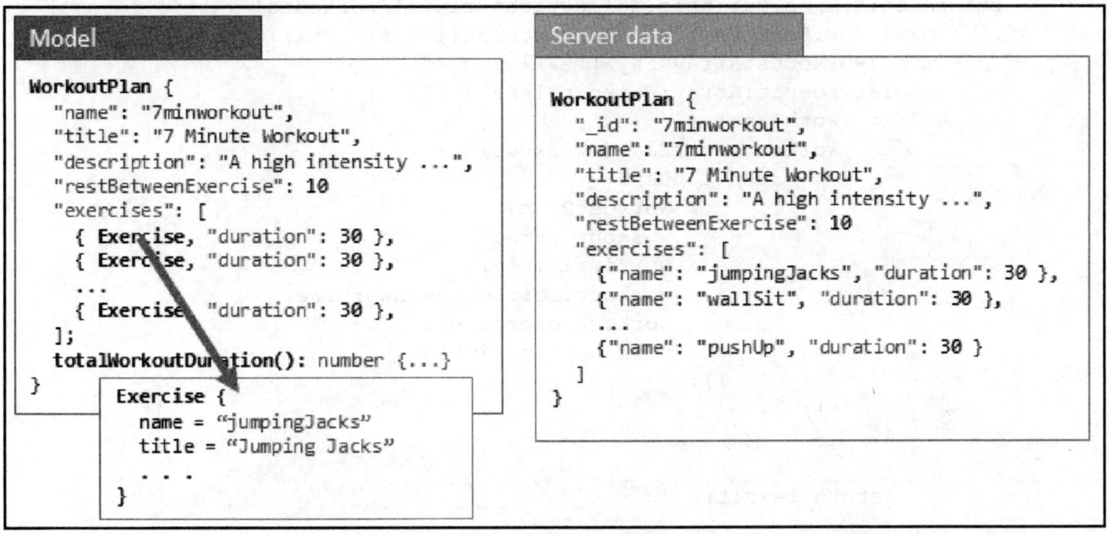

Model

```
WorkoutPlan {
  "name": "7minworkout",
  "title": "7 Minute Workout",
  "description": "A high intensity ...",
  "restBetweenExercise": 10
  "exercises": [
    { Exercise, "duration": 30 },
    { Exercise, "duration": 30 },
    ...
    { Exercise, "duration": 30 },
  ];
  totalWorkoutDuration(): number {...}
}
Exercise {
  name = "jumpingJacks"
  title = "Jumping Jacks"
  . . .
}
```

Server data

```
WorkoutPlan {
  "_id": "7minworkout",
  "name": "7minworkout",
  "title": "7 Minute Workout",
  "description": "A high intensity ...",
  "restBetweenExercise": 10
  "exercises": [
    {"name": "jumpingJacks", "duration": 30 },
    {"name": "wallSit", "duration": 30 },
    ...
    {"name": "pushUp", "duration": 30 }
  ]
}
```

The two major differences between the model and server data are as follows:

- The model defines the `totalWorkoutDuration` method.
- The `exercises` array representation also differs. The `exercises` array of the model contains the full `Exercise` object while the server data stores just the exercise identifier or name.

This clearly means that loading and saving a workout require model mapping.

The way we will do this is by adding a second map to further transform the Observable response object. So far, we have only transformed the response into a plain JavaScript object. The nice thing is that the map request that we just used also returns an Observable, which allows us to chain another map request that transforms the JSON object. With the second map request, we will map that JSON object to the `WorkoutPlan` type in our model.

Let's rewrite the `getWorkouts` method in the `workout-service.ts` file to add the second map:

```
getWorkouts() {
    return this.http.get(this.collectionsUrl + '/workouts' + this.params)
        .map((res:Response) => <WorkoutPlan[]>res.json())
        .map((workouts:Array<any>) => {
            let result:Array<WorkoutPlan> = [];
            if (workouts) {
                workouts.forEach((workout) => {
                    result.push(
                        new WorkoutPlan(
                            workout.name,
                            workout.title,
                            workout.restBetweenExercise,
                            workout.exercises,
                            workout.description
                        ));
                });
            }

            return result;
        })
        .catch(WorkoutService.handleError);
}
```

As previously, our first map transforms the response Observable into an Observable consisting of an array of JavaScript objects. The second map then transforms this Observable into one made up of `WorkoutPlan` objects. Each `WorkoutPlan` object (we have only one at the moment) will then have the `totalWorkoutDuration` method that we need.

Looking at the code for the second map, you can see that we operate on the JSON results from the first method, which is why we are using the `<any>` type. And then we create a typed array of `WorkoutPlans` and iterate through the first array using a fat arrow `forEach` function, assigning each JavaScript object to a `WorkoutPlan` object.

We return the results of these mappings to the callers that subscribe to them, `workouts.component.ts` in this case. The callers do not need to make any changes to the code they use to subscribe to our workouts Observable. Instead, the model mapping can take place at one spot in the application and then be used throughout it.

If you rerun the application, you will see that the total number of seconds now displays properly:

Checkpoint 5.2 in the GitHub repository contains the working implementation for what we have covered so far. The GitHub branch is `checkpoint5.2` (folder: `trainer`).

# Loading exercise and workout data from the server

Just as we fixed the `getWorkouts` implementation in `WorkoutService` earlier, we can implement other get operations for exercise- and workout-related stuff. Copy the service implementation for the `getExercise` and `getWorkout` methods of `WorkoutService` from `workout-builder.component.ts` in the `trainer/src/components/workout-builder` folder in checkpoint 5.2.

The `getWorkout` and `getExercise` methods use the name of the workout/exercise to retrieve results. Every MongoLab collection item has an `_id` property that uniquely identifies the item/entity. In the case of our `Exercise` and `WorkoutPlan` objects, we use the name of the exercise for unique identification. Hence, the `name` and `_id` property of each object always match.

At this point, we will need to add one more import to `workout-service.ts`:

```
import 'rxjs/add/observable/forkJoin';
```

This import brings in the `forkJoin` operator which we will be discussing shortly.

Pay special attention to the implementation for the `getWorkout` method because there is a decent amount of data transformation happening due to the model and data storage format mismatch. This is how the `getWorkout` method now looks:

```
getWorkout(workoutName:string) {
```

```
        return Observable.forkJoin(
            this.http.get(this.collectionsUrl + '/exercises' +
            this.params).map((res:Response) => <Exercise[]>res.json()),
            this.http.get(this.collectionsUrl + '/workouts/' +
            workoutName + this.params).map((res:Response) =>
           <WorkoutPlan>res.json())
        ).map(
            (data:any) => {
                let allExercises = data[0];
                let workout = new WorkoutPlan(
                    data[1].name,
                    data[1].title,
                    data[1].restBetweenExercise,
                    data[1].exercises,
                    data[1].description
                )
                workout.exercises.forEach(
                    (exercisePlan:any) => exercisePlan.exercise =
                     allExercises.find(
                         (x:any) => x.name === exercisePlan.name
                    )
                )
                return workout;
            }
        )
        .catch(WorkoutService.handleError);
    }
```

There is a lot happening inside `getWorkout` that we need to understand.

The `getWorkout` method uses Observable and its `forkJoin` operator to return two Observable objects: one that retrieves the `Workout` and another that retrieves a list of all the `Exercises`. What is interesting about the `forkJoin` operator is that not only does it allow us to return multiple Observable streams, but it also waits until both Observable streams have retrieved their data before further processing the results. In other words, it enables us to stream the responses from multiple concurrent HTTP requests and then operate on the combined results.

Once we have the `Workout` details and the complete list of exercises, we use the map operator (which we saw previously with the code for the `Workouts` list) to update the `exercises` array of the workout to the correct `Exercise` class object. It does this by searching the `allExercises` Observable for the name of the exercise in the `workout.exercises` array returned from the server, and then assigning the matching exercise to the workout services array. The end result is that we have a complete `WorkoutPlan` object with the `exercises` array set up correctly.

These WorkoutService changes warrant fixes in upstream callers too. We have already fixed the lists of exercises in the LeftNavExercises and Exercises components and the workouts in the Workouts component.

Now let's fix the Workout and Exercise components along similar lines. The getWorkout and getExercise methods in the workout services are not directly called by these components but by builder services. Let's now fix the builder services together with the Workout and Exercise components.

# Fixing the builder services

Now that we have WorkoutService set up to retrieve a workout from our remote data store, we have to modify WorkoutBuilderService to be able to retrieve that workout as an Observable. The method that pulls the Workout details is startBuilding. In order to do that, we will break the current startBuilding method into two methods, one for new workouts and one for existing workouts that we have retrieved from the server. Here is the code for new workouts:

```
startBuildingNew(name: string){
    let exerciseArray : ExercisePlan[] = [];
    this.buildingWorkout = new WorkoutPlan("", "", 30, exerciseArray);
    this.newWorkout = true;
    return this.buildingWorkout;
}
```

For existing workouts, we add the following code:

```
startBuildingExisting(name: string){
    this.newWorkout = false;
    return this._workoutService.getWorkout(name);

}
```

We'll let you make the same fixes in ExerciseBuilderService.

# Fixing the Workout and Exercise components

Next, we will update our `Workout` and `Exercise` components to work with the Observable that we are returning from our remote data store. We will fix the `Workout` component and leave it to you to fix the `Exercise` component yourself as it follows a similar pattern. `LeftNavExercises`, used in the workout detail page navigation rendering, is already fixed, so let's jump on to fixing the `Workout` component.

The `Workout` component uses its `ngOnit` life cycle hook to load a new or existing workout. When the route successfully resolves to this component, `ngOnit` uses the injected `WorkoutBuilderService` to load the workout. Here is what the method looks like:

```
ngOnInit() {
    this.sub = this.route.params.subscribe(params => {
        if (!params['id']) {
            this.workout = this.workoutBuilderService.startBuildingNew();
        } else {
            let workoutName = params['id'];
            this.workoutBuilderService.startBuildingExisting(workoutName)
                .subscribe(
                    (data:WorkoutPlan) => {
                        this.workout = <WorkoutPlan>data;
                        if (!this.workout) {
                            this.router.navigate(['/builder/workouts']);
                        } else {
                            this.workoutBuilderService.buildingWorkout =
                            this.workout;
                        }
                    },
                    (err:any) => {
                        if (err.status === 404) {
                            this.router.navigate(['/builder/workouts'])
                        } else {
                            console.error(err)
                        }
                    }
                );
        }
    });
}
```

Firstly, we change the method for a new workout to the `WorkoutBuilderService.startBuildingNew` method. This method creates a new `WorkoutPlan` object and assigns it to a local instance of `WorkoutPlan` that will be used to display the workout on the screen.

Secondly, we change the code for retrieving an existing `WorkoutPlan` to handle the fact that what is now being returned is an Observable. We therefore add the code to subscribe to that Observable and set the results to a local instance of `WorkoutPlan`.

To test the implementation, just load any existing workout detail page, such as *7 Minute Workout* under `/builder/workouts/`. The workout data should load successfully.

The exercise detail page also needs fixing. The `Checkpoint 5.2` file contains the fixed `ExerciseBuilderService` and `Exercise` components that you can copy to load the exercise details; or you can do it yourself and compare the implementation.

# Updating the router guards

As we move on to using Observable types with our data access, we are going to have to make some adjustments to the router guards that we have created for the routes leading to workout and exercise screens. This is because of some timing considerations that come into play when working with Observable types. Put simply, because an Observable is push-based, there is often a slight delay between when our subscription is created and the Observable it is subscribing to is returned. In the case of our remote calls that populate a form or display a list of items, we can manage that delay by simply adding a check for the existence of the item or list before displaying it.

However, no such option exists with our guards which run their checks immediately upon the creation of the subscription. To fix this problem, we need to add in some code to `workout-guard.ts` that ensures that the Observable is resolved before we run the check.

First import the `Observable` from RxJS:

```
import {Observable} from "rxjs/Rx";
```

Next update the `canActivate` method in the `WorkoutGuard` component, as follows:

```
canActivate(route:ActivatedRouteSnapshot,
            state:RouterStateSnapshot):Observable<boolean> {
    let workoutName = route.params['id'];
    return this.workoutService.getWorkout(workoutName)
        .take(1)
        .map(workout => !!workout)
        .do(workoutExists => {
            if (!workoutExists)
            this.router.navigate(['/builder/workouts']);
        })
        .catch(error => {
```

```
            if (error.status === 404) {
                this.router.navigate(['/builder/workouts']);
                return Observable.of(false)
            } else {
                return Observable.throw(error);
            }
        }
    )
```

What we are doing here is using the `take` operator and setting it to `1` in order to stop the Observable subscription when a single result is returned. We then `map` the workout object to a `boolean` (using the JavaScript double not operator) to determine whether it exists. Finally, we use the `do` operator to set the route to false in the event that it does not exist and route the user back to the workouts screen. This gives us the immediate result that we are looking for.

`Checkpoint 5.2` in the GitHub repository contains the working implementation for what we have covered thus far. If you are not using Git, download the snapshot of Checkpoint 5.2 (a ZIP file) from the following GitHub location: `https://github.com/chandermani/angular2byexample/tree/checkpoint5.2`. Refer to the `README.md` file in the `trainer` folder when setting up the snapshot for the first time.

It is now time to fix, create, and update scenarios for the exercises and workouts.

# Performing CRUD on exercises/workouts

When it comes to create, read, update, and delete (CRUD) operations, all save, update, and delete methods need to be converted to the Observable pattern.

Earlier in the chapter, we detailed the endpoint access pattern for CRUD operations in a MongoLab collection. Head back to the *Loading exercise and workout data* section and revisit the access patterns. We need this now as we plan to create/update workouts.

Before we start the implementation, it is important to understand how MongoLab identifies a collection item and what our ID generation strategy is. Each collection item in MongoDB is uniquely identified in the collection using the `_id` property. While creating a new item, either we supply an ID or the server generates one itself. Once `_id` is set, it cannot be changed. For our model, we will use the `name` property of the exercise/workout as the unique ID and copy the name into the `_id` field (hence, there is no auto generation of `_id`). Also remember that our model classes do not contain this `_id` field; it has to be created before saving the record for the first time.

Let's fix the workout creation scenario first.

## Creating a new workout

Taking the bottom-up approach, the first thing that needs to be fixed is `WorkoutService`. Update the `addWorkout` method as shown in the following code:

```
addWorkout(workout:any) {
  let workoutExercises:any = [];
  workout.exercises.forEach((exercisePlan:any) => {
    workoutExercises.push({name: exercisePlan.exercise.name,
duration:exercisePlan.duration})
  });
  let body = {
    "_id": workout.name,
    "exercises": workoutExercises,
    "name": workout.name,
    "title": workout.title,
    "description": workout.description,
    "restBetweenExercise": workout.restBetweenExercise
  };
  return this.http.post(this.collectionsUrl + '/workouts' + this.params,
body)
  .map((res:Response) => res.json())
  .catch(WorkoutService.handleError)
}
```

In `getWorkout`, we had to map data from the server model to our client model; the reverse has to be done here. First, we create a new array for the exercises, `workoutExercises`, and then add to that array a version of the exercises that is more compact for server storage. We only want to store the exercise name and duration in the exercises array on the server (this array is of type `any` because in its compact format it does not conform to the `ExercisePlan` type).

Next, we set up the body of our post by mapping these changes into a JSON object. Note that as part of constructing this object, we set the `_id` property as the name of the workout to uniquely identify it in the database of the workouts collection.

The simplistic approach of using the *name* of the workout/exercise as a record identifier (or `id`) in MongoDB will break for any decent-sized app. Remember that we are creating a web-based application that can be accessed simultaneously by many users. Since there is always the possibility of two users coming up with the same name for a workout/exercise, we need a strong mechanism to make sure that names are not duplicated.
Another problem with the MongoLab REST API is that if there is a duplicate `POST` request with the same `id` field, one will create a new document and the second will update it, instead of the second failing. This implies that any duplicate checks on the `id` field on the client side still cannot safeguard against data loss. In such a scenario, assigning auto generation of the `id` value is preferable.
In standard cases where we are creating entities, unique ID generation is done on the server (mostly by the database). The response to when an entity is created then contains the generated ID. In such a case, we need to update the model object before we return data to the calling code.

Lastly, we call the `post` method of the HTTP module, passing the URL to connect to, an extra query string parameter (`apiKey`), and the data we are sending.

The last return statement should look familiar, as we use Observables to return the workout object as part of the Observable resolution. You need to be sure you add `.subscribe` to the Observable chain in order to make it hot. We'll do that shortly by adding a subscription to the save method of the to the `WorkoutComponent`.

## Updating a workout

Why not try to implement the update operation? The `updateWorkout` method can be fixed in the same manner, the only difference being that the HTTP module's `put` method is required:

```
updateWorkout(workout:WorkoutPlan) {
  let workoutExercises:any = [];
  workout.exercises.forEach((exercisePlan:any) => {
    workoutExercises.push({name: exercisePlan.exercise.name,
    duration:exercisePlan.duration})
  });
  let body = {
    "_id": workout.name,
    "exercises": workoutExercises,
    "name": workout.name,
```

```
        "title": workout.title,
        "description": workout.description,
        "restBetweenExercise": workout.restBetweenExercise
      };
      return this.http.put(this.collectionsUrl + '/workouts/' +
      workout.name + this.params, body)
      .map((res:Response) => res.json())
      .catch(WorkoutService.handleError);
    }
```

The preceding request URL now contains an extra fragment (`workout.name`) that denotes the identifier of the collection item that needs to be updated.

The MongoLab `PUT` API request creates the document passed in as the request body if the document is not found in the collection. While making the `PUT` request, make sure that the original record exists. We can do this by making a `GET` request for the same document first and confirming that we get a document before we update it. We'll leave that for you to implement.

## Deleting a workout

The last operation that needs to be fixed is deleting the workout. Here is a simple implementation where we call the HTTP module's `delete` method to delete the workout referenced by a specific URL:

```
deleteWorkout(workoutName:string) {
  return this.http.delete(this.collectionsUrl + '/workouts/' +
  workoutName + this.params)
  .map((res:Response) => res.json())
  .catch(WorkoutService.handleError)
}
```

## Fixing the upstream code

With that, it's now time to fix the `WorkoutBuilderService` and `Workout` components. The `save` method of `WorkoutBuilderService` now looks like this:

```
save(){
  let workout = this.newWorkout ?
this.workoutService.addWorkout(this.buildingWorkout) :
this.workoutService.updateWorkout(this.buildingWorkout);
  this.newWorkout = false;
  return workout;
```

```
}
```

Most of it looks the same as it was earlier because it is the same! We did not have to update this code because we effectively isolated the interaction with the external server in our `WorkoutService` component.

Finally, the save code for the `Workout` component is shown here:

```
save(formWorkout:any) {
  this.submitted = true;
  if (!formWorkout.valid) return;
  this.workoutBuilderService.save().subscribe(
    success => this.router.navigate(['/builder/workouts']),
    err => console.error(err)
  );
}
```

Here we have made a change so that we now subscribe to the save. As you may recall from our previous discussions, `subscribe` makes an Observable live so that we can complete the save.

And that's it! We can now create new workouts, update existing workouts, and delete them too. That was not too difficult!

Let's try it out. Open the new `Workout Builder` page, create a workout, and save it. Also try to edit an existing workout. Both scenarios should work seamlessly.

Check out `checkpoint 5.3` for an up-to-date implementation if you are having issues running your local copy. If you are not using Git, download the snapshot of Checkpoint 5.3 (a ZIP file) from the following GitHub location: `https://github.com/chandermani/angular2byexample/tree/checkpoint5.3`. Refer to the `README.md` file in the `trainer` folder when setting up the snapshot for the first time.

Something interesting happens on the network side while we make `POST` and `PUT` requests to save data. Open the browser's network log console (*F12*) and see the requests being made. The log looks something like this:

| Name | Method | Status |
|---|---|---|
| 7minworkout?apiKey=[illegible] | OPTIONS | 200 |
| 7minworkout?apiKey=[illegible] | PUT | 200 |
| workouts?apiKey=[illegible] | GET | 200 |

An `OPTIONS` request is made to the same endpoint before the actual `PUT` is done. The behavior that we witness here is termed a **prefight request**. This happens because we are making a cross-domain request to `api.mongolab.com`.

# Using promises for HTTP requests

The bulk of this chapter has focused on how the Angular HTTP client uses Observables as the default for `XMLHttpRequests`. This represents a significant change from the way things used to work. Many developers are familiar with using promises for asynchronous HTTP requests. With that being the case, Angular continues to support promises but just not as the default choice. A developer has to opt for promises in an `XMLHttpRequest` in order to be able to use them.

For example, if we want to use promises with the `getExercises` method in `WorkoutService`, we will first need to import the RxJS `toPromise` operator:

```
import 'rxjs/add/operator/toPromise';
```

Then we will have to restructure the command as follows:

```
getExercises() {
    return this.http.get(this.collectionsUrl + '/exercises' + this.params)
    .toPromise().then((res:Response) => <Exercise[]>res.json())
    .catch(WorkoutService.handleError);
}
```

In order to convert this method to use promises, all we have to do is add `.toPromise()` to the method chain and then replace the call to `.map` with a success parameter, `then`, for the promise. We can leave `catch` as it is.

For upstream components, we just have to switch to handling the return value as a promise rather than an Observable. So, to use promises in this case, we would have to change the code in `Exercises.component.ts` and `LeftNavExercises.component.ts` to first add a new property for the error message:

```
errorMessage: any;
```

Then change the method that is calling `WorkoutService` to the following:

```
ngOnInit() {
  this.workoutService.getExercises()
  .then(exerciseList => this.exerciseList = exerciseList,
   error => this.errorMessage = <any>error
);
}
```

Of course, the ease with which we can substitute promises for Observables in this simple example does not indicate that they are essentially the same. A `then` promise returns another promise, which means that you can create successively chained promises. In the case of an Observable, a subscription is essentially the end of the line and cannot be mapped or subscribed to beyond that point.

If you're familiar with promises, it may be tempting at this stage to stick with them and not give Observables a try. After all, much of what we have done with Observables in this chapter can be done with promises as well. For example, the mapping of two streams of Observables that we did with `getWorkouts` using the Observable's `forkJoin` operator can also be done with the promise's `q, all` function.

However, you would be selling yourself short if you took that approach. Observables open up an exciting new way of doing web development using what is called functional reactive programming. They involve a fundamental shift in thinking that treats an application's data as a constant stream of information to which the application reacts and responds. This shift allows applications to be built with a different architecture that makes them faster and more resilient. Observables are at the core of Angular in such things as event emitters and the new version of `NgModel`.

While promises are a useful tool to have in your toolkit, we encourage you to investigate Observables as you get into developing with Angular. They are part of the forward-looking philosophy of Angular and will be useful in future-proofing both your applications and your skill set.

Check out the `checkpoint 5.3` file for an up-to-date implementation that includes the promises-related code that we covered previously. If you are not using Git, download the snapshot of Checkpoint 5.3 (a ZIP file) from the following GitHub location: `https://github.com/chandermani/angular2byexample/tree/checkpoint5.3`. Refer to the `README.md` file in the `trainer` folder when setting up the snapshot for the first time. Be aware that in the next section, we will be reverting to the use of Observables for this code. This code can be found in the `checkpoint 5.4` file.

# The async pipe

As we have seen with many of the data operations covered in this chapter, there is a fairly common pattern being repeated over and over again. When an Observable is returned from an HTTP request, we convert the response to JSON and subscribe to it. The subscription then binds the Observable output to a UI element. Wouldn't it be nice if we could eliminate this repetitive coding and replace it with a simpler way to accomplish what we are wanting to do?

Not surprisingly, Angular provides us with just the right way to do that. It's called the **async pipe**, and it can be used like any other pipe for binding to an element on the screen. However, the async pipe is a much more powerful mechanism than other pipes. It takes an Observable or a promise as an input and subscribes to it automatically. It also handles the tear down of the subscription for an Observable without necessitating any further lines of code.

Let's look at an example of this in our application. Let's go back to the `LeftNavExercises` component that we were just looking at in the previous section in connection with promises. Note that we have converted this component and the `Exercise` component from promises back to using Observables.

Check out `checkpoint 5.4` file for an up-to-date implementation that includes the conversion of this code to use Observables once again. If you are not using Git, download the snapshot of Checkpoint 5.4 (a ZIP file) from the following GitHub location: `https://github.com/chandermani/angular2byexample/tree/checkpoint5.4`. Refer to the `README.md` file in the `trainer` folder when setting up the snapshot for the first time.

Then make the following changes in `LeftNavExercises`. First, change `exerciseList` from an array of exercises to an Observable of the same type:

```
public exerciseList:Observable<Exercise[]>;
```

Then modify the call to `WorkoutService` to get the exercises to eliminate the subscription:

```
this.exerciseList = this.workoutService.getExercises();
```

Finally, open the template files for each of these components and add the `async` pipe to the `*ngFor` loop, like so:

```
<div *ngFor="let exercise of exerciseList|async|orderBy:'title'">
```

Refresh the page and you will still see the Exercise list displaying. But this time, we have used the `async` pipe to eliminate the need to set up the subscription to the Observable. Pretty cool! This is a nice convenience that Angular has added, and since we have been spending time in this chapter understanding how Observables work with subscriptions, we have a clear idea of what the `async` pipe is now handling for us under the hood.

We'll leave it to you to implement the same change in the `Exercises` component.

It is important to understand the cross-domain behavior of the HTTP request and the constructs that Angular provides to make cross-domain requests.

# Cross-domain access and Angular

Cross-domain requests are requests made for resources in a different domain. Such requests, when originated from JavaScript, have some restrictions imposed by the browser; these are called *same-origin policy* restrictions. Such a restriction stops the browser from making AJAX requests to domains that are different from the script's original source. The source match is done strictly based on a combination of protocol, host, and port.

For our own app, the calls to `https://api.mongolab.com` are cross-domain invocations as our source code hosting is in a different domain (most probably, something like `http://localhost/....`).

There are some workarounds and some standards that help relax/control cross-domain access. We will be exploring two of these techniques as they are the most commonly used ones. They are as follows:

- **JSON with Padding (JSONP)**
- **Cross-Origin Resource Sharing (CORS)**

A common way to circumvent this same-origin policy is to use the JSONP technique.

# Using JSONP to make cross-domain requests

The JSONP mechanism of remote invocation relies on the fact that browsers can execute JavaScript files from any domain irrespective of the source of origin as long as the script is included via the `<script>` tag.

In JSONP, instead of making a direct request to a server, a dynamic `<script>` tag is generated, with the `src` attribute set to the server endpoint that needs to be invoked. This `<script>` tag, when appended to the browser's DOM, causes a request to be made to the target server.

The server then needs to send a response in a specific format, wrapping the response content inside a function invocation code (this extra padding around the response data gives this technique the name JSONP).

The Angular JSONP service hides this complexity and provides an easy API to make JSONP requests. The Plunker link, `http://plnkr.co/edit/ZKAUYeOnlIXau27IWG6V?p=preview`, highlights how JSONP requests are made. It uses the *Yahoo Stock API* to get quotes for any stock symbol.

The Angular JSONP service only supports HTTP `GET` requests. Using any other HTTP request, such as `POST` or `PUT`, will generate an error.

If you look at the Plunker, you will see the familiar pattern for component creation that we have followed throughout this book. We will not go over this pattern again but will highlight a few details that are relevant to using the Angular JSONP service.

First, along with the standard imports you will need to import the `JsonpModule` into `app.module.ts` as follows:

```
. . .
import { JsonpModule }  from '@angular/http';
. . .
@NgModule({
  imports: [
    BrowserModule,
    FormsModule,
    JsonpModule
  ],
. . .
})
```

Next, we need to add the following imports to `get-quote-component.ts`.

```
import { Jsonp, URLSearchParams } from '@angular/http';
import {Observable} from 'rxjs/Observable';
import {Subject} from 'rxjs/Subject';
import 'rxjs/Rx';
```

We are importing `Jsonp` and `URLSearchParams` from the HttpModule and the RxJS Observable as well as `rxjs/Rx`. The latter will bring in the RxJS operators that we need for this example:

As you work with Angular JSONP, it is important to understand that by default, it returns Observables using RxJS. This means that we will have to follow the pattern for subscribing to those Observables and use the RxJS operators to manipulate the results. We can also use the async pipe to streamline these operations.

The next step is to inject JSONP into the constructor:

```
constructor(public jsonp: Jsonp) {}
```

Now we have everything in place for our `getQuote` method. Take a look at the `getQuote` method in the Plunker. We start by defining several variables that we will use to construct our request:

```
getQuote (){
  let url = "https://query.yahooapis.com/v1/public/yql";
  let searchTerm ='select * from yahoo.finance.quote where symbol in
  ("' + this.symbol + '")';
  let env = 'store://datatables.org/alltableswithkeys';
  let params = new URLSearchParams();
  params.set('q', searchTerm); // the user's search value
  params.set('format', 'json');
  params.set('env', env);
  params.set('callback', 'JSONP_CALLBACK');
  this.quote = this.jsonp.get(url, { search: params })
  .map(( res: Response) => res.json());
};
```

We are using the JSONP `get` method to execute the remote call to the Yahoo! quote service. In order to set up that method, we first set the URL for the request. The URL contains the address of the Yahoo! service as well as a rather lengthy query string. The query string contains several required name-value pairs that are needed to make a successful call to the Yahoo! service. These include `q` for the query to be executed, `format` for the format of the response, and `env` for the particular environment we are querying against.

The Angular JSONP service provides a convenient way for us to create this query string. We can construct each parameter individually and then pass them in an array to the `get` method. Angular JSONP will then build the query string for us out of these parameters.

To make a JSONP request, the Angular JSONP service requires us to augment the original URL with an extra query string parameter, `callback=JSONP_ CALLBACK`, verbatim. Internally, the Angular JSONP service then generates a dynamic `script` tag and a function. It then substitutes the `JSONP_CALLBACK` token with the function name generated and makes the remote request.

Open the Plunker and enter symbols such as `GOOG`, `MSFT`, or `YHOO` to see the stock quote service in action. The browser network log for requests looks like this:

```
https://query.yahooapis.com/... & &callback=__ng_jsonp__.__req1.finished
```

Here, `__ng_jsonp__.__req1` is the dynamically generated function. And the response looks like this:

```
__ng_jsonp__.__req1.finished({"query"  ...});
```

The response is wrapped in the callback function. Angular parses and evaluates this response, which results in the invocation of the `__ng_jsonp__.__req1` callback function. Then, this function internally routes the data to our `finished` function callback.

We hope this explains how JSONP works and what the underlying mechanism of a JSONP request is. However, JSONP has its limitations, as given here:

- Firstly, we can make only `GET` requests (which is obvious as these requests originate due to script tags)
- Secondly, the server also needs to implement the part of the solution that involves wrapping the response in a function `callback`, as seen before

- There is always a security risk involved, as JSONP depends on dynamic script generation and injection
- Error handling too is not reliable because it is not easy to determine why a script load failed

Ultimately, we must recognize that JSONP is more of a workaround than a solution. As we moved towards Web 2.0, where mashups became commonplace and more and more service providers decided to expose their API over the Web, a far better solution/standard emerged: CORS.

# Cross-origin resource sharing

Cross-origin Resource Sharing (CORS) provides a mechanism for the web server to support cross-site access control, allowing browsers to make cross-domain requests from scripts. With this standard, a consumer application (such as *Personal Trainer*) is allowed to make some types of requests, termed **simple requests**, without any special setup requirements. These simple requests are limited to `GET`, `POST` (with specific MIME types), and `HEAD`. All other types of requests are termed **complex requests**.

For complex requests, CORS mandates that the request should be preceded by an HTTP `OPTIONS` request (also called a preflight request) that queries the server for HTTP methods allowed for cross-domain requests. And only on successful probing is the actual request made.

You can learn more about CORS from the MDN documentation available at `https://developer.mozilla.org/en-US/docs/Web/HTTP/Access_control_CORS`.

The best part about CORS is that the client does not have to make any adjustment as in the case of JSONP. The complete handshake mechanism is transparent to the calling code and our Angular HTTP client calls work without a hitch.

CORS requires configurations to be made on the server, and the MongoLab servers have already been configured to allow cross-domain requests. So the preceding `POST` requests that we made to the MongoLab to add and update `Exercise` and `Workout` documents all caused the preflight `OPTIONS` request.

# Handling workouts not found

You might recall that in `Chapter 4`, *Building Personal Trainer*, we created the `WorkoutGuard` to prevent navigation to the `WorkoutComponent` if a non-existent workout was in the route parameters. Now we would like to augment this functionality by displaying an error message on the workouts screen, indicating that the workout was not found.

In order to do this, we are going to modify `WorkoutGuard` so that it reroutes to the workouts screen if a workout is not found. To start, add the following child route to `workoutBuilderRoutes` (making sure it precedes the existing workouts route):

```
children: [
  {path: '', pathMatch: 'full', redirectTo: 'workouts'},
  {path: 'workouts/workout-not-found', component: WorkoutsComponent'},
  {path: 'workouts', component: 'WorkoutsComponent'},
   *** other child routes ***
  },
]
```

Next, modify `WorkoutGuard` to redirect to this route in the event that a workout is not found:

```
.do(workoutExists => {
  if (!workoutExists)  this.router.navigate(['/builder/workouts/
  workout-not-found']);
})
```

Then add a `notFound` boolean set to `false` to the variables in the `Workouts` component:

```
public workoutList:Array<WorkoutPlan> = [];
public notFound:boolean = false;
private subscription:any;
```

And, in the ngOnInit method of that component, add the following code to check for the workout-not-found path and set notFound value to true:

```
ngOnInit() {
  if(this.route.snapshot.url[1] && this.route.snapshot.url[1].path ===
  'workout-not-found') this.notFound = true;
  this.subscription = this.workoutService.getWorkouts()
  .subscribe(
    workoutList => this.workoutList = workoutList,
    (err:any) => console.error(err)
  );
}
```

Finally in the Workouts.component.html template add the following div tag above the workout list that will display if the notFound is set to true.

```
<div *ngIf="notFound" class="not-found-msgbox">Could not load the specific
workout!</div>
```

If we find workout-not-found in the path when a user is returned to the Workouts page, then this displays the following message on the screen:

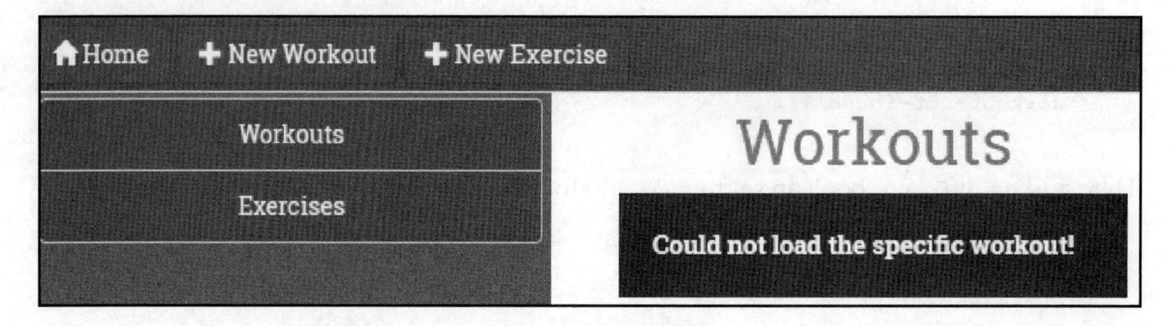


We have fixed routing failure for the Workout Builder page, but the exercise builder page is still pending. Again, we will leave it to you to fix it yourself.

Another major (and pending) implementation is fixing *7 Minute Workout*, as it currently caters to only one workout routine.

# Fixing the 7 Minute Workout app

As it stands now, the *7 Minute Workout* (or *Workout Runner*) app can play only one specific workout. It needs to be fixed to support the execution of any workout plan built using *Personal Trainer*. There is an obvious need to integrate these two solutions. We already have the groundwork done to commence this integration. We've the shared model services and `WorkoutService` to load data, enough to get us started.

Fixing *7 Minute Workout* and converting it into a generic *Workout Runner* roughly involves the following steps:

- Removing the hardcoded workout and exercises used in *7 Minute Workout*.
- Fixing the start page to show all available workouts and allowing users to select a workout to run.
- Fixing the workout route configuration to pass the selected workout name as the route parameter to the workout page.
- Loading the selected workout data using `WorkoutService` and starting the workout.
- And, of course, we need to rename the *7 Minute Workout* part of the app; the name is now a misnomer. I think the complete app can be called *Personal Trainer*. We can remove all references to *7 Minute Workout* from the view as well.

An excellent exercise to try out yourself! And that is why we are not going to walk you through the solution. Instead, go ahead and implement the solution. Compare your implementation with the one available at `checkpoint 5.4`.

It's time to end the chapter and summarize your learning.

# Summary

We now have an app that can do a lot of stuff. It can run workouts, load workouts, save and update them, and track history. And if we look back, we have achieved this with minimal code. We bet that if we were to try this in standard jQuery or some other framework, it would require substantially more effort as compared to Angular.

We started the chapter by providing a *MongoDB* database on *MongoLab* servers. Since MongoLab provided a RESTful API to access the database, we saved some time by not setting up our own server infrastructure.

The first Angular construct that we touched upon was the HTTP client, which is the primary service for connecting to any HTTP backend.

You also learned how the HTTP module uses Observables. For the first time, in this chapter, we created our own Observable and explained how to create subscriptions to those Observables.

We fixed our *Personal Trainer* app so that it uses the HTTP module to load and save workout data. (Note that data persistence for exercises is left for you to complete.) In the process, you also learned about issues surrounding cross-domain resource access. You learned about JSONP, a workaround to circumvent a browser's *same-origin* restrictions, and how to issue JSONP requests using Angular. We also touched upon CORS, which has emerged as a standard when it comes to cross-domain communication.

We have now covered most of the building blocks of Angular, except the big one: Angular directives. We have used directives everywhere but have not created one. The next chapter is exclusively dedicated to Angular directives. We will be creating a number of small directives, such as a remote validator, AJAX button, and a validation cues directive for the *Workout Builder* app. You will also learn how to integrate a jQuery plugin as a directive in Angular.

# 6

# Angular 2 Directives in Depth

**Directives** are everywhere. They are the fundamental building blocks of Angular 2. Each extension to the application has resulted in us creating new **component directives**. These component directives have further consumed **attribute directives** (such as `NgClass` and `NgStyle`) and **structural directives** (such as `NgIf` and `NgFor`) to extend their behavior.

While we have built a number of component directives and a lone attribute directive, there are still some concepts of directive building that are worth exploring. This is especially true for attribute and structural directives, which we are yet to cover in detail.

The topics we will cover in this chapter include the following:

- **Building directives**: We build multiple directives and learn where directives are useful, how they differ from components, and how directives communicate with each other and/or their host component. We explore all directive types, including *component directives, attribute directives*, and *structural directives*.
- **Asynchronous validation**: Angular makes it easy to validate rules that require server interaction and hence are async in nature. We build our first async validator in this chapter.
- **Using renderer for view manipulation**: Renderer allows view manipulation in a platform-agnostic way. We utilize renderer for the busy indicator directive and learn about its API.
- **Host binding**: Host binding allows directives to communicate with their *host element*. This chapter covers how to utilize such bindings for directives.
- **Directive injection**: The Angular DI framework allows directive injection based on where in the HTML hierarchy the directives are declared. We will cover multiple scenarios pertaining to such injections.
- **Working with view children and content children**: Components have the capability to include external view templates into their own view. How to work with the injected content is something we will cover here.

- **Understanding the NgIf platform directive**: We will look under the hood of the `NgIf` platform directive, and try to comprehend the working of *structural directives* such as `NgIf`.
- **View encapsulation of Angular components**: We will learn how Angular uses concepts derived from *web components* to support view and style encapsulation.

Let's start the chapter by reiterating the basic classification of directives.

# Classifying directives

Angular directives integrate the HTML view with the application state. Directives help us manipulate views on application state changes and respond to view updates with little or no manual interaction with the actual DOM.

Depending upon how they affect the view, these directives are further classified into three types.

# Components

**Component directives** or **components** are directives with an encapsulated view. In Angular, when we build UI widgets, we are building components. We have already built a lot of them, such as `WorkoutRunner`, `WorkoutAudio`, `WorkoutBuilder`, and many more!

An important point to realize here is that the view is bound to the *component* implementation and can only work with properties and events defined on the backing component.

# Attribute directives

**Attribute directives**, on the other hand, extend an existing component or HTML element. Consider them as behavioral extensions to these components/elements.

Since directives are behavioral extensions for predefined elements, every directive building exercise involves manipulating the state of the components/elements on which these directives are applied. The `MyAudioDirective` built in `Chapter 3`, *More Angular 2 – SPA, Routing, and Data Flows in Depth*, does the same. The directive wraps the HTML 5 *audio* element (`HTMLAudioElement`) for easy usage. Platform directives such as `ngStyle` and `ngClass` also function in a similar manner.

## Structural directives

**Structural directives**, like attribute directives, do not define their own view. Instead, they work on the *view template* (HTML fragment) provided to them as part of their usage. More often than not, the purpose of a structural directive is to show/hide or clone the template view provided to it. Platform directives such as `NgFor`, `NgIf`, and `NgSwitch` are the prime examples in this category.

Hope this quick refresher on directives is enough to get us started. We'll begin our pursuit by extending the workout builder validations and build an async validator directive.

We are starting from where we left off in `Chapter 5`, *Supporting Server Data Persistence*. The Git branch `checkpoint5.4` can serve as the base for this chapter.
The code is also available on GitHub (`https://github.com/chandermani/angular2byexample`) for everyone to download. Checkpoints are implemented as branches in GitHub.
If you are not using Git, download the snapshot of `checkpoint2.4` (a zip file) from the GitHub location `http://bit.ly/ng2be-checkpoint5-4`.
Refer to the `README.md` file in the `trainer` folder when setting up the snapshot for the first time.
Also remember to update the API key in `services/workout-service.ts` with your own API key.

# Building a remote validator directive

We ended `Chapter 5`, *Supporting Server Data Persistence*, with *Workout Runner* capable of managing workouts in the MongoDB store. Since each workout should have a unique name, we need to enforce the uniqueness constraint. Therefore, while creating/editing a workout, every time the user changes the workout name, we can query MongoDB to verify that the name already exists.

As is the case with any remote invocation, this check too is asynchronous, and hence it requires a *remote validator*. We are going to build this remote validator using Angular's *async validator support*.

**Async validators** are similar to standard custom validators, except that instead of returning a key-value object map or null, the return value of a validation check is a **promise**. This promise is eventually resolved with the validation state being set (if there is an error), or null otherwise (on validation success).

We are going to create a validation directive that does workout name checks. There are two possible implementation approaches for such a directive:

- We can create a directive specifically for unique name validation
- We can create a generic directive that can perform any remote validation

**Validation directives**
While we are building a validation directive here, we could have built a standard custom validator class. The advantage of creating a directive is that it allows us to incorporate the directive in a template-driven form approach, where the directive can be embedded in the view HTML. Or, if the form has been generated using a model (model-driven approach), we can directly use the validator class while creating the `Control` objects.

At first, the requirement of checking duplicate names against a data source (the mLab database) seems to be too a specific requirement and cannot be handled by a generic validator. But with some sensible assumptions and design choices, we can still implement a validator that can handle all types of remote validation, including workout name validation.

The plan is to create a validator that externalizes the actual validation logic. The directive will take the validation function as input. This implies that the actual validation logic is not a part of the validator but a part of the component that actually needs to validate input data. The job of the directive is just to call the function and return the appropriate error keys based on the function's return value.

Let's put this theory into practice and build our remote validation directive, aptly named `RemoteValidatorDirective`.

The companion code base for the following section is Git branch `checkpoint6.1`. You can work along with us or check out the implementation available in the aforementioned folder.
Or if you are not using Git, download the snapshot of `checkpoint6.1` (a zip file) from GitHub location `http://bit.ly/ng2be-checkpoint6-1`. Refer to the `README.md` file in the `trainer` folder when setting up the snapshot for the first time.

## Validating workout names using async validator

Like custom validators, async validators too inherit from the same `Validator` class; but this time, instead of returning an object map, async validators return a `Promise`.

Let's look at the definition of the validator. Create a file, `remote-validator.directive.ts`, in the `workout-builder/shared` folder and add this `RemoteValidatorDirective` implementation:

```
import {provide, Directive, Input} from '@angular/core';
import { NG_ASYNC_VALIDATORS, Validators, Validator, FormControl }
from '@angular/forms';

@Directive({
selector: `[a2beRemoteValidator][ngModel]`,
   providers:[{ provide: NG_ASYNC_VALIDATORS,
 useExisting: RemoteValidatorDirective,
 multi: true }]
  ]
})

export class RemoteValidatorDirective implements Validator {
  @Input("a2beRemoteValidator") validationKey: string;
  @Input("validateFunction")
execute: (value: string) => Promise<boolean>;

  validate(control: FormControl): { [key: string]: any } {
    let value: string = control.value;
    return this.execute(value).then((result: boolean) => {
      if (result) { return null; }
      else {
        let error: any = {};
        error[this.validationKey] = true;
        return error;
      }});
  }
}
```

Since we are registering the validator as a directive instead of registering using a `FormControl` instance (generally used when building forms with a *model-driven approach*), we need the extra provider configuration setting (added in the preceding `@Directive` metadata) by using this syntax:

```
    providers:[{ provide: NG_ASYNC_VALIDATORS,
useExisting: RemoteValidatorDirective,
multi: true }]
```

This statement registers the validator with the existing async validators.

The strange directive selector, `selector:` `` `[a2beRemoteValidator][ngModel]` `` used in the preceding code will be covered in the next section, where we will build a busy indicator directive.

Before we dig into the validator implementation, let's add it to the workout name input. This will help us correlate the behavior of the validator with its usage.

Update the workout name input (`workout.component.html`) with the validator declaration:

```
<input type="text" name="workoutName" ...
  a2beRemoteValidator="workoutname"
[validateFunction]="validateWorkoutName">
```

And add the customary declaration for the validation directive to the workout builder module (`workout-builder.module.ts`):

```
import { RemoteValidatorDirective } from "./shared/remote-
validator.directive";
...
declarations: [WorkoutBuilderComponent,...
RemoteValidatorDirective],
```

The remote validator is referenced in the view as `a2beRemoteValidator`.

**Prefixing the directive selector**
Always prefix your directives with an identifier (`a2be` as you just saw) that distinguishes them from framework directives and other third-party directives.

The directive implementation takes two inputs: the validation key (`validationKey`) used to set the *error key*, and the validation function called to validate the value of the control. Both inputs are annotated with the `@Input` decorator.

The input parameter `@Input("validateFunction") execute: (value: string) => Promise<boolean>;`, binds to a function, not a standard component property. We are allowed to treat the function as a property due to the nature of the underlying language, TypeScript (as well as JavaScript).

When the async validation fires (on a change of `input`), Angular invokes the `validate` function, passing in the underlying `control`. As the first step, we pull the current input value and then invoke the `execute` function with this input. The `execute` function returns a promise, which should eventually resolve to `true` or `false`:

- If it is `true`, the validation is successful, the promise callback function returns `null`.
- If it is `false`, the validation has failed, and an error key-value map is returned. The *key* here is the string literal that we set when using the validator (`a2beRemoteValidator="workoutname"`).

This *key* comes in handy when there are multiple validators declared on the input, allowing us to identify validations that have failed.

Let's add a validation message for this failure too. Add this label declaration after the existing validation `label` for *workout name*:

```
<label *ngIf="name.control.hasError('workoutname')" class="alert alert-
danger">A workout with this name already exists.</label>
```

And then wrap these two labels inside a `div`, as we do for *workout title* error labels.

The `hasError` function checks whether the `'workoutname'` validation key is present.

The last missing piece of this implementation is the actual validation function we assigned when applying the directive (`[validateFunction]="validateWorkoutName"`) but never implemented.

Add the `validateWorkoutName` function to `workout.component.ts`:

```
validateWorkoutName = (name: string): Promise<boolean> => {
if (this.workoutName === name) return Promise.resolve(true);
return this.workoutService.getWorkout(name)
          .toPromise()
          .then((workout: WorkoutPlan) => {
             return !workout;
          }, error => {
               return true;
          });
}
```

Before we explore what the preceding function does, we need to do some more fixes on `WorkoutComponent` class. The `validateWorkoutName` function is dependent on `WorkoutService` to get a workout with a specific name. Let's inject the service in the constructor and add the necessary import in the imports section:

```
import { WorkoutService }  from "../../../services/workout-service";
...
constructor(... private workoutService: WorkoutService) {
```

Let's also update the `ngOnInit` function and convert the local variable `workoutName` into a *class member*. Declare `workoutName` with other class members:

```
private workoutName: string;
```

Change the first statement inside the `else` statement to:

```
this.workoutName = params['id'];
```

And use the same variable later during the `startBuildingExisting` function call.

The first `if` condition in `validateWorkoutName` is for the update scenario. We obviously do not want to validate the existing workout name. The `return Promise.resolve(true);` statement returns a promise that always resolves to `true`.

The reason for defining the `validateWorkoutName` function as an *instance function* (the use of the *arrow operator*) instead of defining it as a standard function (which declares the function on the *prototype*) is the `'this'` scoping issue.

Look at the validator function invocation inside `RemoteValidatorDirective` (declared using `@Input("validateFunction") execute;`):

```
return this.execute(value).then((result: boolean) => { ... });
```

When the function (named `execute`) is invoked, the `this` reference is bound to `RemoteValidatorDirective` instead of the `WorkoutComponent`. Since `execute` is referencing the `validateWorkoutName` function in the preceding setup, any access to `this` inside `validateWorkoutName` is problematic.

This causes the `if (this.workoutName === name)` statement inside `validateWorkoutName` to fail, as `RemoteValiatorDirective` does not have a `workoutName` instance member. By defining `validateWorkoutName` as an instance function, the *TypeScript* compiler *creates a closure* around the value of `this` when the function is defined.

With the new declaration, the `this` inside `validateWorkoutName` always points to the `WorkoutComponent` irrespective of how the function gets invoked.

We can also look at the compiled JavaScript for `WorkoutComponent` to know how the closure works with respect to `validateWorkoutName`. The parts of the generated code that interest us are as follows:

```
function WorkoutComponent(...) {
  var _this = this;
  ...
  this.validateWorkoutName = function (name) {
    if (_this.workoutName === name)
      return Promise.resolve(true);
```

If we look at the validation function implementation, we see that it involves querying *mLab* for a specific workout name. The `validateWorkoutName` function returns `true` when a workout with the same name is not found and `false` when a workout with the same name is found (actually a *promise* is returned).

The `getWorkout` function on `WorkoutService` returns an *observable* but we convert it into a *promise* by calling the `toPromise` function on the observable. We need a promise object as `RemoteValidatorDirective` requires a promise to be returned from the validator function.

The validation directive can now be tested. Create a new workout and enter an existing workout name such as `7minworkout`. See how the validation error message shows up eventually:

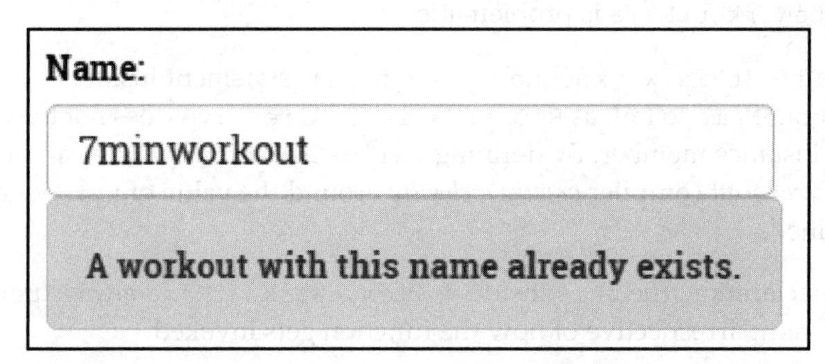


Excellent! It looks great, but there is still something missing. The user is not informed that we are validating the workout name. We can improve this experience.

# Building a busy indicator directive

While the workout name is being validated remotely, we want the user to be aware of the activity in the background. A visual clue around the input box while the remote validation happens should serve the purpose.

Think carefully; there is an input box with an asynchronous validator (which does remote validation) and we want to adorn the input box with a visual clue during validation. Seems like a common pattern to solve? Indeed it is, so let's create another directive!

But before we start, it is imperative to understand that we are not into it alone. The busy indicator directive requires the help of another directive, `NgModel`. We have already used the `NgModel` directive on `input` elements in `Chapter 4`, *Building Personal Trainer*. `NgModel` helps us track the input element state. The following example taken from `Chapter 4`, *Building Personal Trainer*, highlights how `NgModel` helps us validate inputs:

```
<input type="text" name="workoutName" #name="ngModel"  class="form-control"
id="workout-name" ... [(ngModel)]="workout.name" required>
...
<label *ngIf="name.control.hasError('required') && (name.touched ||
submitted)" class="alert alert-danger">Name is required</label>
```

Even the error label for the unique workout name validation done in the previous section employs the same technique of using NgModel to check the validation state.

Let's begin with defining the outline of the directive, along with the decorator metadata. Create a busy-indicator.directive.ts file in the workout-builder/shared folder, and add the following code:

```
import {Directive} from '@angular/core';
import {NgModel} from '@angular/form;

@Directive({
  selector: '[a2beBusyIndicator]',
})
export class BusyIndicatorDirective {
  constructor(private model: NgModel) { }
}
```

The directive selector metadata specifies that the busy indicator will be applied to elements/components having the a2beBusyIndicator attribute.

Add this directive to the declaration section of the workout builder module (workout-builder.module.ts) before proceeding any further.

The only point of interest in the preceding code is Angular DI injecting the NgModel directive associated with the input element. Remember that the NgModel directive is already present on input (workoutname):

```
    <input... name="workoutName" #name="ngModel" [(ngModel)]="workout.name"
...>
```

This is enough to integrate our new directive in the workout view, so let's do it quickly.

Open workout.component.html from workout-builder and add the busy indicator directive to the workout name input:

```
<input type="text" name="workoutName" ... a2beBusyIndicator>
```

Create a new workout or open an existing one to see whether the BusyIndicatorDirective is loaded and the NgModel injection worked fine. This can be easily verified by putting a breakpoint inside the BusyIndicatorDirective constructor.

Angular injects the same NgModel instance into BusyIndicatorDirective that it created when it encountered ngModel on the input HTML.

You may be wondering what happens if we apply this directive on an input element that does not have the `ngModel` attribute, or as a matter of fact on any HTML element/component, such as this:

```
<div a2beBusyIndicator></div>
<input type="text" a2beBusyIndicator>
```

Will the injection work?

Of course not! We can try it on the create workout view. Open `workout.component.html` and add the following `input` above the workout name `input`. Refresh the app:

```
<input type="text" name="workoutName1" a2beBusyIndicator>
```

Angular throws an exception, as follows:

```
EXCEPTION: No provider for NgModel! (BusyIndicatorDirective -> NgModel)
```

How to avoid this? Well, Angular's DI can rescue us here as it allows us to declare an optional dependency.

Remove the `input` control you just added before proceeding further.

# Injecting optional dependencies with the @Optional decorator

Angular has an `@Optional` decorator, which when applied to a constructor argument instructs the Angular *injector* to inject `null` if the dependency is not found.

Hence, the busy indicator constructor can be written as follows:

```
constructor(@Optional() private model: NgModel) { }
```

Problem solved? Not really; as stated previously, we require the `NgModel` directive for `BusyIndicatorDirective` to work. So, while we have learned something new, it is not very useful in the current scenario.

Before proceeding further, remember to revert the `workoutnameinput` to its original state, with `a2beBusyIndicator` applied.

`BusyIndicatorDirective` should only to be applied if there is an `NgModel` directive already present on the element.

The `selector` directive is going to save our day this time. Update the `BusyIndicatorDirective` selector to this:

```
selector: `[a2beBusyIndicator][ngModel]`
```

This selector creates the `BusyIndicatorDirective` only if the combination of `a2beBusyIndicator` with the `ngModel` attribute is present on the element. Problem solved!

It's now time to add the actual implementation.

# Implementation 1 – using renderer

For `BusyIndicatorDirective` to work, it needs to know when the async validation on the `input` fires and when it is over. This information is only available with the `NgModel` directive. `NgModel` has a property, `control`, which is an instance of the `Control` class. It is this `Control` class that tracks the current state of the input, including the following:

- Currently assigned validators (sync and async)
- The current value
- The input element state, such as `pristine`, `dirty`, and `touched`
- The input validation state, which could be any one of `valid`, `invalid`, or `pending` in the case of validation being performed asynchronously
- Events that track when the value changes or the validation state changes.

`Control` seems to be a useful class, and it's the `pending` state that interests us!

Let's add our first implement for `BusyIndicatorDirective` class. Update the class with this code:

```
private subscriptions: Array<any> = [];
ngAfterViewInit() {
this.subscriptions.push(
this.model
          .control.statusChanges
          .subscribe((status: any) => {
              if (this.model.control.pending) {
                this.renderer.setElementStyle(
                  this.element.nativeElement,
"border-width", "3px");
                this.renderer.setElementStyle(
                  this.element.nativeElement,
"border-color", "gray");
              }
              else {
                this.renderer.setElementStyle(
                  this.element.nativeElement,
"border-width", null);
this.renderer.setElementStyle(
                      this.element.nativeElement,
"border-color", null);
              }
        }));
    }
```

Two new dependencies need to be added to the constructor, as we use them in the `ngAfterViewInit` function. Update the `BusyIndicatorDirective` constructor to look like this:

```
constructor(private model: NgControl,
    private element: ElementRef,     private renderer: Renderer) { }
```

And also add imports for `ElementRef` and `Renderer` in `'@angular/core'`.

`ElementRef` is a wrapper object over the underlying HTML element (`input` in this case). The `MyAudioDirective` directive built in `Chapter 3`, *More Angular 2 – SPA, Routing, and Data Flows in Depth,* used `ElementRef` to get hold of the underlying `Audio` element.

The `Renderer` injection deserves a bit of attention. Calling `setElementStyle` is a dead giveaway that `Renderer` is responsible for managing the DOM. But before we delve more deeply into the role of `Renderer`, let's try to understand what the preceding code is doing.

In the preceding code, the `control` property on the model (the `NgModel` instance) defines an event (an `Observable`), `statusChanges`, which we can subscribe to in order to know when the control validation state changes. The available validation states are `valid`, `invalid`, and `pending`.

The subscription checks whether the control state is `pending` or not, and accordingly adorns the underlying element using the `Renderer` API function, `setElementStyle`. We set the `border-width` and `border-color` of the input.

The preceding implementation is added to the `ngAfterViewInit` directive lifecycle hook, which is called after the view has initialized.

Let's try it out. Open the create workout page or the existing *7 Minute Workout*. As soon as you start typing/editing the workout name, the `input` style changes and reverts once the remote validation of the workout name is complete. Nice!

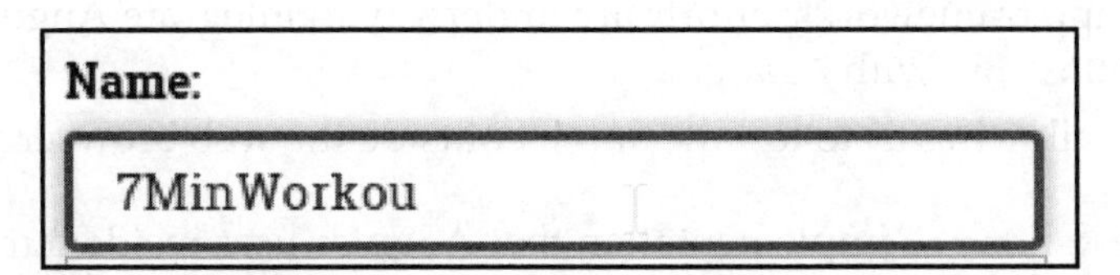


Before moving forward, also add the un-subscription code to the `BusyIndicatorDirective` to avoid memory leak. Add this function to `BusyIndicatorDirective`:

```
ngOnDestroy() {
    this.subscriptions.forEach((s) => s.unsubscribe());
}
```

**Always unsubscribe from observables**
Always remember to unsubscribe from any `Observable/EventEmitter` subscription done in the code to avoid memory leak.

The implementation looks good. The `Renderer` is doing its job. But there are some unanswered questions.

Why not just get hold of the underlying DOM object and use the standard DOM API to manipulate the input styles? Why do we need the *renderer*?

# Angular renderer, the translation layer

One of the primary design goals of Angular 2 was to make it run across environments, frameworks, and devices. Angular enabled this by dividing the core framework implementation into an application layer and a rendering layer. The application layer has the API we interact with, whereas the rendering layer provides an abstraction that the application layer can use without worrying about how and where the actual view is being rendered.

By separating the rendering layer, Angular can theoretically run in various setups. These include (but are not limited to):

- Browser
- Browser main thread and web worker thread, for obvious performance reasons
- Server-side rendering
- Native app frameworks; efforts are underway to integrate Angular 2 with `NativeScript` with `ReactNative`
- Testing, allowing us to test the app UI outside the web browser

The `Renderer` implementation that Angular uses inside our browser is `DOMRenderer`. It is responsible for translating our API calls into browser DOM updates. In fact, we can verify the renderer type by adding a breakpoint in the `BusyIndicatorDirective`'s constructor and seeing the value of `renderer`.

For this precise reason, we avoid direct manipulation of DOM elements inside `BusyIndicatorDirective`. You never know where the code will end up running. We could have easily done this:

```
this.element.nativeElement.style.borderWidth="3px";
```

Instead, we used the `Renderer` to do the same in a platform-agnostic way.

Look at the `Renderer` API function `setElementStyle`:

```
this.renderer.setElementStyle(
          this.element.nativeElement, "border-width", "3px");
```

It takes the element on which the style has to be set, the style property to update, and the value to set. The `element` references the `input` element injected into `BusyIndicatorDirective`.

**Resetting styles**
Styles set by calling `setElementStyle` can be reset by passing a `null` value in the third argument. Check out the `else` condition in the preceding code.

The `Renderer` API has a number of other methods that can be used to set attributes, set properties, listen to events, and even create new views. Whenever you build a new directive, remember to evaluate the `Renderer` API for DOM manipulation.

A more detailed explanation of `Renderer` and its application is available as part of Angular's design documents here:
`http://bit.ly/ng2-render`

We are not done yet! With Angular's awesomeness, we can improve the implementation. Angular allows us to do *host binding* in directive implementation, helping us avoid a lot of boilerplate code.

# Host binding in directives

In the Angular realm, the component/element that a directive gets attached to is termed the **host element**: a container that hosts our directive/component. For the `BusyIndicatorDirective`, the `input` element is the *host*.

While we can use the `Renderer` to manipulate the host (and we did too), the Angular data binding infrastructure can reduce the boilerplate code further. It provides a declarative way to manage directive-host interaction. Using the host binding concepts, we can manipulate an element's *properties* and *attributes* and subscribe to its *events*.

Let's understand each of the host binding capabilities, and at the end, we will fix our `BusyIndicatorDirective` implementation.

# Property binding using @HostBinding

Use **host property binding** to bind a *directive property* to a *host element property*. Any changes to the directive property will be synced with the linked host property during the change detection phase.

We just need to use the `@HostBinding` decorator on the directive property that we want to sync with. For example, consider this binding:

```
@HostBinding("readOnly") get busy() {return this.isbusy};
```

When applied to `input`, it will set the `inputreadOnly` property to `true` when the `isbusy` directive property is `true`.

Note that `readonly` is also an attribute on `input`. What we are referring to here is the input property `readOnly`.

## Attribute binding

**Attribute binding** binds a directive property to a host component attribute. For example, consider a directive with binding like this:

```
@HostBinding("attr.disabled") get canEdit(): string
  { return !this.isAdmin ? "disabled" : null };
```

If applied to input, it will add the `disabled` attribute on `input` when the `isAdmin` flag is `false`, and clear it otherwise. We follow the same attribute binding notation used in the HTML template here too. The attribute name is prefixed with string literal `attr`.

We can do something similar with *class* and *style binding* too. Consider the following line:

```
@HostBinding('class.valid')
   get valid { return this.control.valid; }
```

This line sets up a class binding, and the following line creates a style binding:

```
@HostBinding("style.borderWidth")
   get focus(): string { return this.focus?"3px": "1px"};
```

## Event binding

Lastly, **event binding** is used to subscribe to the events raised by the host component/element. Consider this example:

```
@Directive({ selector: 'button, div, span, input' })
class ClickTracker {
  @HostListener('click', ['$event.target'])
  onClick(element: any) {
    console.log("button", element, "was clicked");
  }
}
```

This sets up a listener on the host event `click`. Angular will instantiate the preceding directive for every *button, div, span,* and *input* on the view and set up the host binding with the `onClick` function. The `$event` variable contains the event data for the event raised, and `target` refers to the element/component that was clicked on.

Event bindings work for components too. Consider the following example:

```
@Directive({ selector: 'workout-runner' })
class WorkoutTracker {
  @HostListener('workoutStarted', ['$event'])
  onWorkoutStarted(workout: any) {
    console.log("Workout has started!");
  }
}
```

With this directive, we track the `workoutStarted` event defined on the `WorkoutRunner` component. The `onWorkoutStarted` function is called when the workout starts, with the details of the started workout.

Now that we understand how these bindings work, we can improve our `BusyIndicatorDirective` implementation.

# Implementation 2 – BusyIndicatorDirective with host bindings

You may have already guessed it! We will use *host property binding* instead of `Renderer` to set styles. Want to give it a try? Go ahead! Clear the existing implementation and try to set up a host binding for the `borderWidth` and `borderColor` style attributes without looking at the following implementation.

This is how the directive will look after the host binding implementation:

```
import {Directive, HostBinding} from '@angular/core';
import {NgModel} from '@angular/forms';

@Directive({ selector: `[a2beBusyIndicator][ngModel]`})
export class BusyIndicatorDirective {
  private get validating(): boolean {
   return this.model.control != null &&
          this.model.control.pending; };

  @HostBinding("style.borderWidth") get controlBorderWidth():
      string { return this.validating ? "3px" : null; };
  @HostBinding("style.borderColor") get controlBorderColor():
      string { return this.validating ? "gray" : null };

  constructor(private model: NgModel) { }
}
```

We have moved the `pending` state check into a directive property called `validating` and then used the `controlBorderWidth` and `controlBorderColor` properties for style binding. This is definitely more succint than our earlier approach! Go test it out.

And if we tell you that this can be done without the need for a custom directive, don't be surprised. This is how we do it, just by using style bindings on the workout name `input`:

```
<input type="text" name="workoutName" ...
[style.borderColor]="name.control.pending ? 'gray' :
null"[style.borderWidth]="name.control.pending ? '3px' : null">
```

We get the same effect!

No, our effort did not go to waste. We did learn about **renderer** and **host binding**. These concepts will come in handy while building directives that provide complex behavior extension instead of just setting element styles.

If you are having a problem with running the code, look at the Git branch `checkpoint6.1` for a working version of what we have done thus far. Or if you are not using Git, download the snapshot of `checkpoint6.1` (a zip file) from `http://bit.ly/ng2be-checkpoint6-1`. Refer to the `README.md` file in the `trainer` folder when setting up the snapshot for the first time.

The next topic that we are going to take up is *directive injection*.

# Directive injection

Go back a few pages and look at the `BusyIndicatorDirective` implementation that uses the renderer, specifically the constructor:

```
constructor(private model: NgModel ...) { }
```

Angular automatically locates the `NgModel` directive created for the current element and injects it into `BusyIndicatorDirective`. This is possible because both directives are declared on the same *host element*.

The good news is that we can influence this behavior. Directives created on a parent HTML tree or child tree can also be injected. The next few sections talk about how to inject directives across the component tree, a very handy feature that allows cross-directive communication for directives that have a *common lineage* (in a view).

We will use Plunker (`http://bit.ly/ng2be-directive-tree`) to demonstrate these concepts.

To start with, look at the Plunker file `app.component.ts`. It has three directives: `Relation`, `Acquaintance`, and `Consumer` and this view hierarchy is defined:

```
<div relation="grand-parent" acquaintance="jack">
    <div relation="parent">
      <div relation="me" consumer>
        <div relation="child-1">
          <div relation="grandchild-1"></div>
        </div>
        <div relation="child-2"></div>
      </div>
    </div>
</div>
```

In the next few sections, we describe the various ways in which we can inject the different `relation` and `Acquaintance` directives into the `consumer` directive. Check out the browser console for the injected dependencies that we log during the `ngAfterViewInit` lifecycle hook.

# Injecting directives defined on the same element

Constructor injection by default supports injecting directives defined on the same element. The constructor function just needs to declare the directive type variable that we want to inject:

```
variable:DirectiveType
```

The `NgModel` injection we did in `BusyIndicatorDirective` falls under this category. If the directive is not found on the current element, the Angular DI will throw an error, unless we mark the dependency `@Optional`.

**Optional dependency**
The `@Optional` decorator is not limited to directive injection. It's there to mark any type of dependency optional.

From the plunk example, the first injection injects the `Relation` directive with the me attribute (`relation="me"`) into the consumer directive:

```
constructor(private me:Relation ...
```

# Injecting directive dependency from the parent

Prefixing a constructor argument with the `@Host` decorator instructs Angular to search for the dependency on the *current element, its parent,* or its *parents* until it reaches the component boundaries (a component with the directive present somewhere in its view hierarchy). Check the second `consumer` injection:

```
constructor(..., @Host() private myAcquaintance:Acquaintance
```

This statement injects the `Acquaintance` directive instance declared two levels up the hierarchy.

Like the `@Option` decorator described previously, the usage of `@Host()` too is not limited to directives. Angular service injection also follows the same pattern. If a service is marked with `@Host`, the search stops at the host component. It does not continue further up the component tree.

The `@Skipself` decorator can be used to skip the current element for a directive search.

From the plunk example, this injection injects the `Relation` directive with the parent attribute (`relation="parent"`) into consumer:

```
@SkipSelf() private myParent:Relation
```

# Injecting a child directive (or directives)

If there is a need to inject directive(s) defined on nested HTML into a parent directive/ component, there are four decorators that can help us:

- `@ViewChild/@ViewChildren`
- `@ContentChild/@ContentChildren`

As these naming conventions suggest, there are decorators to inject a single child directive or multiple children directives:

To understand the significance of `@ViewChild/@ViewChildren` versus `@ContentChild/@ContentChildren`, we need look at the what view and content children are, a topic that we will take up soon. But for now, it's enough to understand that view children are part of a component's own view and content children are external HTML injected into the component's view.

Look how, in Plunket, the `ContentChildren` decorator is used to inject the child `Relation` directive into `Consumer`:

```
@ContentChildren(Relation) private children:QueryList<Relation>;
```

Surprisingly, the data type of the variable `children` is not an array but a custom class- `QueryList`. The `QueryList` class is not a typical array but a collection that is kept up to date by Angular whenever dependencies are added or removed. This can happen if the DOM tree is created/destroyed when using structural directives such as `NgIf` or `NgFor`. We will also talk more about `QueryList` in the coming sections.

You may have observed that the preceding injection is not a constructor injection as were the earlier two examples. This is for a reason. The injected directive(s) will not be available until the underlying component/element's content has initialized. For this precise reason, we have the console.log statements inside the `ngAfterViewInit` lifecycle hook. We should only access the content children post this lifecycle hook execution.

The preceding sample code injects in all three child `relation` objects into the `consumer` directive.

# Injecting descendant directive(s)

The standard `@ContentChildren` decorator (or as a matter of fact `@ViewChildren` too) only injects the immediate children of a directive/component and not its descendants. To include all its descendants, we need to provide an argument to `Query`:

```
@ContentChildren(Relation, {descendants: true}) private
allDescendents:QueryList<Relation>;
```

Passing the `descendants: true` parameter will instruct Angular to search for all descendants.

If you look at the plunker log, the preceding statement injects in all four descendants.

The Angular DI, while it seems simple to use, packs a lot of functionality. It manages our services, components, and directives and provides us with the right stuff in the right place at the right time. Directive injection in components and other directives provides a mechanism for directives to communicate with each other. Such injections allow one directive to access the public API (public functions/properties) of another directive.

It's now time to explore something new. We are going to build an Ajax button component that allows us to inject an external view into the component, a process also known as **content transclusion**.

# Building an Ajax button component

When we save/update an exercise or workout, there is always the possibility of duplicate submission (or duplicate `POST` requests). The current implementation does not provide any feedback as to when the save/update operation started and when it completed. The user of an app can knowingly or unknowingly click on the **Save** button multiple times due to the lack of visual clues.

Let's try to solve this problem by creating a specialized button—an *Ajax button* that gives some visual clues when clicked on and also stops duplicate Ajax submissions.

The button component will work on these lines. It takes a function as input. This input function (input parameter) should return a promise pertaining to the remote request. On clicking on the button, the button internally makes the remote call (using the input function), tracks the underlying promise, waits for it to complete, and shows some busy clues during this activity. Also, the button remains disabled until the remote invocation completes to avoid duplicate submission.

The companion code base for the following section is git branch `checkpoint6.2`. You can work along with us, or check out the implementation available in the branch.
Or if you are not using Git, download the snapshot of `checkpoint6.2` (a zip file) from the GitHub location `http://bit.ly/ng2be-checkpoint6-2`. Refer to the `README.md` file in the `trainer` folder when setting up the snapshot for the first time.

Let's create the component outline to make things clearer. Create a file, `ajax-button.component.ts`, inside the `workout-builder/shared` folder and add the following component outline:

```
import {Component, Input } from '@angular/core';

@Component({
  selector: 'ajax-button',
  template: `<button [attr.disabled]="busy"
              class="btn btn-primary">
            <span [hidden]="!busy">
            <span class="glyphicon
                glyphicon-refresh spin"></span>
            </span>
            <span>Save</span>
          </button>`
})
export class AjaxButtonComponent {
  busy: boolean = null;
  @Input() execute: any;
  @Input() parameter: any;
}
```

The component (`AjaxButtonComponent`) takes two property bindings, `execute` and `parameter`. The `execute` property points to the function that is invoked on the Ajax button click. The `parameter` is the data that can be passed to this function.

Look at the usage of the `busy` flag in the view. We disable the button and show the spinner when the `busy` flag is set. Let's add the implementation that makes everything work. Add this code to the `AjaxButtonComponent` class:

```
@HostListener('click', ['$event'])
onClick(event: any) {
    let result: any = this.execute(this.parameter);
    if (result instanceof Promise) {
      this.busy = true;
      result.then(
        () => { this.busy = null;},
  (error:any) => { this.busy = null; });
    }
}
```

We set up a *host event binding* to the click event on `AjaxButtonComponent`. Anytime the `AjaxButtonComponent` component is clicked on, the `onClick` function is invoked.

The `HostListener` import needs to be added to the `'@angular/core'` module.

The `onClick` implementation calls the input function with a lone parameter as `parameter`. The result of the invocation is stored in the `result` variable.

The `if` condition checks whether the `result` is a `Promise` object. If yes, the `busy` indicator is set to `true`. The button then waits for the promise to get resolved, using the `then` function. Irrespective of whether the promise is resolved with *success* or *error*, the busy flag is set to `null`.

The reason the busy flag is set to `null` and not `false` is due to this attribute binding `[attr.disabled]="busy"`. The `disabled` attribute will not be removed unless `busy` is `null`. Remember that in HTML, `disabled="false"` does not enable the button. The attribute needs to be removed before the button becomes clickable again.

Suppose there is still confusion about this line:

```
let result: any = this.execute(this.parameter);
```

Then you need to look at how the component is used. Open `workout.component.html` and replace the `Save` button HTML with the following:

```
<ajax-button [execute]="save" [parameter]="f"></ajax-button>
```

We pass the `Workout.save` function binds to `execute` and `parameter`; it takes the `FormControl` object `f`.

We need to change the `save` function in the `Workout` class to return a promise for `AjaxButtonComponent` to work. But before proceeding, open `workout.component.ts`, add `AjaxButtonComponent`, import it, and reference it in the `directives` array.

Then change the `save` function implementation to the following:

```
save = (formWorkout: any): Promise<any> => {
        this.submitted = true;
        if (!formWorkout.valid) return;
        let savePromise =
        this.workoutBuilderService.save().toPromise();
        savePromise.then(
           (data) => this.router.navigate(['/builder/workouts']),
           (err) => console.error(err)
        );
           return savePromise;
}
```

The `save` function now returns a *promise* that we build by calling the `toPromise` function on the *observable* returned from the call to `workoutBuilderService.save()`.

Also add `AjaxButtonComponent` to the workout builder module's (`workout-builder.module.ts`) declarations section.

Make note of how we define the `save` function as an *instance function* (with the use of the arrow operator) to create a closure over *this*. It's something we did earlier too while building the *remote validator directive*.

Time to test our implementation! Refresh the application and open the create/edit workout view. Click on the **Save** button and see the Ajax button in action:

We started this section with the aim of highlighting how external elements/components can be transcluded into a component. Let's do it now!

# Transcluding external components/elements into a component

From the very start, we need to understand what **transclusion** means. And the best way to understand this concept would be to look at an example.

No component that we have built thus far has borrowed content from outside. Not sure what this means?

Consider the preceding `AjaxButtonComponent` example:

```
<ajax-button [execute]="save" [parameter]="f"></ajax-button>
```

What if we change the `ajax-button` usage to the following?

```
<ajax-button [execute]="save" [parameter]="f">Save Me!</ajax-button>
```

Will the `Save Me!` text show up on the button? It will not, try it!

The `AjaxButtonComponent` component already has a template, and it rejects the content we provide in the preceding declaration. What if we can somehow make the content (`Save Me!` in the preceding example) load inside the `AjaxButtonComponent`? This act of injecting an external view fragment into the component's view is what we call **transclusion**, and the framework provides the necessary constructs to enable transclusions.

It's time to introduce two new concepts, *content children* and *view children*.

# Content children and view children

To define it succinctly, the HTML structure that a component defines internally (using `template/templateUrl`) is the **view children** of the component. However, the HTML view provided as part of the component usage added to the host element (like `<ajax-button>`), defines the **content children** of the component.

By default, Angular does not allow *content children* to be embedded as we saw before. The `Save Me!` text was never emitted. We need to explicitly tell Angular where to emit the *content children* inside the *component view template*. To understand this concept, let's fix the `AjaxButtonComponent` view. Open `ajax-button.component.ts` and update the view template definition to:

```
`<button [attr.disabled]="busy" class="btn btn-primary">
<span [hidden]="!busy">
<ng-content select="[data-animator]"></ng-content>
   </span>
   <ng-content select="[data-content]"></ng-content>
 </button>`
```

The two `ng-content` elements in the preceding view define the *content injection locations,* where the content children can be injected/transcluded. The `selector` property defines the *CSS selector* that should be used to locate the content children.

It starts to make more sense as soon as we fix the `AjaxButtonComponent` usage in `workout.component.html`. Change it to the following:

```
<ajax-button [execute]="save" [parameter]="f">
   <span class="glyphicon glyphicon-refresh spin" data-animator>
   </span><span data-content>Save</span>
</ajax-button>
```

The `span` with `data-animator` is injected into the `ng-content` with the `select=[data-animator]` property and the other `span` (with the `data-content` attribute) is injected into the second `ng-content` declaration.

Refresh the application again and try to save a workout. While the end result is the same, the resultant view is a combination of multiple view fragments: one part for component definition (*view children*) and another part for component usage (*content children*).

The following diagram highlights this difference for the rendered `AjaxButtonComponent`:

```
<ajax-button [execute]="save" [parameter]="f">
   <span class="glyphicon glyphicon-refresh
        spin" data-animator></span>
   <span data-content>Save</span>
</ajax-button>
```

Component Use

Content injected

```
<button [attr.disabled]="busy" ...>
  <span [hidden]="!busy">
    <ng-content select="[data-animator]">
    </ng-content>
  </span>
  <ng-content select="[data-content]">
  </ng-content>
</button>
```

Component View

View Rendered

```
<ajax-button>
  <button class="btn btn-primary">
    <span>
      <span class="glyphicon glyphicon-refresh spin"
        data-animator="">
      </span>
    </span>
    <span data-content="">Save</span>
  </button>
</ajax-button>
```

Save

The `ng-content` can be declared without the `selector` attribute. In such a scenario, the complete content defined inside the component tag is injected.

*Content injection* into an existing component view is a very powerful concept. It allows the component developer to provide extension points that the component consumer can readily consume and customize the behavior of the component, that too in a controlled manner.

The content injections that we defined for the `AjaxButtonComponent` allow the consumer to change the busy indicator animation and the button content, keeping the behavior of the button intact.

Angular's advantages do not end here. It has the capability to inject *content children* and *view children* into the *component code/implementation*. This allows the component to interact with its content/view children and control their behavior too.

# Injecting view children using @ViewChild and @ViewChildren

In `Chapter 3`, *More Angular 2 – SPA, Routing, and Data Flows in Depth,* we made use of something similar, *view children injection*. To refresh what we did, let's look at the relevant parts of the `WorkoutAudioComponent` implementation.

The view definition looked like this:

```
<audio #ticks="MyAudio" loop
  src="/static/audio/tick10s.mp3"></audio>
<audio #nextUp="MyAudio"
  src="/static/audio/nextup.mp3"></audio>
<audio #nextUpExercise="MyAudio"
  [src]="'/static/audio/' + _nextupSound"></audio>
// Some other audio elements
```

And the injection looked like the following:

```
@ViewChild('ticks') private _ticks: MyAudioDirective;
@ViewChild('nextUp') private _nextUp: MyAudioDirective;
@ViewChild('nextUpExercise') private _nextUpExercise: MyAudioDirective;
```

The directive (`MyAudioDirective`) associated with the `audio` tag was injected into the `WorkoutAudio` implementation using the `@ViewChild` decorator. The parameters passed to `@ViewChild` are the *template variable* names (such as `tick`) used to locate the element in the view definition. The `WorkoutAudio` component then used these audio directives to control the audio playback for *7 Minute Workout*.

While the preceding implementation injects `MyAudioDirective`, even child components can be injected. For example, instead of using the `MyAudioDirective`, suppose we build a `MyAudioComponent`, something like this:

```
@Component({
  selector: 'my-audio',
  template: '<audio ...></audio>',
})
export class MyAudioComponent {
  ...
}
```

We can then use it instead of the `audio` tag:

```
<my-audio #ticks loop
  src="/static/audio/tick10s.mp3"></my-audio>
```

The injection would still work.

What happens if there is more than one directive/component of the same type defined on the component view? Use the `@ViewChildren` decorator. It allows you to query injections of one type. The syntax for the use of `@ViewChildren` is as follows:

```
@ViewChildren(directiveType) children: QueryList<directiveType>;
```

This injects all the view children of type `directiveType`. For the `WorkoutAudio` component example stated previously, we can use the following statement to get hold of all `MyAudioDirective`'s:

```
@ViewChildren(MyAudioDirectives) private all: QueryList<MyAudioDirectives>;
```

The `ViewChildren` decorator can also take a list of comma-separated selectors (*template variable names*) instead of type. For example, to select multiple `MyAudioDirective` instances in the `WorkoutAudio` component, we can use this:

```
@ViewChildren('ticks, nextUp, nextUpExercise, halfway, aboutToComplete')
private all: QueryList<MyAudioDirective>;
```

The `QueryList` class is a special class provided by Angular. We introduced `QueryList` in the *Injecting descendant directive(s)* section earlier in the chapter. Let's explore `QueryList` further.

# Tracking injected dependencies with QueryList

For components that require multiple components/directives to be injected (using either `@ViewChildren` or `@ContentChildren`), the dependency injected is a `QueryList` object.

The `QueryList` class is a *read-only collection* of injected components/directives. Angular keeps this collection in sync based on the current state of the user interface.

Consider, for example, the `WorkoutAudio` directive view. It has five instances of `MyAudioDirective`. Hence, for the following statement, `all.length` will be *five*:

```
@ViewChildren(MyAudioDirective) private all: QueryList<MyAudioDirective>;
```

While the preceding example does not highlight the syncing part, Angular can track components/directives being added or removed for the view. This happens when we use content generation directives such as `ngFor`.

Take this hypothetical template for example:

```
<div *ngFor="let audioData of allAudios">
  <audio [src]=" audioData.url"></audio>
</div>
```

The number of `MyAudioDirective` directives injected here equals the size of the `allAudios` array. During the program's execution, if elements are added to or removed from the `allAudios` array, the directive collection is also kept in sync by the framework.

While the `QueryList` class is not an array, it can be iterated over (as it implements the **ES6 iterable interface**) using the `for (var item in queryListObject)` syntax. It also has some other useful properties, such as `length`, `first`, and `last`, that can come in handy. Check out the framework documentation (`http://bit.ly/ng2-querylist-class`) for more details.

From the preceding discussion, we can conclude that `QueryList` saves the component developer a lot of boilerplate code that would be required if tracking had to be done manually.

**View children access timing**
View children injections are not available when the component/directive initializes.
Angular makes sure that the view children injections are available to the component no later than the `ngAfterViewInit` lifecycle event. Make sure you access the injected components/directives only when (or after) the `ngAfterViewInit` event has fired.

Let's now look at content children injection, which is almost similar, except for a few minor differences.

## Injecting content children using @ContentChild and @ContentChildren

Angular allows us to inject *content children* too, using a parallel set of attributes: `@ContentChild` to inject a specific content child and `@ContentChildren` to inject content children of a specific type.

If we look back at the usage of `AjaxButtonComponent`, its content children spans can be injected into `AjaxButtonComponent` by doing this:

```
@ContentChild('spinner') spinner:ElementRef;
@ContentChild('text') text:ElementRef;
```

This can also be done by adding template variables onto the corresponding spans in `workout.component.html`:

```
<span class="glyphicon glyphicon-refresh spin"
data-animator #spinner></span><span data-content #text>Save</span>
```

While the preceding injection is `ElementRef`, it could have been a component too. Had we defined a component for spinner such as:

```
<ajax-button>
<busy-spinner></busy-spinner>
...
</ajax-button>
```

We could have injected it too using:

```
@ContentChild(BusySpinner) spinner: BusySpinner;
```

The same holds good for directives too. Any directive declared on `AjaxButtonComponent` can be injected into the `AjaxButtonComponent` implementation. For the preceding case, since it is standard HTML, we injected `ElementRef`, a wrapper that Angular creates for any HTML element.

Like *view children*, Angular makes sure that the content children references are bound to the variables injected before the `ngAfterContentInit` lifecycle event.

While we are on the subject of injecting dependencies, let's talk about some variations around *injecting services into components*.

# Dependency injection using viewProvider

We are already familiar with the mechanism of DI registration in Angular, where we register a dependency at the global level by adding it to any module declaration:

Or we can do it at a component level using the `providers` property on the `@Component` decorator:

```
providers:[WorkoutHistoryTracker, LocalStorage]
```

Just to avoid confusion, we are now talking about injecting dependencies other than directive/component objects. Directives/components are registered in the `declarations` array of a module before they can be injected using decorator hints such as `@Query`, `@ViewChild`, `@ViewChildren`, and a few others.

Dependencies registered at the component level are available for its *view children* and *content children* and their descendants.

Before we proceed, we hope that the distinction between *view* and *content children* is crystal clear to everyone. If in doubt, refer to the *Content children and view children* section again.

Let's take an example from `Chapter 4`, *Building Personal Trainer*. The `WorkoutBuilderService` service was registered at the app level in the workout builder module (`WorkoutBuilderModule`):

```
providers: [ExerciseBuilderService, ...
            WorkoutBuilderService]);
```

This allows us to inject `WorkoutBuilderService` across the app in order to build workouts and while running workouts. Instead, we could have registered the service at the `WorkoutBuilderComponent` level since it is the parent of all workout/exercise creation components, something like this:

```
@Component({
    template: `...`
    providers:[ WorkoutBuilderService ]
})
export class WorkoutBuilderComponent {
```

This change would disallow injecting `WorkoutBuilder` in `WorkoutRunner` or any component related to workout execution.

What if the `WorkoutBuilderService` service is registered at the app level as well as at the component level (as shown in the preceding example)? How does the injection happen?
From our experience, we know that Angular will inject a different instance of the `WorkoutBuilderService` service into `WorkoutBuilderComponent` (and its descendants), while other parts of the application (*Workout runner*) will get the global dependency. Remember **hierarchical injectors**!

Angular does not stop here. It provides some further scoping of dependencies using the `viewProviders` property. The `viewProviders` property, available on the `@Component` decorator, allows the registering of dependencies that can be injected only in the view children.

Let's consider the `AjaxButtonComponent` example again, and a simple directive implementation called `MyDirective`, to elaborate on our discussion:

```
@Directive({
  selector: '[myDirective]',
})
export class MyDirective {
  constructor(service:MyService) { }
  ...
}
```

The `MyDirective` class depends upon a service, `MyService`.

To apply this directive to the *button element* in the `AjaxButtonComponent` template, we need to register the `MyService` dependency too (assuming that `MyService` has not been registered globally):

```
@Component({
  selector: 'ajax-button',
  template:` <button [attr.disabled]="busy" ...
             myDirective>
             ...
             <button>`
  providers:[MyService],
...
```

Since `MyService` is registered with `AjaxButtonComponent`, `MyDirective` can be added to its content children too. Hence the `myDirective` application on *spinner HTML* will also work (the code in `workout.component.html`):

```
<span class="glyphicon glyphicon-refresh spin"
  data-animator #spinner myDirective></span>
```

But changing the `providers` property to `viewProviders` :

```
viewProviders:[MyService]
```

Will fail the `MyService` injection for the `AjaxButtonComponent`'s content children (the `span` in the preceding code), with a DI error in the console.

Dependencies registered with `viewProviders` are invisible to its content children.

This dependency scoping for *view* and *content children* may not seem useful at first sight, but it does have its benefits. Imagine we are building a reusable component that we want to package and deliver to developers for consumption. If the component has a service dependency that it prepackages too, we need to be extra cautious. If such a component allows *content injection* (content children), the dependent service is widely exposed if *provider-based* registration is used on the component. Any content children can get hold of the service dependency and use it, leading to undesirable consequences. By registering the dependency using `viewProvider`, only the component implementation and its child views have access to the dependency, providing the necessary layer of encapsulation.

Yet again, we are amazed by the flexibility and level of customization the DI framework provides. While it may be intimidating for starters, once we start building more and more components/directives with Angular, we will always find areas where these concepts make our implementation simpler.

Let's shift our focus to the third classification of directives: *structural directives*.

# Understanding structural directives

While we will often be using structural directives, such as `NgIf` and `NgFor`, there is seldom a need to creating a structural directive. Think carefully. If we need a new view, we create a *component*. If we need to extend an existing element/component, we use a *directive*. Whereas the most common use of structural directives is to clone a piece of a view (also called a *template view*) and then, based on some conditions:

- Either inject/destroy these templates (`NgIf` and `NgSwitch`)
- Or duplicate these templates (`NgFor`)

Any behavior implemented using structure directives will inadvertently fall into either of these two categories.

Given this fact, instead of building our own structural directive, let's look at the source code of the `NgIf` implementation.

This is the complete implementation for `NgIf`:

```
@Directive({selector: '[ngIf]', inputs: ['ngIf']})
export class NgIf {
  private _prevCondition: boolean = null;

  constructor(private _viewContainer: ViewContainerRef,
              private _templateRef: TemplateRef) {}

  set ngIf(newCondition /* boolean */) {
    if (newCondition && (isBlank(this._prevCondition)
                        || !this._prevCondition)) {
      this._prevCondition = true;
      this._viewContainer
          .createEmbeddedView(this._templateRef);
    }
    else if (!newCondition && (isBlank(this._prevCondition)
                               || this._prevCondition)) {
      this._prevCondition = false;
      this._viewContainer.clear();
    }
  }
}
```

No magic here, just simple structural directive that checks a boolean condition to create/destroy the view!

`prevCondition` is tracking the last value of the `ngIf` expression. The extra checks with `prevCondition` are there to make sure that the template add/remove logic runs only when the expression being watched (`newCondition`) actually flips. Nothing happens when both `newCondition` and `prevCondition` are either `true` or `false`.

It's not difficult to understand how the directive works. What needs to be detailed are the two new injections, `ViewContainerRef (_viewContainer)` and `TemplateRef (_TemplateRef)`.

## TemplateRef

The `TemplateRef` class (`_templateRef`) stores the reference to the template the structural directive is referring to. Remember the discussion on structural directives from `Chapter 2`, *Building Our First App – 7 Minute Workout*? All structural directives take a template HTML that they work on. When we use a directive such as `NgIf`:

```
<h3 *ngIf="currentExercise.exercise.name=='rest'">
  ...
</h3>
```

Angular internally translates this declaration to the following:

```
<template [ngIf]="currentExercise.exercise.name=='rest'">
  <h3> ... </h3>
</template>
```

This is the template that structural directives work with, and `_templateRef` points to this template.

The `template` is an HTML5 tag used to hold content that we do not want to be rendered in the browser. The default style set for the `template` tag is `display:none`.
If we manually add template tags in an Angular view, the Angular templating engine will replace them with empty `<script></script>` tags.

The other injection is `ViewContainerRef`.

# ViewContainerRef

The `ViewContainerRef` class points to the container where templates are rendered. This class has a number of handy methods for managing views. The two functions that `NgIf` implementation uses, `createEmbeddedView` and `clear`, are there to add and remove the template HTML.

The `createEmbeddedView` function takes the template reference (again injected into the directive) and renders the view.

The `clear` function destroys the element/component already injected and clears the view container. Since every component and its children referenced inside the template (`TemplateRef`) are destroyed, all the associated bindings also cease to exist.

Structural directives have a very specific area of application. Still, we can do a lot of nifty tricks using the `TemplateRef` and `ViewContainerRef` classes.

We can implement a structural directive that, depending on the user role, shows/hides the view template.

Consider this example of a hypothetical structural directive, `forRoles`:

```
<button *forRoles="admin">Admin Save</button>
```

The `forRoles` directive will not render the button if the user does not belong to the *admin* role. The core logic would look something like the following:

```
if(this.loggedInUser.roles.indexOf(this.forRole) >=0){
      this.viewContainer.createEmbeddedView(this.templateRef);
}
else {
      this.viewContainer.clear();
}
```

The directive implementation will need some sort of service that returns the logged-in user's details. We will leave the implementation for such a directive to the readers.

What the `forRoles` directive does can also be done using `NgIf`:

```
<button *ngIf="loggedInUser.roles.indexOf('admin')>=0">Admin Save</button>
```

But the `forRoles` directive just adds to the template's readability with clear intentions.

A fun application of structural directives may involve creating a directive that just duplicates the template passed to it. It would be quite easy to build one; we just need to call `createEmbeddedView` twice:

```
ngOnInit() {
  this.viewContainer.createEmbeddedView(this._templateRef);
  this.viewContainer.createEmbeddedView(this._templateRef);
}
```

Another fun exercise!

The `ViewContainerRef` class also has some other functions that allow us to inject *components*, get the number of embedded views, reorder the view, and so on and so forth. Look at the framework documentation for `ViewContainerRef` (`http://bit.ly/view-container-ref`) for more details.

That completes our discussion on structural directives and it's time to start something new!

The components that we have built thus far derive their styles (CSS) from the common *bootstrap style sheet* and some custom styles defined in `app.css`. Angular has much more to offer in this area. A truly reusable component should be completely self-contained, in terms of both behavior and user interface.

# Component styling and view encapsulation

A longstanding problem with web app development is the lack of encapsulation when it comes to DOM element behavior and styles. We cannot segregate one part of the application HTML from another through any mechanism.

In fact, we have too much power at our disposal. With libraries like jQuery and powerful *CSS selectors*, we can get hold of any DOM element and change its behavior. There is no distinction between our code and any external library code in terms of what it can access. Every single piece of code can manipulate any part of the rendered DOM. Hence, the encapsulation layer is broken. A badly written library can cause some nasty issues that are hard to debug.

The same holds true for CSS styling too. Any UI library implementation can override global styles if the library implementation wants to do so.

These are genuine challenges that any library developer faces when building reusable libraries. Some emerging web standards have tried to address this issue by coming up with concepts such as **web components**.

**Web components**, in simple terms, are reusable user interface widgets that encapsulate their *state, style, user interface,* and *behavior*. Functionality is exposed through well-defined APIs, and the user interface parts are encapsulated too.

The *web component* concept is enabled through four standards:

- HTML templates
- Shadow DOM
- Custom elements
- HTML imports

For this discussion, the technology standard we are interested in is **Shadow DOM.**

# Overview of Shadow DOM

**Shadow DOM** is like a parallel DOM tree hosted inside a component (*an HTML element, not to be confused with Angular components*), hidden away from the main DOM tree. No part of the application has access to this shadow DOM other than the component itself.

It is the implementation of the Shadow DOM standard that allows view, style, and behavior encapsulation. The best way to understand Shadow DOM is to look at HTML5 `video` and `audio` tags.

Have you ever wondered how this `audio` declaration:

```
<audio src="/static/audio/nextup.mp3" controls></audio>
```

Produces the following?

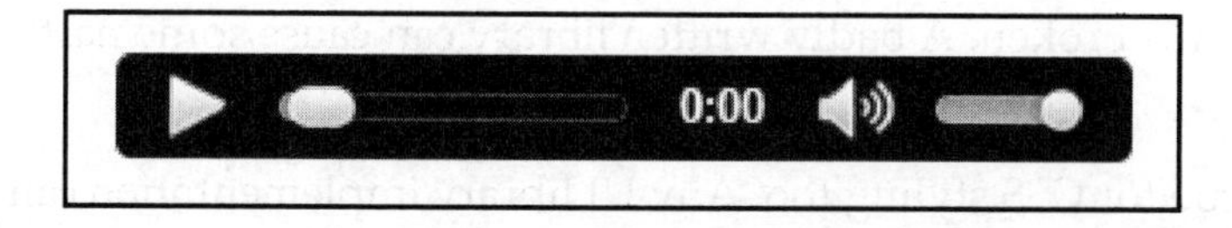

It is the browser that generates the underlying Shadow DOM to render the audio player. Surprisingly, we can even look at the generated DOM! Here is how we do it:

- Take the preceding HTML, create a dummy HTML page, and open it in Chrome.
- Then open the developer tools window (F12). Click on the **Setting** icon in the top left.
- On the **General** settings, click on the checkbox, as highlighted in the following screenshot, to enable the inspection of Shadow DOM:

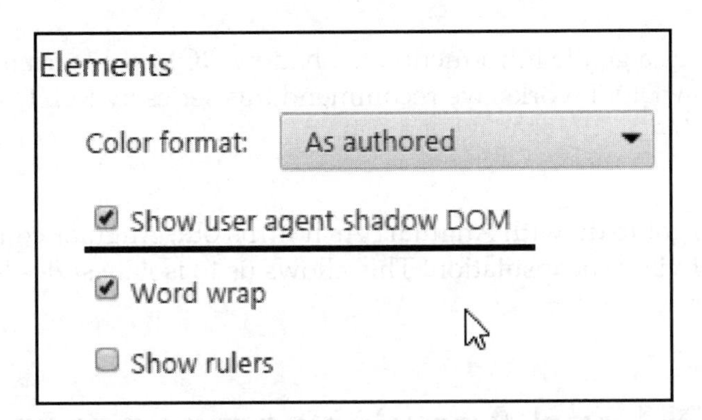


Refresh the page, and if we now inspect the generated `audio` HTML, the Shadow DOM shows up:

```
▼<audio controls src="/static/audio/nextup.mp3">
  ▼#shadow-root (user-agent)
   ▼<div pseudo="-webkit-media-controls">
     ▼<div pseudo="-webkit-media-controls-overlay-enclosure">
       ►<input type="button" style="display: none;">…</input>
      </div>
     ▼<div pseudo="-webkit-media-controls-enclosure">
       ▼<div pseudo="-webkit-media-controls-panel">
         ►<input type="button" pseudo="-webkit-media-controls-play-button">…</input>
         ►<input type="range" step="any" pseudo="-webkit-media-controls-timeline" max="0.972">…</input>
          <div pseudo="-webkit-media-controls-current-time-display" style>0:00</div>
          <div pseudo="-webkit-media-controls-time-remaining-display" style="display: none;">0:00</div>
         ►<input type="button" pseudo="-webkit-media-controls-mute-button">…</input>
         ►<input type="range" step="any" max="1" pseudo="-webkit-media-controls-volume-slider" style>…</input>
         ►<input type="button" pseudo="-webkit-media-controls-toggle-closed-captions-button" style="display: none;">…</input>
         ►<input type="button" style="display: none;">…</input>
         ►<input type="button" pseudo="-webkit-media-controls-fullscreen-button" style="display: none;">…</input>
        </div>
      </div>
    </div>
</audio>
```

Under `shadow-root`, there is a whole new world that the other part of the page and script do not have access to.

In the Shadow DOM realm, **shadow-root** (`#shadow-root` in the preceding code) is the root node for the generated DOM, hosted inside a **shadow host** (in this case the `audio` tag).
When the browser renders this element/component, what gets rendered is the content from the *shadow root* and not the *shadow host*.

From this discussion, we can conclude that Shadow DOM is a parallel DOM created by the browser that encapsulates the *markup*, *style*, and *behavior* (DOM manipulation) of an HTML element.

This was a gentle introduction to Shadow DOM. To learn more about how Shadow DOM works, we recommend this series by Rob Dodson: `http://bit.ly/shadow-dom-intro`

But what has all this got to do with Angular? As it turns out, Angular components too support some sort of view encapsulation! This allows us to isolate styles for Angular components too.

# Shadow DOM and Angular components

To understand how Angular employs the concept of Shadow DOM, we will first have to learn about styling Angular components.

When it comes to styling the apps built as part of this book, we have taken a conservative approach. Be it *Workout Builder* or the *Workout Runner* (*7 Minute Workout*) app, all the components that we built derive their style from *bootstrap CSS* and from custom styles defined in `app.css`. No component has defined its own style.

While this adheres to the standard practices of web app development, sometimes we do need to deviate. This is especially true when we are building self-contained, packaged, and reusable components.

Angular allows us to define styles specific to a component by using the `style` (for inline style) and `styleUrl` (external style sheet) properties on the `@Component` decorator. Let's play around with the `style` property and see what Angular does.

We will use the AjaxButtonComponent implementation as our playground for the next exercise. But before doing that, let's look at the AjaxButtonComponent HTML as it stands now. The HTML tree for AjaxButtonComponent looks like this:

```
▼<ajax-button>
  ▼<button class="btn btn-primary">
    ▼<span hidden>
       <span class="glyphicon glyphicon-refresh spin" data-animator></span>
     </span>
     <span data-content>Save</span>
   </button>
 </ajax-button>
```

Let's override some styles using the styles property:

```
@Component({
  ...
  styles:[`
    button {
      background: green;
    }`]
})
```

The preceding *CSS selector* sets the background property to green for all HTML buttons. Save the above style and refresh the work builder page. The button style has been updated. No surprises here? Not true, there are some! Look at the generated HTML:

```
▼<ajax-button _nghost-eaq-14>
  ▼<button class="btn btn-primary" _ngcontent-eaq-14>
    ▼<span _ngcontent-eaq-14 hidden>
       <span class="glyphicon glyphicon-refresh spin" data-animator></span>
     </span>
     <span data-content>Save</span>
   </button>
 </ajax-button>
```

There are some new attributes added to a number of HTML elements. And where have the recently defined styles landed? At the very top, inside the `head` tag:

```
<style>button[_ngcontent-eaq-14] {
      background: green;
   }</style>
```

The style defined in the `head` section has an extra scope with the `_ngcontent-eaq-14` attribute (the attribute name may differ in your case). This scoping allows us to style `AjaxButtonComponent` independently and it cannot override any global styles.

The styles that, by definition, should have affected the appearance of all the buttons in the application have had no effect. Angular has scoped these styles.

This scoping makes sure that the component styles do not mess with the already defined style, but the reverse is not true. Global styles will still affect the component unless overridden in the component itself.
Angular does the same even if we use the `styleUrls` property. Suppose we had embedded the same CSS in an external CSS file and used this:
`styleUrls:['static/css/ajax-button.css']`
Angular would have still in-lined the styles into the `head` section, by fetching the CSS, parsing it, and then injecting it.

This scoped style is the result of Angular trying to emulate the Shadow DOM paradigm. The styles defined on the component never leak into the global styles. All this awesomeness without any effort!

If you are building components that define their own styles and want a degree of isolation, use the component's `style/styleUrl` property instead of using the old-school approach of having a common CSS file for all styles.

We can further control this behavior by using a `@Component` decorator property called **encapsulation**. The API documentation on this property mentions:

> **encapsulation : ViewEncapsulation**
> *Specify how the template and the styles should be encapsulated. The default is***ViewEncapsulation.Emulated** *if the view has styles, otherwise* **ViewEncapsulation.None**.

As we can see, as soon as we set the style on the component, the encapsulation effect is `Emulated`. Otherwise, it is `None`.

If we explicitly set `encapsulation` to `ViewEncapsulation.None`, the scoping attributes are removed and the styles are embedded in the head section as normal styles.

And then there is a third option, `ViewEncapsulation.Native`, in which Angular actually creates Shadow DOM for the components view. Set the `encapsulation` property on the `AjaxButtonComponent` implementation to `ViewEncapsulation.Native`, and now look at the rendered DOM:

```
▼<ajax-button>
  ▼#shadow-root (open)
    <style>
        button {
          background: green;
        }
        </style>
    ▼<button class="btn btn-primary">
      ▶<span hidden>…</span>
      <span data-content>Save</span>
    </button>
  </ajax-button>
```

The `AjaxButtonComponent` now has a shadow DOM! This also implies that the complete styling of the button is lost (style derived from bootstrap CSS) and the button needs to now define its own style.

Angular goes to great lengths to make sure that the components we develop can work independently and are reusable. Each component already has its own template and behavior. In addition to that, we can also encapsulate component styles, allowing us to create robust, standalone components.

This brings us to the end of the chapter, and it's time to wrap up the chapter with what we've learned.

# Summary

As we conclude this chapter, we now have a better understanding of how directives work and how to use them effectively.

We started the chapter by building a `RemoteValidatorDirective`, and learned a lot about Angular's support for *asynchronous validations*.

Next in line was `BusyIndicatorDirective`, again an excellent learning ground. We explored the **renderer** service, which allows component view manipulation in a platform-agnostic way. We also learned about **host bindings**, which let us bind to a host element's *events, attributes,* and *properties*.

Angular allows directives declared across the view lineage to be injected within the lineage. We dedicated a few sections to understanding this behavior.

The third directive (component) that we created was `AjaxButtonComponent`. It helped us understand the critical difference between *content children* and *view children* for a component.

We also touched upon structural directives, where we explored the `NgIf` platform directive.

Lastly, we looked at Angular's capabilities in terms of view encapsulation. We explored the basics of Shadow DOM and learned how the framework employs the Shadow DOM paradigm to provide view plus style encapsulation.

The next chapter is all about testing Angular apps, a critical piece in the complete framework offering. The Angular framework was built with testability in mind. The framework constructs and the tooling support make automated testing in Angular easy. More about this in the next chapter....

# 7
# Testing Personal Trainer

Unless you are a superhero who codes perfectly, you need to test what you build. Also, unless you have loads of free time to test your application again and again, you need some test automation.

When we say Angular was built with testability in mind, we really mean it. It has a strong **Dependency Injection** (**DI**) framework, some good mock constructs, and awesome tools that make testing in an Angular app a fruitful endeavor.

This chapter is all about testing and is dedicated to testing what we have built over the course of this book. We test everything from components to pipes, services, and our app directives.

The topics we cover in this chapter include:

- **Understanding the big picture**: We will try to understand how testing fits into the overall context of Angular app development. We will also discuss the types of testing Angular supports, including unit and **end-to-end** (**E2E**) testing.
- **Overview of tools and frameworks**: We cover the tools and frameworks that help in both unit and end-to-end testing with Angular. These include **Karma** and **Protractor**.
- **Writing unit tests**: You will learn how to do unit testing with Angular using Jasmine and Karma inside a browser. We will unit-test what we have built in the last few chapters. This section also teaches us how to unit-test various Angular constructs, including pipes, components, services, and directives.
- **Creating end-to-end tests**: Automated end-to-end tests work by mimicking the behavior of the actual user through browser automation. You will learn how to use Protractor combined with WebDriver to perform end-to-end testing.

Let the testing begin!

The code for this chapter can be found at `checkpoint7.1`. It is available on GitHub (`https://github.com/chandermani/angular2byexample`) for everyone to download. Checkpoints are implemented as branches in GitHub.
If you are not using Git, download the snapshot of `checkpoint7.1` (a ZIP file) from this GitHub location: `https://github.com/chandermani/angular2byexample/archive/checkpoint7.1.zip`. Refer to the `README.md` file in the `trainer` folder when setting up the snapshot for the first time.

# The need for automation

The size and complexity of apps being built for the Web are growing with each passing day. The plethora of options that we now have to build web apps is just mind-boggling. Add to this the fact that the release cycles for products/apps have shrunk drastically from months to days, or even multiple releases per day! This puts a lot of burden on software testing. There is too much to be tested. Multiple browsers, multiple clients and screen sizes (desktop and mobile), multiple resolution, and so on.

To be effective in such a diverse landscape, automation is the key. *Automate everything that can be automated* should be our mantra.

# Testing in Angular

The Angular team realized the importance of testability and hence created a framework that allowed easy testing (automated) for apps built on it. The design choice of using DI constructs to inject dependencies everywhere helped. This will become clear as the chapter progresses and we build a number of tests for our apps. However, before that, let's understand the types of testing that we target when building apps on this platform.

# Types of testing

There are broadly two forms of testing that we do for a typical Angular app:

- **Unit testing**: Unit testing is all about testing a component in isolation to verify the correctness of its behavior. Most of the dependencies of the component under test need to be replaced with mock implementations to make sure that the unit tests do not fail due to failure in a dependent component.
- **End-to-end testing**: This type of testing is all about executing the application like a real end user and verifying the behavior of the application. Unlike unit testing, components are not tested in isolation. Tests are done against a running system in real browsers, and assertions are done based on the state of the user interface and the content displayed.

Unit testing is the first line of defense against bugs, and we should be able to iron out most issues with code during unit testing. But unless E2E is done, we cannot confirm that the software is working correctly. Only when all the components within a system interact in the desired manner can we confirm that the software works; hence, E2E testing becomes a necessity.

Who writes unit and E2E tests and when are they written are important questions to answer.

# Testing – who does it and when?

Traditionally, E2E testing was done by the **Quality Assurance** (**QA**) team and developers were responsible for unit-testing their code before submitting. Developers too did some amount of E2E testing but the overall the E2E testing process was manual.

With the changing landscape, modern testing tools, especially on the web front, have allowed developers to write automated E2E tests themselves and execute them against any deployment setup (such as development/stage/production). Tools such as Selenium, together with WebDrivers, allow easy browser automation, thus making it easy to write and execute E2E tests against real web browsers.

A good time to write E2E scenario tests is when the development is complete and ready to be deployed.

When it comes to unit testing, there are different schools of thought around when a test should be written. A *TDDer* writes tests before the functionality is implemented. Others write tests when the implementation is complete to confirm the behavior. Some write while developing the component itself. Choose a style that suits you, keeping in mind that the earlier you write your tests, the better.

We are not going to give any recommendations, nor are we going to get into an argument over which one is better. Any amount of unit tests is better than nothing.
Our personal preference is to use the *middle approach*. With TDD, we feel that the test creation effort at times is lost as the specifications/requirements change. Tests written at the start are prone to constant fixes as the requirement changes.
The problem with writing unit tests at the end is that our target is to create tests that pass according to the current implementation. The tests that are written are retrofitted to test the implementation whereas they should test the specifications.
Adding tests somewhere in the middle works best for us.

Let's now try to understand the tooling and technology landscape available for Angular testing.

## The Angular testing ecosystem

Look at the following diagram to understand the tools and frameworks that support Angular testing:

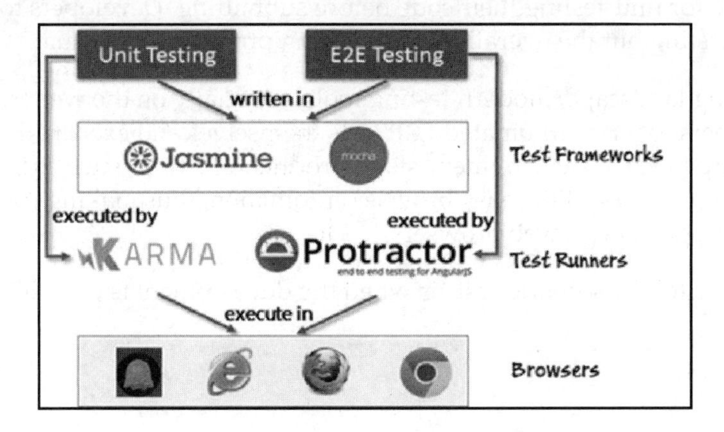

As we can see, we write our tests using unit testing libraries such as Jasmine or Mocha.

At the moment, the Angular testing library works by default with Jasmine. However, the Angular team has indicated that they have made the framework more generic so that you can use other testing libraries such as Mocha with it. The Angular documentation has not yet been updated to include how to do this.

These tests are executed by either Karma or Protractor depending on whether we are writing unit or integration tests. These test runners in turn run our tests in a browser such as Chrome, Firefox, or IE, or headless browsers such as PhantomJS. It is important to highlight that not only E2E but also unit tests are executed in a real browser.

Except for browsers, the complete Angular testing setup is supported by the awesome libraries and framework made available through the **Node.js** ecosystem. Some libraries such as Jasmine have standalone versions too, but we will stick to Node.js packages.

All the tests in this chapter are written using Jasmine (both unit and integration tests). Karma will be our test runner for unit tests and Protractor for E2E tests.

# Getting started with unit testing

The ultimate aim of unit testing is to test a specific piece of code/component in isolation to make sure that the components work according to the specification. This reduces the chances of failures/bugs in the component when integrated with other parts of the software. Before we start writing tests, there are some guidelines that can help us write good and maintainable tests:

- One unit should test one behavior. For obvious reasons, testing one behavior per unit test makes sense. A failing unit test should clearly highlight the problem area. If multiple behaviors are tested together, a failed test requires more probing to assert what behavior was violated.
- Dependencies in a unit test should be mocked away. Unit testing, as the name suggests, should test the unit and not its dependencies.
- Unit tests should not change the state of the component being tested permanently. If it does happen, other tests may get affected.

- The order of execution of unit tests should be immaterial. One unit test should not be dependent on another unit test to execute before it. This is a sign of a brittle unit test. It may also mean that the dependencies are not mocked.
- Unit tests should be fast. If they are not fast enough, developers will not run them. This is a good reason to mock all dependencies such as database access, remote web service call, and others in a unit test.
- Unit tests should try to cover all code paths. Code coverage is a metric that can help us assess the effectiveness of unit tests. If we have covered all positive and negative scenarios during testing, the coverage will indeed be higher. A word of caution here: high code coverage does not imply that the code is bug-free, but low coverage clearly highlights a lack of areas covered in unit tests.
- Unit tests should test both positive and negative scenarios. Just don't concentrate on positive test cases; all software can fail, and hence unit testing failure scenarios are as important to test as success scenarios.

These guidelines are not framework-specific but give us enough ammunition for writing good tests. Let's begin the process of unit testing by setting up the components required for it.

# Setting up Karma for unit testing

Since the complete test automation infrastructure is supported using Node, this needs to be installed first. Follow the instructions on the Node website (`https://nodejs.org`) and get Node installed locally.

Node comes with a package manager called **Node Package Manager** (**NPM**) that is used to install all other components (**packages** in the Node world) required for testing.

Start by installing Karma from the command line. Navigate to the root of your application code base (in the `trainer` folder) and install Karma using this command:

```
npm install karma --save-dev
```

To use Karma from the command line, we need to install its command-line interface:

```
npm install -g karma-cli
```

The Karma version against which the code was tested is 0.13.22. The **karma-cli** version was 1.0.1 .
To install a specific version of a package, suffix the package name with `@version`, for example, `npm install karma@0.13.22 --save-dev`.

This completes the Karma installation and it's now time to configure the test runner. Configuring Karma is all about setting up its configuration file so that it contains enough details for it to run our scripts and test them. Create a `tests` folder in the root (in the `trainer` folder), navigate to it, and start the Karma configuration setup with the following command:

```
karma init
```

This starts a command-line wizard that guides us through the options available, including the test framework, folders to watch, and other such settings. Once the wizard is complete, it generates a `karma.conf.js` file. Instead of using the generated configuration file, copy the `karma.conf.js` file from the companion code base at `checkpoint7.1/tests` to your local `tests` folder.

The `karma init` wizard installs some packages based on our selection. If we skip the wizard, the `karma-chromelauncher` and `karma-jasmine` packages need to be installed manually for unit testing.

Next, we will install Jasmine. So execute the following command:

```
npm install jasmine-core --save-dev --save-exact
```

## The Karma configuration files

Open `karma.conf.js` in the `tests` directory. It contains settings that affect the tests we run. We will not be covering each and every configuration supported by Karma here, but will focus on configurations that are unique and/or required for our test setup. Refer to the Karma documentation (`http://karma-runner.github.io/1.0/config/configuration-file.html`) to understand more about the various Karma configuration options.

First, we'll need to set the base path for resolving file locations. In our case, because we have put Karma in a `tests` folder just below the root of the application, we will use a relative URL that will point to that location:

```
basePath: '../',
```

We then set our framework to Jasmine:

```
frameworks: ['jasmine'],
```

The next Karma configuration is the `files` array. Files can be referenced either by their file path or by using patterns. In our case, we are referencing the first ten files using file paths. These are the files for the polyfills, SystemJS and zone.js. When Karma first starts it will load all these files in the browser using a `<script>` tag.

For the rest, we are using patterns because this approach allows us to set the `included` and `watched` properties. The `included` property determines whether Karma will initially load the file. If set to `true`, Karma will load these files in a `<script>` tag. If not, Karma will not load the file, so that some other mechanism must be found to do that. The `watched` property indicates whether a file will be watched for changes.

As you look through the file patterns in our configuration, you will notice files you have seen before that are used for building our application such as RxJS, and Angular itself. We have set the `included` property for all but one of these files to `false`.

The one file pattern where the `included` property has been set to `true` is one we have not seen before: `karma-test-shim.js`:

```
{pattern: 'tests/karma-test-shim.js', included: true, watched: true},
```

This pattern references an additional file that we have added to the testing folder to handle the special requirements for running Angular tests in Karma and it will be loaded by Karma when it first starts. We will discuss that file shortly.

The `watched` property in all our files (except for RxJS and the files that support debugging) is set to true, which means that Karma will watch these files and rerun our tests if any of them are changed.

The next file pattern in the array after `karma-test-shim.js` is a little different from the rest:

```
{pattern: 'dist/**/*.js', included: false, watched: true},
```

This pattern contains a path to our application files (the code that we have been building throughout the earlier chapters), as well as a path to the test files we will create.

But note that it is pointing to the `dist` directory that holds the JavaScript files that have been generated by the TypeScript transpiler and not the TypeScript files themselves. This stands to reason because Karma, of course, is a JavaScript test runner (and will not itself transpile our TypeScript files). This in turn means we will need an initial transpilation step in order to get our tests running.

Getting back to the files where we have set the `included` property to false, how do we load them if we do not use Karma? As you know, we are using SystemJS to load our modules and Karma does not know anything about SystemJS. If Karma were to run our test files before the related modules are loaded, our tests would fail. This means that for these files, we need to run SystemJS and load our modules before Karma runs. Setting the `included` property to false means that Karma will not run these files automatically.

But if the config file sets the `included` property for these files to `false`, how and when will we run the tests in these files? The answer to that question brings us to the Karma test shim file.

## The Karma test shim file

Open `karma-test-shim.js` in the `tests` directory. This file addresses the challenges of using Karma with Angular and SystemJS.

The Karma test shim file currently uses the SystemJS loader but according to the Angular team it
could be adapted for something else such as Web Pack.

The Karma test shim file does three things, which we'll discuss now.

First, it cancels Karma's synchronous start:

```
__karma__.loaded = function() {};
```

Then it adds the configuration for SystemJS:

```
System.config({
    baseURL: '/base'
});

System.config(
    {
        paths: {
            // paths serve as alias
            'npm:': 'node_modules/'
        },
        map: {
            'app': 'dist',
            '@angular/core': 'npm:@angular/core/bundles/core.umd.js',
...[other Angular modules] ...
            // angular testing umd bundles
            '@angular/core/testing': 'npm:@angular/core/bundles/core-
testing.umd.js',
...[other Angular testing modules] ...
            // other libraries
            'rxjs': 'npm:rxjs',
            'angular2-modal': 'npm:angular2-modal',
        },
        packages: {
            'app': {
                defaultExtension: 'js'
            },
            'rxjs': {
                defaultExtension: 'js'
            }
        }
    });
```

This should already be familiar. After all, we are already doing this kind of configuration for SystemJS in the `system.js.config` file in the root of our application. What this configuration does is set up the path mapping for the modules that will be used by Karma including the Angular testing modules. We have modified the file so that it points to the right locations in our application for this mapping.

It's important to understand that this configuration is for using SystemJS with our test runner and not for use with our underlying application.

Finally, it uses SystemJS with a promise to import our modules. If the promise resolves successfully, it starts up Karma, and if not, throws an error:

```
Promise.all([
    System.import('@angular/core/testing'),
    System.import('@angular/platform-browser-dynamic/testing')
]).then(function (providers) {
    var testing = providers[0];
    var testingBrowser = providers[1];
testing.TestBed.initTestEnvironment(testingBrowser.BrowserDynamicTestingMod
ule,
        testingBrowser.platformBrowserDynamicTesting());

}).then(function() {
    // Finally, load all spec files.
    // This will run the tests directly.
    return Promise.all(
        allSpecFiles.map(function (moduleName) {
            return System.import(moduleName);
        }));
}).then(__karma__.start, __karma__.error);
```

Note that we are importing `@angular/platform-browser-dynamic/testing` and passing parameters that are specific to testing browser-based applications to the `initTestEnvironment` method of `TestBed`. We will be discussing `TestBed` in detail later in this chapter.

With this file in place, we are now ready to begin using Karma for Angular testing. While this configuration has been somewhat tedious, the good news is that once we have completed configuring Karma, Angular makes it easy to create and run unit tests, as we will see soon.

# Organization and naming of our test files

To unit-test our app, the plan is to create one test (such as `workout-runner.spec.ts`) file for each TypeScript file that we have in our project.

Naming the test files with the name of the file under test plus `.spec` is a convention that is used by developers who test with Jasmine. It is also used to facilitate the mapping of files to tests in the configuration steps we outlined previously.

This test file will contain the unit test specification for the corresponding component, as shown in the following screenshot (taken in the Karma debugger when running our unit tests):

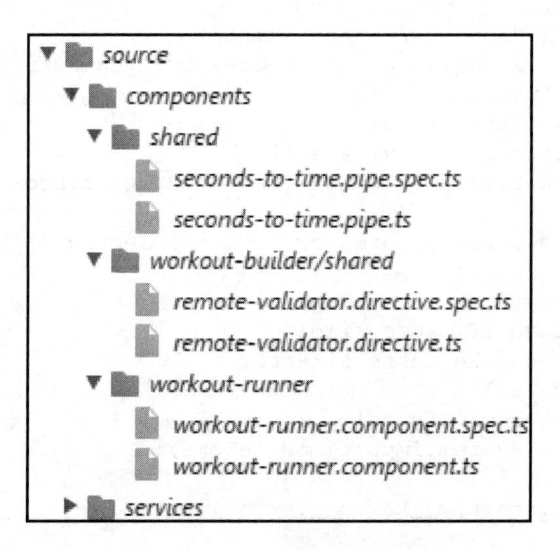


# Unit-testing Angular applications

Over the course of this book, we have built components that cover every construct available in Angular. We have built components, pipes, a few services, and finally some directives too. All of these are testable in unit tests.

Just to get the hang of unit testing with Jasmine, let's test the smallest and easiest component first: the pipe.

## Unit-testing pipes

Pipes are the easiest to test as they have minimum or zero dependencies on other constructs. The `SecondsToTimePipe` that we created for *Workout Runner* (the *7 Minute Workout* app) has no dependencies and can be easily unit-tested.

Look at the Jasmine framework documentation to understand how to write unit tests using Jasmine. We are using Jasmine 2.0 for our unit tests ( http://jasmine.github.io/2.0/introduction.html). Jasmine has some of the best documentations available and the overall framework is very intuitive to use. We strongly recommend that you head over to the Jasmine site and get yourself familiar with the framework before you proceed.

Add a seconds-to-time.pipe.spec.ts file to the trainer/src/components/shared folder and add this unit test to it:

```
import {SecondsToTimePipe} from "./seconds-to-time.pipe";
describe('SecondsToTime pipe', () => {
    let pipe:SecondsToTimePipe;
    beforeEach(() => {
        pipe = new SecondsToTimePipe();
    });
    it('should convert integer to time format', () => {
        expect(pipe.transform(5)).toEqual('00:00:05');
        expect(pipe.transform(65)).toEqual('00:01:05');
        expect(pipe.transform(3610)).toEqual('01:00:10');
    });
});
```

Let's take a look at what we are doing here in our test file.

Not surprisingly, we import the SecondsToTimePipe, which we are going to test. This is just like the imports we have used elsewhere in our TypeScript classes. Notice that we use a relative path to the file in which it is located './seconds-to-time.pipe'. In Angular, this means to look for the component to test in the same directory as the test itself. As you recall, this is the way we set up our file structure: putting our tests in the same directory as the file under test.

In the next line, we start using Jasmine syntax. First, we wrap the test in a describe function that identifies the test. The first parameter of this function is a user-friendly description of the test; in this case, it is the SecondsToTime pipe. For the second parameter, we pass a lambda (fat arrow) function that will contain our test. After setting up a local variable to hold the pipe, we call Jasmine's beforeEach function and use this to inject an instance of our pipe.

Since the `beforeEach` function runs before every test that is in our `describe` function, we can use it for common code that will run in each of our tests. In this case, it is not strictly necessary since there is only one test in our `describe` function. But it is a good idea to get into the habit of using it for common setup scenarios, as we will see going forward.

Next, we call Jasmine's `it` function and pass it a title, along with three calls to Jasmine's `expect` function (Jasmine's name for assertions). These are all self-explanatory.

It is not necessary to explicitly import these Jasmine functions in our test.

## Running our test files

Now it's time to run our tests using Karma. As you recall, in order to do that, we first have to transpile our files from TypeScript to JavaScript. To do this, we will simply start up our application itself in a terminal window by calling:

```
gulp play
```

The gulp file for the application will transpile our TypeScript files to JavaScript and then watch for changes in these files.

Next, we need to run Karma and we do that by executing the following command in a separate terminal window in the `trainer` folder:

```
karma start tests/karma.conf.js
```

We should then see this output in the terminal window:

```
19 03 2016 14:28:47.012:INFO [karma]: Karma v0.13.22 server started at http://localhost:9876/
19 03 2016 14:28:47.031:INFO [launcher]: Starting browser Chrome
19 03 2016 14:28:49.617:INFO [Chrome 49.0.2623 (Windows 8 0.0.0)]: Connected on socket /#Awjpf2kJKARV0f9mAAAA with id 49661790
Chrome 49.0.2623 (Windows 8 0.0.0): Executed 1 of 1 SUCCESS (0.007 secs / 0.003 secs)
```

The last line shows that our test passed successfully. To make sure that it is reporting the correct pass/fail results, let's make a change in the test to cause one of the expectations to fail. Change the time in the first expectation to 6 seconds rather than 5, like so:

```
expect(pipe.transform(5, [])).toEqual('00:00:06');
```

We get the following error message:

```
Chrome 49.0.2623 (Windows 8 0.0.0) SecondsToTime pipe should convert integer to time format FAILED
        Expected '00:00:05' to equal '00:00:06'.
```

What's nice about this error message is that it combines the `describe` and `it` descriptions into a complete sentence that provides a clear summary of the error. This shows how Jasmine allows us to write readable tests so that someone who is new to our code can quickly understand any problems that may arise in it. The next line shows us which expectation was not met, what was expected, and what the actual results were that did not meet this expectation.

We also get a lengthy stack trace below this message and a final line that shows the overall results of our tests:

```
Chrome 49.0.2623 (Windows 8 0.0.0): Executed 1 of 1 (1 FAILED) ERROR (0.012 secs / 0.005 secs)
```

One thing you'll notice is that when we make the change to our test, we do not have to rerun Karma. Instead, it watches for any changes in our files and related tests and immediately reports success or failure whenever we make a change.

Pretty cool! Let's undo the last change that we made and put the test back into a passing state.

To sum up, we'll be taking the following multi-step approach to executing all our tests. First, we use our gulp script to convert the TypeScript files to JavaScript. Then we call Karma to run our tests against those JavaScript files. The following diagram sets forth these steps:

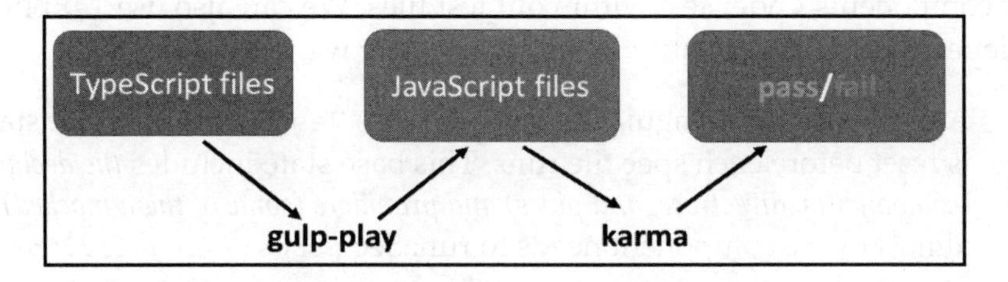


We will not be repeating the description of these steps in the tests we cover moving forward. So, be sure to follow them with every test we are exploring in this section. Now let's move on to unit-testing components.

# Unit-testing components

Testing Angular components is more complicated than testing simple pipes or services. That is because Angular components are associated with views and also usually have more dependencies than services, filters, or directives.

## Angular testing utilities

Because of their complexity, Angular has introduced utilities that enable us to test our components more easily. These testing utilities include the `TestBed` class (which we previously used to initialize our tests) and several helper functions in `@angular/core/testing`.

`TestBed` has a `createComponent` method that returns a `ComponentFixture` containing several members and methods, including:

- `debugElement`: For debugging a component
- `componentInstance`: For accessing the component properties and methods
- `nativeElement`: For accessing the view's markup and other DOM elements
- `detectChanges`: For triggering the component's change detection cycle

`ComnponentFixture` also contains methods for overriding the view, directives, bindings, and providers of a component. Going forward, we will be using `TestBed` throughout the rest of our tests.

`TestBed` has a method called `configureTestingModule` that we can use to set up our testing as its own module. This means we can bypass the initial bootstrap process and compile our components under test within out test files. We can also use `TestBed` to specify additional dependencies and identify the providers that we will need.

According to the Angular documentation, `TestBed` has a base state that is reset before each spec file runs. This base state includes *the declarables (components, directives, and pipes) and providers (some of them mocked)* that almost every component needs to run. See `https://angular.io/docs/ts/latest/guide/testing.html#!#atu-intro`.

## Managing dependencies in our tests

Components in Angular integrate the view with everything else. Due to this, components normally have more dependencies compared to any of the services, filters, or directives.

Notwithstanding the fact that our unit tests focus on the code within the component itself, we still need to account for these dependencies in our tests or else the tests will fail (we skipped the dependency setup for pipe testing as it did not have external dependencies).

Two approaches exist for handling these dependencies: inject them into our component or create a mock or fake for them that we can use in our tests. If a dependency is simple enough, we can just inject an instance of it into our test class. However, if the dependency is significantly complicated, especially if it has dependencies of its own and/or makes remote server calls, then we should be mocking it. The Angular testing library provides the tools for us to do that.

The component that we plan to test in this section is the `WorkoutRunner` component. Located inside `trainer/src/components/workout-runner/`, this is the component that runs a specific workout.

# Unit-testing WorkoutRunnerComponent

With this background, let's get started unit testing `WorkoutRunnerComponent`.

First, add a new file, `workout-runner-component.spec.ts`, with the following imports:

```
import { inject, fakeAsync, async, tick, TestBed, ComponentFixture} from
'@angular/core/testing';
import { NO_ERRORS_SCHEMA }          from '@angular/core';
import {Router} from '@angular/router';
import {Observable} from "rxjs/Rx";

import {WorkoutHistoryTracker} from '../../services/workout-history-
tracker';
import {WorkoutRunnerComponent} from './workout-runner.component';
import {WorkoutService} from '../../services/workout-service';
import {Exercise, WorkoutPlan, ExercisePlan} from "../../services/model";
import {SecondsToTimePipe} from "../shared/seconds-to-time.pipe";
```

These imports identify the test utilities (and things such as `Router` and `Observable` from `RxJS`) that we will be using in our tests along with the types and dependencies our component requires. We'll discuss these dependencies in a moment. One import that looks different from the others is the one that imports `NO_ERRORS_SCHEMA` from `@angular/core`. We will use this import to ignore elements in the component that we will not be testing. Again, we will discuss that further in a moment.

One more thing to note with the imports is that `@angular/core/testing` is a part of the core module and not in a separate testing module. This is a common pattern with imports for Angular testing. For example, when we get to testing HTTP, you will see that we are importing from `@angular/http/testing`.

## Setting up component dependencies

Next, we need to establish our component's dependencies and determine whether we need to inject or mock them. If we look at the code for the `WorkoutRunner` component, we see that there are three dependencies being injected into our component:

- `WorkoutHistoryTracker`: This is a component that has some behavior attached to it. So we definitely want to mock it.
- `Router`: We'll have to mock this too in order to isolate `WorkoutRunner` from the rest of the application and prevent our test from trying to navigate away from the `WorkoutRunner` view.
- `WorkoutService`: This is a service that we will use to make an HTTP call to retrieve our workouts. We will mock this service as well since we don't want to be making a call to an external system within our test.

## Mocking dependencies – workout history tracker

Angular allows us to mock our dependencies in a straightforward manner using simple classes. Let's start with mocking `WorkoutHistoryTracker`. To do that, add the following class just after the imports:

```
class MockWorkoutHistoryTracker {
    startTracking() {}
    endTracking() {}
    exerciseComplete() {}
}
```

We do not need to mock the entire `WorkoutHistoryTracker` class but only the methods that `WorkoutRunner` will be calling. In this case, those methods are `startTracking()`, `endTracking()`, and `exerciseComplete()`. We have made these methods empty because we do not need anything returned from them in order to test `WorkoutRunner`. Now we can inject this dummy implementation into `WorkoutRunner` wherever it is looking for `WorkoutHistoryTracker`.

## Mocking dependencies – workout service

In `Chapter 5`, *Supporting Server Data Persistence,* we extended the workout service to make a remote call to retrieve the data that populates a workout. For unit-testing the workout runner, we will want to replace that call with a mock implementation that returns some static data that we can use to run the test. So we will add a third mock class, as follows:

```
class MockWorkoutService {
    sampleWorkout = new WorkoutPlan(
         "testworkout",
         "Test Workout",
          40,
          [
              new ExercisePlan(new Exercise( "exercise1", "Exercise 1",
              "Exercise 1 description",  "/image1/path",
             "audio1/path"), 50),
              new ExercisePlan(new Exercise( "exercise1", "Exercise 2",
             "Exercise 2 description",  "/image2/path",
             "audio2/path"), 30),
              new ExercisePlan(new Exercise( "exercise1", "Exercise 3",
             "Exercise 3 description",  "/image3/path",
             "audio3/path"), 20)
          ],
          "This is a test workout"
    );
    getWorkout(name: string) {
        return Observable.of(this.sampleWorkout);
    }
    totalWorkoutDuration(){
        return 180;
    };
    export class MockRouter {
    navigate = jasmine.createSpy('navigate');
    }
}
```

Notice that the `getWorkout` method is returning an `Observable`. Otherwise the class is self-explanatory.

## Mocking dependencies – router

As with `WorkoutHistoryTracker` and `WorkoutService`, we also will be using mocking to handle the dependency that we have on the Angular router. But here we will be taking a slightly different approach. We will assign a jasmine spy to a navigate method on our mock. This will be sufficient for our purposes because we only want to make sure that the router's navigate method is being called with the appropriate route (`finished`) as a parameter. The jasmine spy will allow us to do that as we will see later.

## Configuring our test using TestBed

Now that we have our imports and dependencies out of the way, let's get started with the tests themselves. We begin by adding a Jasmine `Describe` function that will wrap our tests, followed by setting two local variable using `let`: one for `fixture` and the other for `runner`:

```
describe('Workout Runner', () =>{
    let fixture:any;
    let runner:any;
```

Next we'll add a `beforeEach` function that sets up our test configuration:

```
beforeEach( async(() =>{
    TestBed
        .configureTestingModule({
            declarations: [ WorkoutRunnerComponent, SecondsToTimePipe ],
            providers: [
                {provide: Router, useClass: MockRouter},
                {provide: WorkoutHistoryTracker ,useClass:
                MockWorkoutHistoryTracker},
                {provide: WorkoutService ,useClass: MockWorkoutService}
            ],
            schemas: [ NO_ERRORS_SCHEMA ]
        })
        .compileComponents()
        .then(() => {
            fixture = TestBed.createComponent(WorkoutRunnerComponent);
            runner = fixture.componentInstance;
        });
}));
```

The `beforeEach` method executes before each test, which means that we will only have to set this up once in our test file. Inside `beforeEach`, we add an `async` call. This is required because of the asynchronous `compileComponents` method we are calling.

The Angular documentation indicates that the `async` function *arranges for the tester's code to run in a special* `async` *test zone that hides the mechanics of asynchronous execution, just as it does when passed to an* `it` *test*. For more information refer to `https://angular.io/docs/ts/latest/guide/testing.html#!#async-in-before-each`. We'll discuss this in more detail shortly.

Let's go through each method call in the order they are executed. The first method, `configureTestingModule`, allows us to build on the base configuration of the testing module and add things such as imports, declarations (of the components, directives, and pipes we will be using in our test), and providers. In the case of our test, we are first adding declarations for the workout runner, our component under test, and the `SecondsToTimePipe`:

```
declarations: [ WorkoutRunnerComponent, SecondsToTimePipe ],
```

Then we add three providers for our `Router`, `WorkoutHistoryTracker`, and `WorkoutService`:

```
providers: [
{provide: Router, useClass: MockRouter},
{provide: WorkoutHistoryTracker ,useClass: MockWorkoutHistoryTracker},
{provide: WorkoutService ,useClass: MockWorkoutService}
],
```

For each of these providers, we set the `useClass` property to our mocks instead of the actual components. Now, anywhere in our test, when the `WorkoutRunner` requires any of these components, the mock will be used instead.

The next configuration may seem a bit mysterious:

```
schemas: [ NO_ERRORS_SCHEMA ]
```

This setting allows us to bypass the errors we would otherwise get regarding the custom elements associated with two components that we are using in the component's template: `ExerciseDescriptionComponent` and `VideoPlayerComponent`. At this point, we don't want to be testing these components within the test for the `WorkoutRunnerComponent`. Instead, we should be testing them separately. One thing to be aware of, however, when you use this setting is that it will suppress all schema errors related to elements and attributes in the template of the component under test; so it may hide other errors that you do want to see.

When you set up a test using `NO_ERRORS_SCHEMA`, you are creating what is called a shallow test, one that does not go deeper than the component you are testing. Shallow tests allow you to reduce complexities in the templates within the component you are testing and reduce the need for mocking dependencies.

The final steps in the configuration of our test are to compile and instantiate our components:

```
.compileComponents()
.then(() => {
    fixture = TestBed.createComponent(WorkoutRunnerComponent);
    runner = fixture.componentInstance;
});
```

As mentioned previously, we are using an `async` function in our `beforeEach` method because this is required when we call the `compileComponents` method. This method call is asynchronous and we need to use it here because our component has an external template that is specified in a `templateUrl`. This method compiles that external template and then inlines it so that it can be used by the `createComponent` method (which is synchronous) to create our component fixture. This component fixture in turn contains a `componentInstance–WorkoutRunner`. We then assign both the `fixture` and the `componentInstance` to local variables.

As mentioned previously, the `async` function we are using creates a special `async` test zone in which our tests will run. You'll notice that this function is simplified from normal `async` programming and lets us do things such as using the `.then` operator without returning a promise.

You can also compile and instantiate test components inside individual test methods. But the `beforeEach` method allows us to do it once for all our tests.

Now that we have configured our test, let's move on to unit-testing `WorkoutRunner`.

# Starting unit testing

Starting from the loading of workout data to transitioning of exercises, pausing workouts, and running exercise videos, there are number of aspects of the `WorkoutRunner` that we can test. The `workout.spec.ts` file (available in the `components/workout-runner` folder under `trainer/src`) contains a number of unit tests that cover the preceding scenarios. We will pick up some of those tests and work through them.

To start with, let's add a test case that verifies that the workout starts running once the component is loaded:

```
it('should start the workout', () => {
expect(runner.workoutTimeRemaining).toEqual(runner.workoutPlan.totalWorkout
Duration());
    expect(runner.workoutPaused).toBeFalsy();
});
```

This test asserts that the total duration of the workout is correct and the workout is in the running state (that is, not paused).

Assuming that the `autoWatch` property of `karma.conf.js` is true, saving this test automatically triggers the test execution. But this test fails (check the Karma console). Strange! All the dependencies have been set up correctly but still the second expect function of the `it` block fails as it is undefined.

We need to debug this test.

# Debugging unit tests in Karma

Debugging unit tests in Karma is easy as the tests are run in the browser. We debug tests as we debug the standard JavaScript code. And since our Karma configuration has added mappings from our TypeScript files to our JavaScript files, we can debug directly in TypeScript.

When Karma starts, it opens a specific browser window to run the tests. To debug any test in Karma, we just need to click on the **Debug** button available at the top of the browser window.

There is one window opened by Karma and one when we click on **Debug**; we can use the original window too for testing, but the original window is connected to Karma and does a live reload. Also, the script files in the original window are timestamped, which changes whenever we update the test and hence requires us to put in a breakpoint again to test.

Once we click on **Debug**, a new tab/window opens with all the tests and other app scripts loaded for testing. These are scripts that were defined during the Karma configuration setup in the `karma.conf.js` files section.

To debug the preceding failure, we need to add breakpoints at two locations. One should be added inside the test itself and the second one inside `Workoutcomponent`, where it loads the workout and assigns the data to the appropriate local variables.

Perform the following steps to add a breakpoint in Google Chrome:

1. Open the Karma debug window/tab by clicking on the **Debug** button on the window loaded by Karma when it started.
2. Press the F12 key to open the developer console.
3. Go to the **Sources** tab and the TypeScript files for your application will be located in the `source` folder.
4. We can now put breakpoints at the required locations just by clicking on the line number. This is the standard mechanism to debug any script. Add breakpoints at the locations highlighted here:

components
shared
workout-builder/shared
workout-runner
seconds-to-time.pipe.spec.ts
seconds-to-time.pipe.ts
workout-runner.component.spec.ts
workout-runner.component.ts

*workout-runner.component.spec.ts*

```
67        expect(fixture.componentInstance instanceof WorkoutRunnerComponent).toBe(true, 'should create Work
68    });
69
70    it('should start the workout', () => {
71        expect(runner.workoutTimeRemaining).toEqual(runner.workoutPlan.totalWorkoutDuration());
72        expect(runner.workoutPaused).toBeFalsy();
73    });
74
75    it('should start the first exercise', () => {
76        spyOn(runner, 'startExercise').and.callThrough();
77        runner.ngOnInit();
78        runner.ngDoCheck():
```

components
shared
workout-builder/shared
workout-runner
seconds-to-time.pipe.spec.ts
seconds-to-time.pipe.ts
workout-runner.component.spec.ts
workout-runner.component.ts
services
debug.html
(no domain)

*workout-runner.component.ts*

```
35    ngOnInit() {
36        this.getWorkout(this.workoutName);
37    }
38
39    ngDoCheck():any {
40        if (!this.dataLoaded) {
41            this.start();
42        }
43    }
57    start() {
58        if(this.workoutPlan)
59        {
60            this.dataLoaded = true;
61            this.restExercise = new ExercisePlan(new Exercise("rest", "Relax!", "Relax a bit", "rest.png")
62            this.tracker.startTracking();
63            this.workoutTimeRemaining = this.workoutPlan.totalWorkoutDuration();
64            this.currentExerciseIndex = 0;
65            this.startExercise(this.workoutPlan.exercises[this.currentExerciseIndex]);
66            this.workoutStarted.emit(this.workoutPlan);
67        }
```

5. We refresh the **Debug** page (the one we opened when we clicked on the **Debug** button). The breakpoint in `workout-runner.ts` is never hit, causing the test to fail.

   What we overlooked is that the code that we were trying to reach is within the `start` method of `workout-runner`, and the `start` method is not being called in the constructor. Instead it is called in `ngDoCheck` after the data for the workout has been loaded through a call to the `getWorkout` method in `ngOnInit`. Add calls to `ngOnInit` and `ngDoCheck` in your test, like so:

```
it('should start the workout', () => {
    runner.ngOnInit();    runner.ngDoCheck();
    expect(runner.workoutTimeRemaining).toEqual(
          runner.workoutPlan.totalWorkoutDuration());
    expect(runner.workoutPaused).toBeFalsy();
});
```

6. Save the change and Karma will run the test again. This time it will pass.

As the number of tests grows, unit testing may require us to concentrate on a specific test or a specific suite of tests. Karma allows us to target one or more tests by prepending `f` to the existing `it` block; that is, `it` becomes `fit`. If Karma finds tests with `fit`, it only executes those tests. Similarly, a specific test suite can be targeted by prepending `f` to the existing `describe` block: `fdescribe`. Also, if you prepend `x` to an `it` block, making it `xit`, then that block will be skipped.

Let's continue unit-testing the component!

# Unit-testing WorkoutRunner continued...

What other interesting things can we test? We can test whether the first exercise has started. We add this test to `workout.spec.ts` after the one we just added:

```
it('should start the first exercise', () => {
    spyOn(runner, 'startExercise').and.callThrough();
    runner.ngOnInit();
    runner.ngDoCheck();
    expect(runner.currentExerciseIndex).toEqual(0);
    expect(runner.startExercise).toHaveBeenCalledWith(
    runner.workoutPlan.exercises[runner.currentExerciseIndex]);
    expect(runner.currentExercise).toEqual(
    runner.workoutPlan.exercises[0]);
```

```
});
```

The second `expect` function in this test is interesting. It uses a Jasmine feature: spies. Spies can be used to verify method invocations and dependencies.

## Using Jasmine spies to verify method invocations

A spy is an object that intercepts every call to the function it is spying on. Once the call is intercepted, it can either return fixed data or pass the call to the actual function being invoked. It also records the call invocation details that can be used later in `expect` as we did in the preceding test.

Spies are very powerful and can be used in a number of ways during unit testing. Look at the documentation on spies at `http://jasmine.github.io/2.0/introduction.html#section-Spies` to learn more about them.

The second `expect` function verifies that the `startExercise` method was called when the workout started (`toHaveBeenCalledWith`). It is also asserting the correctness of the parameters passed to the function. The second `expect` statement asserts the behavior using a spy, but we first need to set up the spy to make this assert work.

In this case, we are using the spy to mock a call to the `startExercise` method. We can use the spy to determine whether the method has been called and with what parameters, using Jasmine's `toHaveBeenCalledWith` function.

Look at the Jasmine documentation for the `toHaveBeenCalled` and `toHaveBeenCalledWith` functions to learn more about these assert functions.

Here, the method is being called with the current `Exercise` as a parameter. Since the previous `expect` confirms that this is the first exercise, this `expect` confirms that a call to start that first exercise was executed.

There are a couple of things to note here. First, you have to be careful to put the setup for `spyOn` prior to calling `ngOnInit`. Otherwise, the spy will not be *spying* when the `startExercise` method is called and the method invocation will not be captured.

Second, since the spy is a mock, we will normally not be able to verify anything within the `startExercise` method. This is because the method itself is being mocked. This means that we cannot actually verify that the `currentExercise` property has been set, since that is being done inside the mocked method. However, Jasmine allows us to chain the spy with `.and.callThrough`, which will mean that in addition to tracking the calls to the method, it will delegate to the actual implementation. This then allows us to test that the `currentExercise` has also been set correctly inside the `startExercise` method.

## Using Jasmine spies to verify dependencies

While we just used a spy to verify the call to a method within our class, Jasmine spies are also useful in mocking calls to external dependencies. But why test calls to our external dependencies at all? After all, we are trying to limit our testing to the component itself!

The answer is that we mock a dependency to make sure that the dependency does not adversely affect the component under test. From a unit testing perspective, we still need to make sure that these dependencies are called by the component being tested at the right time with the correct input. In the Jasmine world, spies help us assert whether dependencies were invoked correctly.

If we look at the `WorkoutRunner` implementation, we emit a message with the details of the workout whenever the workout starts. An external dependency, `WorkoutHistoryTracker`, subscribes to this message/event. So let's create a spy and confirm that `WorkoutHistoryTracker` started when the workout started.

Add this `it` block after the preceding one:

```
it("should start history tracking", inject([WorkoutHistoryTracker],
(tracker: WorkoutHistoryTracker) => {
     spyOn(tracker, 'startTracking');
     runner.ngOnInit();
     runner.ngDoCheck();
     expect(tracker.startTracking).toHaveBeenCalled();
 }));
```

Within the `it` block, we add a spy on the `tracker`, a local instance of the `WorkoutHistoryTracker`. Then we use the spy to verify that the `startTracking` method of that dependency has been called. Simple and expressive!

You may recall that we are using `MockHistoryWorkoutTracker` here; it contains a mock, a `startTracking` method that is empty and returns nothing. That is fine because we are not testing the `WorkoutHistoryTracker` itself but just the method invocation on it being made by the `WorkoutRunner`. This test shows how useful it is to be able to combine mocks with spies to fully test the inner workings of the `WorkoutRunner`, separately and apart from its dependencies.

## Testing event emitters

Examining the code for the `WorkoutRunner`, we see that it sets up several event emitters that look like the following one for `workoutStarted`:

```
@Output() workoutStarted: EventEmitter<WorkoutPlan> = new
EventEmitter<WorkoutPlan>();
```

The Angular documentation describes an event emitter as *an output property that fires events to which we can subscribe with an event binding*. In `Chapter 2`, *Building Our First App – 7 Minute Workout*, we described in detail how event emitters are used in WorkoutRunner. So we have a good understanding of what they do. But how do we unit-test our event emitters and determine that they are firing events in the way we expect?

It's actually pretty easy to do. If we remember that an event emitter is an `Observable Subject` to which we can subscribe, we realize that we can simply subscribe to it in our unit test. Let's revisit our test that verifies that a workout is starting and add the highlighted code to it:

```
it('should start the workout', () => {
    runner.workoutStarted.subscribe((w: any) => {
      expect(w).toEqual(runner.workoutPlan);    });
    runner.ngOnInit();
    runner.ngDoCheck();
    expect(runner.workoutTimeRemaining).toEqual(
    runner.workoutPlan.totalWorkoutDuration());
    expect(runner.workoutPaused).toBeFalsy();
});
```

We injected the `WorkoutService` and added a subscription to the `WorkoutStarted` event emitter and an expectation that checks to see whether the property is emitting a `WorkoutPlan` when the event is triggered. The subscription is placed before `ngOnInit` because that is the method that results in the `workoutStarted` event being triggered, and we need to have our subscription in place before that happens.

## Testing interval and timeout implementations

One of the interesting challenges for us is to verify that the workout progresses as time elapses. The `Workout` component uses `setInterval` to move things forward with time. How can we simulate time without actually waiting?

The answer is the Angular testing library's `fakeAsync` function, which allows us to run otherwise asynchronous code synchronously. It does this by wrapping the function to be executed in a `fakeAsync` zone. It then supports using synchronous timers within that zone and also allows us to simulate the asynchronous passage of time with `tick()`.

For more information about `fakeAsync`, see the Angular documentation at `https://angular.io/docs/ts/latest/guide/testing.html#!#async`, under the heading *The fakeAsync function*.

Let's see how we can use the `fakeAsync` function to test the timeout and interval implementations in our code. Add the following test to `workout-runner.spec.ts`:

```
it('should increase current exercise duration with time', fakeAsync(() => {
    runner.ngOnInit();
    runner.ngDoCheck();
    expect(runner.exerciseRunningDuration).toBe(0);
    tick(1000);
    expect(runner.exerciseRunningDuration).toBe(1);
    tick(1000);
    expect(runner.exerciseRunningDuration).toBe(2);
    TestHelper.advanceWorkout(7);
    expect(runner.exerciseRunningDuration).toBe(10);
    runner.ngOnDestroy();
}));
```

In addition to injecting `WorkoutRunner`, we first wrap the test in `fakeAsync`. Then we add a call to the `WorkoutRunner`'s `ngOnit()` method. This kicks off the timers for the exercises within `WorkoutRunner`. Then within the test, we use the `tick()` function set at various durations to test the operation of the timer for an exercise, and make sure that it continues running for the duration that we expected it to run. Using `tick()` allows us to *fast forward* through the code and avoid having to wait for the exercise to complete over several seconds as it would if we ran the code asynchronously.

You'll notice that there is a helper method, `advanceWorkout`, being used here. This takes care of an anomaly that seemed to exist if the parameter passed to tick was anything other than `1000`.

At the end, we call the `WorkoutRunner`'s `ngOnDestroy()` method to clear any pending timers.

Let's try another similar test. We want to make sure that the WorkoutRunner is correctly transitioning from one exercise to the next. Add the following test to `workout-runner.ts`:

```
it("should transition to next exercise on one exercise complete",
fakeAsync(() => {
    runner.ngOnInit();
    runner.ngDoCheck();
    let exerciseDuration = runner.workoutPlan.exercises[0].duration;
    TestHelper.advanceWorkout(exerciseDuration);
    expect(runner.currentExercise.exercise.name).toBe('rest');
    expect(runner.currentExercise.duration).toBe(
    runner.workoutPlan.restBetweenExercise);
    runner.ngOnDestroy();
}));
```

Again we wrap the test in `fakeAsync` and call `runner.ngOnInit` to start the timer. Then we grab the duration of the first exercise and use the `tick()` function to advance the timer 1 second beyond the duration of that exercise. Next, we test the expectation that we are now in the rest exercise and thus have transitioned from the first exercise.

## Testing workout pause and resume

When we pause a workout, it should stop and the time counter should not lapse. To check this, add the following time test:

```
it("should not update workoutTimeRemaining for paused workout on
    interval lapse", fakeAsync(() => {
    runner.ngOnInit();
    runner.ngDoCheck();
    expect(runner.workoutPaused).toBeFalsy();
    tick(1000);
    expect(runner.workoutTimeRemaining).toBe(
    runner.workoutPlan.totalWorkoutDuration() - 1);
    runner.pause();
    expect(runner.workoutPaused).toBe(true);
    tick(1000);
    expect(runner.workoutTimeRemaining).toBe(
```

```
        runner.workoutPlan.totalWorkoutDuration() - 1);
        runner.ngOnDestroy();
    }));
```

The test starts with verifying the state of the workout as not paused, advances the time for 1 second, pauses it, and then verifies that the time of `workoutTimeRemaining` does not change after the pause.

# Unit-testing services

Unit testing of services is not much different from unit-testing components. Once we get the hang of how to set up a component and its dependencies (mostly using mocks), it becomes a routine affair to apply that learning to testing services. More often than not, the challenge is to set up the dependencies for the services so that testing can be done effectively.

Things are a little different for services that make remote requests (using either `http` or `jsonp`). There is some setup required before we can test such services in isolation.

We will target `WorkoutService` and write some unit tests for it. Since this service makes remote requests to load workout data, we will explore how to test such a service with a mock HTTP backend. Angular provides us with a `MockBackend` and `MockConnection` for doing that.

# Mocking HTTP request/response with MockBackend

When testing services (or, as a matter of fact, any other Angular construct) that make remote requests, we obviously do not want to make actual requests to a backend to check the behavior. That does not even qualify for a unit test. The backend interaction just needs to be mocked away. Angular provides a service for precisely that: `MockBackend`! Using `MockBackend`, we intercept HTTP requests, mock actual responses from the server, and assert endpoints invocation too.

Create a new file named `workout-service.spec.ts` and add the following import statements at the top of the file:

```
import {addProviders, fakeAsync, inject, tick} from
'@angular/core/testing';
import {BaseRequestOptions, Http, Response, ResponseOptions} from
'@angular/http';
import {MockBackend, MockConnection} from '@angular/http/testing';
import {WorkoutService} from './workout-service';
import {WorkoutPlan} from "./model";
```

In addition to the imports from the `testing` module, we are importing both the `http` module and the `MockBackend` and `MockConnection` from the `http/testing` module. We are also importing the `WorkoutService` and `WorkoutPlan` that we will be testing.

Once we have the imports in place, we will begin creating the test with the Jasmine `describe` statement that wraps our tests, and will set several local variables:

```
describe('Workout Service', () => {
    let collectionUrl:string = "...[mongo connnection url]...";
    let apiKey:string = "...[mongo key]...";
    let params:string = '?apiKey=' + apiKey;
    let workoutService:WorkoutService;
    let mockBackend:MockBackend;
```

In addition to creating local variables for `WorkoutService` and `MockBackend`, you'll also notice that we are setting local variables for our Mongo connection. To be clear, we are not setting these variables in order to make a remote call to Mongo but instead to test that the connection properties are being set properly.

The next step is set up the providers and dependency injection for our tests. To handle the providers, add the following to the test file:

```
beforeEach(() => {
    addProviders([
        MockBackend,
        BaseRequestOptions,
        {
            provide: Http,
            useFactory: (backend:MockBackend,
            options:BaseRequestOptions) => {
                return new Http(backend, options);
            },
            deps: [MockBackend, BaseRequestOptions]
        },
        WorkoutService
    ])
```

```
});
```

Not surprisingly, we are adding `MockBackEnd` and `WorkoutService` as providers. At the same time, we are also adding `BaseRequestOptions` from the http module. And then we are adding a provider for HTTP that uses a factory with the `MockEnd` and `BaseRequestOptions`. This factory will then return an `Http` service that is using the `MockBackend`. So now we can make an HTTP call from our tests that will not be a remote call, but instead will use the `MockBackEnd` to fake that call.

To complete the setup for our tests, we add the following to inject dependencies into each of our tests:

```
beforeEach(inject([WorkoutService, MockBackend], (service:WorkoutService,
backend:MockBackend) => {
    workoutService = service;
    mockBackend = backend
}));
```

With this setup in place, we are now in a position to create tests for the workout-service that avoid our making a remote call. We'll start with a simple test that makes sure that the `workoutService` loads:

```
it("should load Workout service", () => {
    expect(workoutService).toBeDefined();
});
```

While this test may seem trivial, it is important to place it here because it acts as a check to make sure that we have set up our configuration correctly.

Next, we'll move to testing several of the methods in the workout-service. First, we will make sure that it returns all workouts when the get Workouts method is called. To do that, add the following test:

```
it("should return all workout plans", fakeAsync(() => {
    let result:any;
    mockBackend.connections.subscribe((connection:MockConnection) => {
      expect(connection.request.url).toBe(collectionUrl + "/workouts" +
      params);
      let response = new ResponseOptions({body: '[{ "name": "Workout1",
      "title": "workout1" }, { "name": "Workout1", "title": "workout1"
      }]'});
        connection.mockRespond(new Response(response));
    });
    workoutService.getWorkouts().subscribe((response:Response) => {
        result = response;
    });
```

```
        expect(result.length).toBe(2);
        expect(result[0] instanceof WorkoutPlan).toBe(true);
    }));
```

Notice that we are using `fakeAsync` to synchronously run what would otherwise be an asynchronous HTTP call.

Note that there would be a problem with this approach if we were making a true HTTP call using XHR. See `https://github.com/angular/angular/issues/8280`. But here, we are not making a real XHR call in our mock.

Because the `Http` module returns `RxJS` observables, we are also using the pattern of subscribing to those observables. You should be used to seeing this pattern from our coverage of observables in `Chapter 5`, *Supporting Server Data Persistence*. Specifically, we are subscribing to the `connections` property of the `mockBackEnd` and passing in a `MockConnection` as our connection. After confirming that our connection has been set properly, we construct a `response` containing two `workouts`. Then we call the `mockRespond` method on our `connection` and return the `response` that we have constructed. The net result is that we have been able to construct a testable response for our service that avoids making a remote call.

The final step in the process is to set the workout's `getWorkouts` method to subscribe to that `response`, and then add the appropriate `expect` statements to confirm that we are getting back the correct number of `workouts` in the `response` and that the first `workout` is of type `WorkoutPlan`.

We'll follow the same pattern of using `fakeAsync` with our `mockBackend` to build additional tests that confirm that we are able to do the following:

- Return a `workout` plan with a specific name
- Map `exercises` correctly within the `getWorkout` method

You can review these tests in the code for checkpoint 7.1. But one thing to note is that in both these tests we have the following conditional logic within the `connections.subscribe` method:

```
if (connection.request.url === collectionUrl + "/workouts/Workout1" +
params) {
    let response = new ResponseOptions({
        body: '{ "name" : "Workout1" , "title" : "Workout 1" , "exercises"
: [ { "name" : "exercise1" , "duration" : 30}]}'
    });
    connection.mockRespond(new Response(response));
} else {
    connection.mockRespond(new Response(
        new ResponseOptions({
            body: [{name: "exercise1", title: "exercise 1"}]
        })));
}
```

This may seem a little confusing at first until we realize that with the `getWorkout` method we are actually making two `Http` calls: one to retrieve `workout` and one to retrieve all `exercises`. As you recall from `Chapter 5`, *Supporting Server Data Persistence,* we are doing that in order to create a fuller description of each `exercise` that is included within `workout`. So what we are doing here is checking to make sure that we construct a `response` made up of a `workout` for the call that is retrieving a `workout` and a `response` made up of an `exercise` for the other call.

With that, we are finished with testing our service.

Next, we need to learn how to test directives. The next section is dedicated to understanding the challenges in directive testing and how to overcome them.

# Unit-testing directives

No other Angular constructs that we have tested so far do not involve any UI interaction. But directives, as we know, are a different beast. Directives are all about enhancing a component's view and extending the behavior of HTML elements. While testing directives, we cannot ignore the UI connections, and hence directive testing may not strictly qualify as unit testing.

The good thing about directive testing is that its setup process is not as elaborate as that for services or components. The pattern to follow while unit-testing directives is as follows:

1. Take an HTML fragment containing the directive markup.
2. Compile and link it to a mock component.
3. Verify that the generated HTML has the required attributes.
4. Verify the changes if the directive created changes the state.

## The TestBed class

As mentioned previously, Angular provides the `TestBed` class to facilitate this kind of UI testing. We can use it to dig into the markup in a component's view and check for DOM changes that are triggered by events. Armed with this tool, let's get started with the testing of our directives. In this section, we are going to test `remoteValidator`.

This will be a good time to revisit the directives that we built in the previous chapter. Also, keep the code handy for the tests we will create in the following sections.

## Testing remote validator

Let's start with unit-testing the `remoteValidatorDirective`. Just to refresh our memory, `remoteValidatorDirective` validates an input against remote rules. It does so by calling a component method that returns a promise. If the promise is resolved with success, the validation passes; otherwise, the validation fails. The `[validateFunction]` attribute provides the link between the DOM and the component's method that checks for the duplication.

Similar to our other test files, we add a `remote-validator.directive.spec.ts` file in the workout builder folder. Refer to the file in checkpoint 7.1 for the imports, which we will not cover at this point.

Just below the import statements, add the following component definition:

```
@Component({
  template: `
  <form>
  <input type="text" name="workoutName" id="workout-name"
  [(ngModel)]="workoutName" a2beBusyIndicator
  a2beRemoteValidator="workoutname"
```

```
    [validateFunction]="validateWorkoutName">
    </form>
    `
})
export class TestComponent {
    workoutName: string;

    constructor() {
        this.workoutName = '7MinWorkout';
    }
    validateWorkoutName = (name: string): Promise<boolean> => {
        return Promise.resolve(false);
    }
}
```

This component looks a lot like the components we set up in our other tests to mock dependencies. Here, however, it is serving a slightly different purpose; it is acting as a host container for the directive that we will be testing. Using this minimal component, lets us avoid having to load the actual host for this directive, which is the `Workout` component.

One thing to notice here is that we have set up a method for `validateWorkoutName` that will be called by our directive. It is essentially a stub that just returns a resolved `Promise` of `false`. Remember we are not concerned with how this method handles its validation but with verifying that the directive calls it and returns the correct result, either `true` or `false`.

Next, we set up the describe statement for our test suite by adding the following code, which injects `RemoteValidatorDirective` into our tests:

```
describe('RemoteValidator', () => {
    let fixture: any;
    let comp: any;
    let debug: any;
    let input: any;

    beforeEach(async(() => {
        TestBed.configureTestingModule({
            imports: [ FormsModule ],
            declarations: [ TestComponent, RemoteValidatorDirective ]
        });
        fixture = TestBed.createComponent(TestComponent);
        comp = fixture.componentInstance;
        debug = fixture.debugElement;
        input = debug.query(By.css('[name=workoutName]'));
    }));
```

As you can see, we are setting up local variables for `fixture`, its `componentInstance`, and `debugElement`. We are also using `by.css` (which we will see more of in our end-to-end tests) along with the query method on `debugElement` to extract the `workoutName` input from our component. We'll be using these to delve into the rendered HTML in our directive.

Now we are ready to write our individual tests. First, we'll write a test to confirm that we have been able to load `RemoteValidatorDirective`. So add the following code:

```
it("should load the directive without error", fakeAsync(() => {
    expect(input.attributes.a2beRemoteValidator).toBe('workoutname',
'remote validator directive should be loaded.')
}));
```

What's interesting about this test is that using the `debugElement`, we have been able to drill-down into the attributes of the input tag in our host component and find our validator, confirming that it has indeed been loaded. Also notice the use of `fakeAsync`, which we discussed in connection with unit testing. Using it makes it possible for us to write our tests in a synchronous fashion and avoid the complications that would otherwise exist with trying to manage the asynchronous rendering of our host component. Next, we'll write two tests to confirm that our validator is working properly. The first test will make sure that an error is created if remote validation fails (that is, a workout with the same name as the one we are using is found). Add the following code for that test:

```
it('should create error if remote validation fails', fakeAsync(() => {
    spyOn(comp, 'validateWorkoutName').and.callThrough();
    fixture.detectChanges();
    input.nativeElement.value = '6MinWorkout';
    tick();
    let form: NgForm = debug.children[0].injector.get(NgForm);
    let control = form.control.get('workoutName');
    expect(comp.validateWorkoutName).toHaveBeenCalled();
    expect(control.hasError('workoutname')).toBe(true);
    expect(control.valid).toBe(false);
    expect(form.valid).toEqual(false);
    expect(form.control.valid).toEqual(false);
    expect(form.control.hasError('workoutname',
    ['workoutName'])).toEqual(true);
}));
```

Again, we are using `fakeAsync` to eliminate the challenges we would otherwise have with the async behavior associated with the rendering and execution of our `remoteValidatorDirective`. Next, we add a spy to track the invocation of the `validateWorkoutName` method. We also set the spy to call through to our method because in this case, we are expecting it to return false. The spy is being used to verify that our method has indeed been invoked. Next, we set `fixture.detectChanges`, which triggers a change detection cycle. We then set the value of our input and call tick, which will, we hope, trigger the response we are expecting from our remote validator. We then grab the form encapsulating our input tag using the injector that is available from the child element array of the debug element. From there, we extract the form control for our input box. Then we run several expectations that confirm that an error has been added both to our control and to the form and that both are now in an invalid state. The next test is the mirror opposite of this test and checks for a positive:

```
it('should not create error if remote validation succeeds', fakeAsync(() =>
{
    spyOn(comp,' validateWorkoutName').and.returnValue(
    Promise.resolve(true));
    fixture.detectChanges();
    input.nativeElement.value = '6MinWorkout';
    tick();
    let form: NgForm = debug.children[0].injector.get(NgForm);
    let control = form.control.get('workoutName');
    expect(comp.validateWorkoutName).toHaveBeenCalled();
    expect(control.hasError('workoutname')).toBe(false);
    expect(control.valid).toBe(true);
    expect(form.control.valid).toEqual(true);
    expect(form.valid).toEqual(true);
    expect(form.control.hasError('workoutname',
['workoutName'])).toEqual(false);
}));
```

Other than changing the expectations, the only change we are making from the previous test is setting up our spy to return a value of true. Unit-testing our `remoteValidatorDirective` shows how powerful the `TestBed` utilities are in testing our UI and the elements and behaviors associated with it.

# Getting started with E2E testing

Automated **E2E** testing is an invaluable asset if the underlying framework supports it. As the size of an app grows, automated E2E testing can save a lot of manual effort.

Without automation, it's just a never-ending battle to make sure that the app is functional. However, remember that in an E2E setup, not everything can be automated; automation may require a lot of effort. With due diligence, we can offload a sizable amount of manual effort but not everything.

The process of E2E testing of a web-based application is about running the application in a real browser and asserting the behavior of the application based on the user interface state. This is how an actual user does testing.

Browser automation holds the key here, and modern browsers have become smarter and more capable in terms of supporting automation. Selenium tools for browser automation are the most popular option out there. Selenium has the WebDriver (`https://www.w3.org/TR/webdriver/`) API that allows us to control the browser through the automation API that modern browsers natively support.

The reason behind bringing up Selenium WebDriver is that the Angular E2E testing framework/runner **Protractor** also uses **WebDriverJS**, which is a JavaScript binding of WebDriver on Node. These language bindings (like the preceding JavaScript binding) allow us to use the automation API in the language of our choice.

Let's discuss Protractor before we start writing some integration tests for our app.

# Introducting Protractor

Protractor is the de facto test runner for E2E testing in Angular. Protractor uses Selenium WebDriver to control a browser and simulate user actions.

Protractor supersedes an earlier E2E framework known as **AngularJS Scenario Runner**. Karma had a plugin that allowed Karma to execute E2E tests.

A typical Protractor setup has the following components:

- A test runner (Protractor)
- A Selenium server
- A browser

We write our test in Jasmine and use some objects exposed by Protractors (which is a wrapper over WebDriverJS) to control the browser.

When these tests run, Protractor sends commands to the Selenium server. This interaction happens mostly over HTTP.

The Selenium server, in turn, communicates with the browser using the *WebDriver Wire Protocol,* and internally the browser interprets the action commands using the browser driver (such as *ChromeDriver* in the case of Chrome).

It is not that important to understand the technicalities of this communication, but we should be aware of the E2E testing setup. Check out the article from the Protractor documentation at `http://angular.github.io/protractor/#/infrastructure` to learn more about this flow.

Another important thing to realize when using Protractor is that the overall interaction with the browser or the browser control flow is asynchronous in nature and promise-based. Any HTML element action, whether `sendKeys`, `getText`, `click`, `submit`, or any other, does not execute at the time of invocation; instead the action is queued up in a control flow queue. For this precise reason, the return value of every action statement is a promise that gets resolved when the action completes.

To handle this *asynchronicity* in Jasmine tests, Protractor patches Jasmine, and therefore assertions like these work:

```
expect(element(by.id("start")).getText()).toBe("Select Workout");
```

They work despite the `getText` function returning a promise and not the element content.

At the time of writing this book, Protractor supports Jasmine version 2.5.2.

With this basic understanding of how Protractor works, let's set up Protractor for end-to-end testing.

# Setting up Protractor for E2E testing

To install Protractor globally, run this command in the console:

```
npm install -g protractor
```

This installs two command-line tools: Protractor and webdriver-manager. Run the following command to make sure that Protractor is set up correctly:

```
protractor --version
```

All E2E tests have been verified against Protractor 4.0.9.web.

webdriver-manager is a helper tool to easily get an instance of a running Selenium server. Before we start the Selenium server, we need to update the driver binaries with the following call:

```
webdriver-manager update
```

Finally, run this command to start the Selenium server:

```
webdriver-manager start
```

Protractor tests send requests to this server to control a local browser. E2E testing can be verified by checking the status of the server at `http://localhost:4444/wd/hub` (the default location).

Protractor also needs to be configured, like Karma, and has a configuration file. We copy the `protractor.config.js` file from the `tests` folder under `chapter7/ checkpoint1/` to our local `tests` folder.

The Protractor configuration file we just added contains four settings that we want to make sure are configured according to our local app setup, and these include the following:

| Key | Description |
|---|---|
| `Specs` | The location of the specification files (the E2E test files). The current assigned value `['e2e/*.js']` should work. |
| `baseUrl` | The base URL where the app is running. Change the server name and port to match your local setup. Navigate to the URL to make sure that the app is running. |
| `seleniumAddress` | The base URL where the Selenium server is running. Unless you have reconfigured the Selenium server settings, the default value should work. |
| `useAllAngular2AppRoots` | Set this to `true`. This lets Protractor know that we are no longer using earlier versions of Angular. |

The configuration file documentation on the Protractor website (`https://github.com/angular/protractor/blob/master/lib/config.ts`) contains details on other supported configurations.

That is enough to start testing with Protractor.

# TypeScript configuration

As with all the other examples in this book, we will be writing our tests in TypeScript. This requires a couple more configuration steps.

First, copy the `tsconfig.json` file from checkpoint 7.1 into the `trainer` folder. You'll notice that this file now has some additional global dependencies for WebDriver and angular-protractor:

```
{
  "globalDependencies": {
    "core-js": "registry:dt/core-js#0.0.0+20160602141332",
    "jasmine": "registry:dt/jasmine#2.2.0+20160621224255",
    "angular-protractor": "registry:dt/angular-
     protractor#1.5.0+20160425143459",    "selenium-webdriver":
     "registry:dt/selenium-webdriver#2.44.0+20160317120654"
  }
}
```

Next, from the command line in the `trainer` folder, run the following:

```
typings install
```

This will install the type definitions for Protractor and web driver that we will be using when in our tests.

Now let's begin writing and executing some tests.

# Writing E2E tests for the app

Let's start in a simple manner and test our app start page (**#/start**). This page has some static content, a workout listing section with search capabilities, and the ability to start a workout by clicking on any workout tile.

All our E2E tests will be added to the `e2e` folder under `tests`.

Add a new file called `workout-runner.spec.ts` to the `e2e` folder under `tests`.

At the top of the file add the following reference:

```
/// <reference path="../../typings/index.d.ts"/>
```

This brings in the type definitions we just installed. Next add the following code:

```
describe("Workout Runner", () => {
describe("Start Page", () => {
    beforeEach(() => {t
        browser.get("");
    });
    it("should load the start page.", () => {
        expect(browser.getTitle()).toBe("Personal Trainer");
        expect(element(by.id("start")).getText()).toBe("Select Workout");
    });
});
});
```

Since we are writing this test in TypeScript, we have to introduce a transpilation step here. Therefore, in a new command window, navigate to the to the `e2e` folder under `tests` and type the following:

```
tsc workout-runner.e2e.ts -w
```

You should see the following message:

```
message TS6042: Compilation complete. Watching for file changes.
```

The `-w` switch means that TypeScript will recompile the tests in this file as we change them. So we will not need to run this command again.

# Executing our E2E tests

Before we execute our first test, we have to make sure that the Selenium server is running (`webdriver-manager start`) and the app is running (run `gulp play` from the command line in the `trainer` folder).

Now from the command line in the `trainer` folder, run the following and see the browser dance to your tune:

```
protractor tests/protractor.conf.js
```

Protractor will open the browser. It will then navigate to the start page; wait for the page, the scripts, and the framework to load; and then perform the test. It finally logs the results of the test in the console. That is pretty awesome!

Let's walk through this simple test.

The first interesting piece is inside the `beforeEach` block. The browser object is a global object exposed by Protractor and is used to control the browser-level actions. Underneath, it is just a wrapper around WebDriver. The `browser.get("")` method navigates the browser to start the app page, every time, before the start of the test.

The actual test verifies that the title of the page is correct. It also checks whether some random content is present on the page.

The preceding test employs two new globals, `element` and `by`, that are made available by Protractor:

- `element`: This function returns an `ElementFinder` object. The primary job of `ElementFinder` is to interact with the selected element. We will be using the `element` function to select `ElementFinder` extensively in our tests.

Refer to the documentation at `http://www.protractortest.org/#/locators#actions` to learn more about element manipulation API support. Functions such as `getText()` are actually defined on `WebElement` but are always accessed using `ElementFinder`. As the documentation suggests, `ElementFinder` can be treated as `WebElement` for most purposes. For more information, you can refer to `http://www.protractortest.org/#/locators#behind-the-scenes-elementfinders-versus-webelements`.

- `by`: This object is there to locate elements. It has functions that create `locators`. In the preceding test, a locator is created to search for elements with `id=start`. There are a number of locators that can be used to search for a specific element. These include by class, by ID, and by css. (At this time by model and by binding are not yet supported.) Refer to the Protractor documentation on locators at `http://angular.github.io/protractor/#/locators` to learn about the supported locators.

Just to reiterate what we discussed earlier, `getTitle()` and `getText()` in the preceding test do not return the actual text but a Promise; we can still assert on the text value.

This simple test highlights another salient feature of Protractor. It automatically detects when the Angular app is loaded and when data is available for testing. There are no ugly hacks to delay testing (using `timeouts`) that may otherwise be required in standard E2E testing scenarios.

Remember, this is an *SPA*; full-page browser refresh does not happen, so it is not that simple to determine when the page is loaded and when the data that is rendered for AJAX calls is available. Protractor makes it all possible.

Protractor may still timeout while trying to assess whether the page is available for testing. If you are hitting timeout errors with Protractor, this article from the Protractor documentation can be really helpful (`http://www.protractortest.org/#/timeouts`) for debugging such issues.

# Setting up backend data for E2E testing

Setting up backend data for E2E testing is a challenge, irrespective of the E2E framework we employ for testing. The ultimate aim is to assert the behavior of an application against some data, and unless the data is fixed, we cannot verify the behavior that involves getting or setting data.

One approach to setting up data for E2E tests is to create a test data store specifically for E2E tests with some seed data. Once the E2E tests are over, the data store can be reset to its original state for future testing. For *Personal Trainer*, we can create a new database in MongoLab dedicated exclusively to E2E testing.

This may seem a lot of effort, but it is necessary. Who said E2E testing is easy! In fact, this challenge is there even if we do manual testing. For a real app, we always have to set up data stores/databases for every environment, whether *dev*, *test*, or *production*.

In this case, we will continue to use our existing backend but go ahead and add another workout that we will use for testing. Name this workout `1minworkout` and give it a title of `1 Minute Workout`. Add two exercises to the workout: Jumping Jacks and Wall Sit. Set the duration of each exercise to 15 seconds and the rest time to 1 second.

We have deliberately kept our new workout short so that we can complete our end-to-end testing of this workout within the normal timeouts provided by Protractor.

# More E2E tests

Let's get back to testing the workout search features on the start page. With the addition of 1 Minute Workout, we now have two workouts and we can assert search behaviors against these.

If you have added other workouts to the backend, just adjust the numbers in this test accordingly.

Add this test after the existing test in `workout-runner.spec.ts`:

```
it("should search workout with specific name.", () => {
    var filteredWorkouts = element.all(by.css(".workout.tile"));
    expect(filteredWorkouts.count()).toEqual(2);
    var searchInput = element(by.css(".form-control"));
    searchInput.sendKeys("1 Minute Workout");
    expect(filteredWorkouts.count()).toEqual(1);
expect(filteredWorkouts.first().element(by.css(".title")).getText()).toBe("
1 Minute Workout");
});
```

The test uses `ElementFinder` and `Locator API` to look for elements on the page. Check the second line of the test. We are using the `element.all` function together with the `by.css` locator to do a multi-element match on all elements on the screen that are using the `.workout.tile` CSS class. This gives us a list of workouts against which the next line asserts the element count of 3.

The test then gets hold of the search input using the `element` function along with the `by.css` locator to do a single element match for an element using the `.form-contol` CSS class. We then use the `sendKeys` function to simulate data entry in the search input.

The last two expect operations check for the count of elements in our list, which after the search should be 1. Also, they check whether the correct workout is filtered based on a div tag using the `title` CSS class that is a child of the element that contains our workout. This last expect statement highlights how we can chain element filtering and get hold of child elements in HTML.

There is one more test associated with the start page that we should add. It tests the navigation from the start page to the workout runner screen. Add this code for that test:

```
it("should navigate to workout runner.", () => {
    var filteredWorkouts = element.all(by.css(".workout.tile"));
    filteredWorkouts.first().click();
    expect(browser.getCurrentUrl()).toContain("/workout/1minworkout");
});
```

This test uses the `click` function to simulate clicking on a workout tile, and then we use the `browser.getCurrentUrl` function to confirm that the navigation is correct.

Run the test again (`protractor tests/protractor.conf.js`) and once again observe the magic of browser automation as the tests run one after another.

Can we automate E2E testing for *Workout Runner*? Well, we can try.

# Testing WorkoutRunner

One of the major challenges with testing WorkoutRunner is that everything is time-dependent. With unit testing, at least we were able to mock the interval, but not anymore. Testing exercise transitions and workout completion is definitely difficult.

However, before we tackle this problem or try to find an acceptable workaround, let's digress and learn about an important technique to manage E2E testing: page objects!

# Using page objects to manage E2E testing

The concept of page objects is simple. We encapsulate the representation of page elements into an object so that we do not have to litter our E2E test code with `ElementFinder` and `locators`. If any page element moves, we just need to fix the page object.

Here is how we can represent our Workout Runner page:

```
class WorkoutRunnerPage{
    pauseResume: any;
    playButton: any;
    pauseButton: any;
    exerciseTitle: any;
    exerciseDescription: any;
    exerciseTimeRemaining; any;

    constructor(){
        this.pauseResume =  element.all(by.id('pause-overlay'));
        this.playButton = element.all(by.css('.glyphicon-play'));
        this.pauseButton = element.all(by.css('.glyphicon-pause'));
        this.exerciseTitle = element.all(by.css(
        '.workout-display-div h1')).getAttribute('value');
        this.exerciseDescription = element.all(by.id(
        'description-panel')).getAttribute('value');
        this.exerciseTimeRemaining = element.all(by.css(
        '.workout-display-div h4')).getAttribute('value');
    }
};
```

This page object now encapsulates many of the elements that we want to test. By organizing the element selection code in one place, we increase the readability and hence the maintainability of E2E tests.

Now add the Workout Runner page object to the top of the test file. We'll use it in a test for the workout runner. Add the following new describe block containing the first of our workout runner tests:

```
describe("Workout Runner page", () => {
    beforeEach(() => {
        browser.get("#/workout/1minworkout");
    });
    it("should load workout data", () => {
        var page = new WorkoutRunnerPage();
        page.pauseResume.click();
        expect(page.exerciseTitle).toBe['Jumping Jacks'];
        expect(page.exerciseDescription).toBe["A jumping jack or
        star jump, also called side-straddle hop is a physical
        jumping exercise."];
    });
});
```

The test verifies that the workout is loaded and the correct data is shown. We make full use of the page object that we defined earlier. Run the test and verify that it passes.

Let's get back to the challenge of testing code based on `interval` or `timeout`. First, we'll add a test that confirms a click event on the screen, when the pause button is pushed:

```
it("should pause workout when paused button clicked", () => {
    let page = new WorkoutRunnerPage(),
        timeRemaining;
    page.pauseResume.click();
    expect(page.playButton.count()).toBe(1);
    expect(page.pauseButton.count()).toBe(0);
    page.exerciseTimeRemaining.then((time)=> {
        timeRemaining = time;
        browser.sleep(3000);
    });
    page.exerciseTimeRemaining.then((time)=> {
        expect(page.exerciseTimeRemaining).toBe(timeRemaining);
    });
});
```

What is interesting here is that we use the `browser.sleep` function within a promise to verify that the exercise time remaining is the same before and after the button is clicked. We are again using our `WorkoutRunner` page object to make the test more readable and understandable.

Next, add the following test to the current test suite:

```
it("should transition exercise when time lapses.", () => {
    var page = new WorkoutRunnerPage();
    browser.sleep(15000);
    page.pauseResume.click();
    expect(page.exerciseTitle).toBe["Relax!"];
    expect(page.exerciseDescription).toBe["Relax a bit!"];
    //expect(page.videos.count()).toBe(0);
});
```

This test checks whether the exercise transition happened. It does so by adding a `browser.sleep` function for 15 seconds and then checking from the UI state whether the exercise-related content of `Rest` is visible. The problem with this test is that it is not very accurate. It can confirm that the transition is happening but cannot confirm that it happened at the right time.

A plausible explanation for this behavior is in the way Protractor works. Before Protractor can start a test, it first waits for the page to load. If the test involves any action (such as `getText`), it again waits till Angular synchronizes the page. During page synchronization, Angular waits for any pending HTTP requests or any timeout-based operations to complete before it starts the test. As a result, when the `browser. sleep` function is invoked and when the browser actually goes to sleep cannot be predicted with great accuracy.

We can disable this synchronization behavior by setting `browser.ignoreSynchronization` to `true`, but we should avoid this as much as possible. If we set it to `true`, the onus is on us to determine when the page content is available for making assertions.

The bottom line is that the *Workout Runner* app workflow is indeed difficult to test. Compared to *Workflow Runner*, other things are far easier to test as we saw with the start page testing.

It's time now to wrap up the chapter and summarize our learning.

# Summary

We do not need to reiterate how important unit- and E2E-testing are for any application. The way the Angular framework has been designed makes testing the Angular app easy. In this chapter, we covered how to write unit tests and E2E tests using libraries and frameworks that target Angular.

For unit testing, we used Jasmine to write our tests and Karma to execute them. We tested a number of filters, components, services, and directives from *Personal Trainer*. In the process, you learned about the challenges and the techniques used to effectively test these types.

For E2E testing, the framework of choice was Protractor. We still wrote out tests in Jasmine but the test runner this time was Protractor. You learned how Protractor automates E2E testing using Selenium WebDriver, as we did some scenario testing for the *Start* and *Workout Runner* pages.

If you have reached this point, you are getting closer to becoming a proficient Angular developer. The next chapter reinforces this with more practical scenarios and implementations built using Angular. We will touch upon important concepts in the last chapter of this book; these include multilingual support, authentication and authorization, communication patterns, performance optimizations, and a few others. You certainly do not want to miss them!

# 8
# Some Practical Scenarios

With seven chapters under our belt, it should feel nice. What you have learned thus far is a direct consequence of the apps we have built in the last few chapters. I believe you now have an adequate understanding of the framework, how it works, and what it supports. Armed with this knowledge, as soon as we start to build some decent size apps, there are some common problems/patterns that will invariably surface, such as these:

- How to authenticate the user and control his/her access (authorize)?
- How to make sure that the app is performing enough?
- My app requires localized content. What do I do?
- What tools can I use to expedite app development?
- I have an Angular 1 app. How do I migrate it?

And some more!

In this chapter, we will try to address such common scenarios and provide some working solutions and/or prescriptive guidance to handle such use cases.

The topics we will cover in this chapter include:

- **Angular seed projects**: You will learn how some seed projects in Angular can help us when starting a new engagement.
- **Authenticating Angular applications**: This is a common requirement. We look at how to support cookie- and token-based Authentication in Angular.
- **Angular performance**: A customary performance section is a must as we try to detail what makes Angular 2 performant and things you can do to make your apps faster.
- **Migrating Angular 1 apps to Angular 2**: Angular 1 and Angular 2 are altogether different beasts. In this chapter, you will learn how to gradually migrate an Angular 1 app to Angular 2.

Let's start from the beginning!

# Building a new app

Imagine a scenario here: we are building a new application and given the super awesomeness of the Angular framework, we have unanimously decided to use Angular. Great! What next? Next is the mundane process of setting up the project.

Although a mundane activity, it's still a critical part of any engagement. Setting up a new project typically involves:

- Creating a standard folder structure. This is at times influenced by the server framework (such as *RoR*, *ASP.Net*, *Node.js*, and others).
- Adding standard assets to specific folders.
- Setting up the build, which in case we are developing an Angular 2-based web application includes:
    - Compiling/transpiling content if using TypeScript
    - Configuring the Module loader
    - Dependency management in terms of framework and third-party components
    - Setting up unit/E2E testing
- Configuring builds for different environments such as dev, test, and production. Again, this is influenced by the server technology involved.
- Code bundling and minification.

There is a lot of stuff to do.

What if we can short-circuit the overall setup process? This indeed is possible; we just need a **seed project** or a **starter site**.

# Seed projects

As we write this book, Angular 2 has just surfaced. There are a number of *seed projects* that can get us started in no time. Some seed projects integrate the framework with a specific backend and some only dictate/provide Angular-specific content. Some come preconfigured with vendor-specific libraries/frameworks (such as *LESS*, *SASS*, *Bootstrap*, and *FontAwesome*) whereas others just provide a plain vanilla setup.

Some of the notable seed projects worth exploring are as follows:

- **Angular 2 Webpack Starter** (`http://bit.ly/ng2webpack`): This seed repo serves as an Angular 2 starter for anyone looking to get up and running with Angular 2 and TypeScript fast. It uses Webpack (module bundler) to build our files and assist with boilerplate. It is a complete build system with a substantial number of integrations.
- **Angular 2 Seed** (`http://bit.ly/ng2seed`): Another seed project similar to Angular 2 Webpack starter. This seed project uses gulp for build automation, and the module bundler system is not as advanced as Webpack.
- **angular-cli** (`http://bit.ly/ng2-cli`): This is a command-line tool created by the Angular team that not only sets up a seed project for us but also has scaffolding capabilities. We can generate boilerplate components, directives, pipes, and services.

These projects provide a head start when building with Angular.

If the app is tied to a specific backend stack, we have two choices, which are as follows:

- Use one of these seed projects and integrate it with the backend manually.
- Find a seed project/implementation that does it for us. Angular 2 is relatively new, but there is a good chance that such seed projects will come up over time.

This discussion cannot be complete without mentioning tools that go a step further. They not only come with a seed implementation but also have scaffolding capabilities that make our lives easier.

# Seed and scaffolding tools

Two of the noteworthy to mention in this space are **Yeoman** and **angular-cli**.

# Yeoman

Yeoman (`http://yeoman.io/`) is a suite of tools targeted toward web application development. It defines a workflow for building modern web applications. It consists of:

- **yo**: This is a scaffolding tool used to generate code on-the-fly
- **Grunt/Gulp**: The de facto choices when it comes to building systems on Node

- **Bower/npm**: Bower is a package manager for the Web and works similarly to npm

The scaffolding component of Yeoman is quite interesting. *yo*, as it is named, uses the concept of a generator to achieve scaffolding.

Scaffolding is the process of generating a code skeleton that can be built upon. Using scaffolding, we can save some initial effort and provide some guidance around how the overall structure of any coding artefact should look.

Generators in Yeoman are used to set up the initial seed project, and later for individual script generation too. Since Yeoman is not targeted specifically towards Angular, there are generators for various client and server stacks. There are a number of generators available for Angular 2 with varied configurations.

Check out http://bit.ly/yogenerators for an exhaustive list of generators supported on Yeoman! You need to filter down to Angular 2 generators and select something that works for you. Feel free to experiment with these generators and see what best fits your needs.

## angular-cli

A scaffolding tool officially endorsed by the Angular team, **angular-cli** works in the same way as Yeoman but specifically targets Angular 2. Let's try it out.

To install angular-cli, from the command line, run:

```
npm install -g angular-cli
```

And then run this set of commands to generate the initial artefacts and build the setup for a new project:

```
ng new PROJECT_NAME
cd PROJECT_NAME
ng serve
```

Open the browser at http://localhost:4200/, and we have a running app!

Let's check its scaffolding capabilities too.

To generate a component, we just need to invoke this command:

```
ng generate component home
```

It generates a boilerplate implementation for our component *Home*. The tool creates a `home` folder inside the `src/app` folder, and adds a number of files related to the component. Not only does this save us some mouse clicks and typing effort, but also the generated code adheres to the best practices outlined by the Angular community. A nice start! Go check out the generated code to understand this structure.

We just saw component generation using angular-cli, but this tool can also generate *directives, pipes, services,* and *routes*. Look at the tool documentation on the GitHub site `http://bit.ly/ng2-cli`.

Time to look at the most touted area in Angular, **performance**. Let's understand what makes Angular run fast and what switches/knobs are available with Angular to improve the overall app performance.

# Angular 2 performance

Angular 2 has been designed with performance in mind. Every part of the framework, starting from the framework footprint, initial load time, memory utilization, change detection plus data binding, and DOM rendering, has been tweaked or is being tweaked for better performance.

The next few sections are dedicated to understanding how performant Angular is and the tricks it uses to achieve some impressive performance gains.

## Byte size

*Byte size* of the framework is a good starting point for performance optimization. While the world is moving towards a high-speed Internet, a sizable population among us is on a slow connection and are using their mobile to connect to the Web. We may not think too much about few KB here or there, but it does matter!

While the byte size of Angular 2 out-of-the-box is bigger than Angular 1, there are techniques that can drastically reduce the size of an Angular 2 bundle.

To start with, the standard techniques of *minification* and *gzipping* can reduce this gap substantially. And with Angular 2, we can do some nifty tricks with *module bundler/loaders* to reduce the Angular 2 bundle size even more.

**Tree shaking** may be a quirky name for a process, but it literally does what it says! As we build apps using TypeScript (or ES2015), containing *modules* and *exports*, a module bundler such as *Rollup* (`http://rollupjs.org`) can perform static code analysis on such code, determine what parts of the code are never used, and remove them before bundling the release bits. Such module bundlers, when added to the app's build process, can analyze the framework bit, any third-party library, and the app code to remove any dead code before creating bundles. *Tree shaking can result in enormous size reduction as you don't bundle framework bits that you don't use.*

One of the biggest framework pieces that can be removed from the framework bundle is the *compiler*. Yes, you read right, it's the compiler!

For curious readers, the compiler is the single biggest framework piece contributing a whopping *500 KB+* to the Angular bundle (In Angular 2 *v2.0.0*).

Using tree shaking together with **Ahead-of-Time** (**AoT**) compilation, we can just get rid of the Angular compiler (in the browser) altogether.

With AoT compilation, the view templates (HTML) are compiled beforehand on the server side. This compilation again is done as part of the app's build process where a server version of Angular 2 compiler (a node package) compiles every view in the application.

With all the templates compiled, there is no need to send the Angular compiler bits to the client side at all. Tree shaking can now just get rid of the compiler and create a far slimmer framework package.

Read more about AoT in the framework documentation available at `http://bit.ly/ng2-aot`.

## Initial load time and memory utilization

The initial load time for any web app with a full-fledged framework is typically slow. This effect is more pronounced on mobile devices, where the JavaScript engine may not be as powerful as a desktop client. For a better user experience, it becomes imperative that the framework initial load time is optimized, especially for mobile devices.

Out of the box, **Angular 2 is five times faster than Angular 1** when it comes to initial load time and re-rendering the view. These numbers will get better as the Angular team evolves the framework.

Further, AoT compilation too can improve the initial load time of the application as a time-consuming activity (view compilation) is not required.

The same holds good for memory utilization. Angular 2 fares better here too, and things will get even better with future releases.

If you are planning to switch to Angular 2, this is something that you should look forward to: a performant framework built for the future.

The next three performance improvements that we are going to talk about have been made possible because of a single architectural decision: *the creation of a separate renderer layer.*

# The Angular rendering engine

The biggest disadvantage of Angular 1 was that the framework was tied to the browser DOM. The directives, the binding, and the interpolations all worked against the DOM.

With Angular 2, the biggest architectural change that came in was a separate rendering layer. Now an Angular 2 app has two layers:

- **Application layer**: This is the layer our code resides in. It uses an abstraction build over the renderer layer to interact with it. The Renderer class we saw in `Chapter 6`, *Angular 2 Directives in Depth*, is the interface that we use to interact with the rendering layer.
- **Rendering layer**: This layer is responsible for translating requests from the application layer into rendered components, and reacting to user input and view updates.

The default renderer implementation for the renderer is `DomRenderer`, which runs inside the browser. But there are other rendering abstractions too and we will discuss them in the following section.

# Server-side rendering

**Prerendering** on the server side is yet another technique for improving the initial load time of an Angular 2 app. This technique is really helpful on mobile devices, as it improves the perceived load time considerably.

Server-side rendering takes care of the initial page load before client-side rendering kicks in (and handles view rendering henceforth).

In such a scenario, when the user requests for a view/page, a piece of software on the server generates a fully materialized HTML page with data pre-bound to the view and sends it to the client along with a small script. The app view hence is immediately rendered, ready for interaction. While the framework loads in the background, the small script that was sent along the first time captures all user inputs and makes them available to the framework, allowing it to replay the interactions once it is loaded.

Angular Universal, as it is touted, allows rendering and sharing of the view both on the server- and client-side.

Server-side rendering is only made possible because of separation of the rendering layer as described previously. The initial view is generated by a renderer implementation on the server, named `ServerDomRenderer`. There is a Node.js plugin (`http://bit.ly/ng2-universal-node`) that can be used in a number of node web frameworks such as *Express*, *Hapi*, *Sail*, and others. Efforts are also ongoing to have a rendered implementation for other popular server platforms such as .NET and PHP.

Look at the Angular design docs for Angular Universal (`http://bit.ly/ng2-universal-design`) and the embedded YouTube videos at the top of the design doc to learn more about server-side rendering.

Performance is not the only benefit with server-side rendering. As it turns out, search indexers too like pre-rendered HTML content. Server-side rendering is really useful in areas such as **search engine optimization (SEO)** and deep linking, which allows easy content sharing.

# Offloading work to a web worker

Offloading work to a **web worker** is a neat idea, again made possible due to the separation of the rendering layer from the application layer.

*Web workers* provide a mechanism for running scripts in background threads. These threads can execute work that does not involve the browser DOM. Be it a CPU-intensive task or a remote XHR invocation, all can be delegated to web workers.

In today's world, CPUs with multiple cores are the norm, but JavaScript execution is still single-threaded. There is a need for a standard/mechanism to utilize these idle cores for our apps. Web worker fits the bill perfectly, and since most modern browsers support them, we all should be writing code that utilizes web workers.

Sadly that's not happening. Web workers are still not mainstream and there are good reasons for that. Web workers impose a good number of restrictions on what is allowed and what is not. These limitations include:

- **No direct access to DOM**: Web workers cannot directly manipulate the DOM. In fact, web workers do not have access to multiple globals such as *window* and *document*, and others are not available on the web worker thread. This severely limits the number of use cases where a web worker can be utilized.
- **Browser support**: Web workers are only available for modern/evergreen browsers (IE 10+).
- **Inter process communication**: Web workers do not share memory with your main browser process and hence need to communicate with the main thread (UI thread) only through *message passing* (serialized data). Adding to that, the message passing mechanism is asynchronous in nature, adding another layer of complexity to the communication model.

Clearly, web workers are hard to use.

Angular 2 tries to alleviate these limitations by integrating the web worker usage into the framework itself. It does that by running the complete application in the web worker thread, except the rendering part.

The framework takes care of the communication between the application code running inside the web worker, and the renderer running inside the main UI thread. From a developer's perspective, there are no visible differences.

This is again made possible due to the separation of the renderer layer in Angular 2. The following diagram shows the layers that run on the app main thread and what runs inside the web worker:

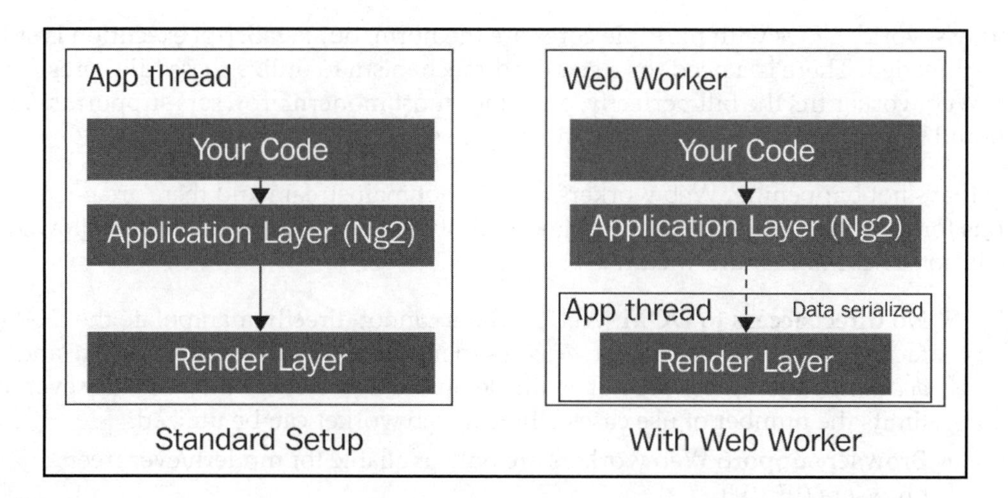

Go through at this talk (`http://bit.ly/yt-ng2-web-worker`) from Jason Teplitz to learn about what web workers have to offer.

# Performant mobile experience

Angular's rendering abstraction again opens up a host of integration avenues, especially on the mobile platform. Rather than running an app on a mobile browser, Angular renderers can be created that can tap the device's native UI capabilities.

Two notable projects in this domain are renderers for platforms:

- **ReactNative** (`http://bit.ly/rnative`): A renderer for ReactNative (`http://bit.ly/ng2-rnative`). It allows writing of Angular 2 apps using ReactNative's rendering capabilities.
- **NativeScript** (`https://www.nativescript.org/`): Angular and NativeScript teams have collaborated to create a renderer for NativeScript (`http://bit.ly/ng2-native-script`).

App platforms such as *ReactNative* and *NativeScript* already do a superb job of providing JavaScript-based APIs for the native mobile platforms (iOS and Android), allowing us to utilize a single code base, with a familiar language. Angular renderers take things a step further. With Angular integration, a good amount of code can be shared across browsers and mobile devices. Things may only differ in terms of view templates and view-related services such as dialogs, popups, and others.

Look at the documentation for the respective renderers to understand how they work and the features they support.

Next up on the line, we have framework improvements in terms of *change detection*.

# Change detection improvements

One of the major performance improvements in Angular 2 over Angular 1 is in how *change detection* works in Angular 2. Angular 2 change detection out of the box is insanely fast, and it can be tweaked further for even better results.

The next few sections talk about Angular change detection in depth. It's an important topic to understand when building anything at scale. It also helps us debug scenarios where it may seem that change detection is not working as advertised.

Let's start the discussion by understanding what change detection is and why it is important.

## Change detection

Angular's *data binding engine* does a great job of binding the view with the model data (component data). These are live bindings where Angular keeps the view in sync with model changes. Any time the model changes, the binding engine re-renders parts of the view that are dependent on the model. To manage this view-model synchronization, Angular needs to know when the model changed and what changed exactly. This is what **change detection** is all about. During app execution, Angular frequently does what we call **change detection runs** to determine what changed.

If you are from Angular 1, *change detection run* is roughly equivalent to **digest cycles**, except that in Angular 2 there are no cycles.

While this problem of keeping the model and view in sync may sound simple, it's a tough nut to crack. Unlike the component tree, the interconnection between multiple models can be complex. Changes in one component model can trigger changes in multiple component models. Furthermore, these interconnections may have cycles. A single model property could be bound to multiple views. All these complex scenarios need to be managed using a robust change detection infrastructure.

In the next few sections, we explore how the Angular change detection infrastructure works, when change detection triggers, and how can we influence change detection behavior in Angular.

## Change detection setup

It all starts with Angular setting up change detectors for every component rendered on the view. Since every Angular app is a hierarchy of components, these change detectors are also set up in the same hierarchy. The following diagram highlights the **change detector hierarchy** of the *Workout Builder* app at a point in time:

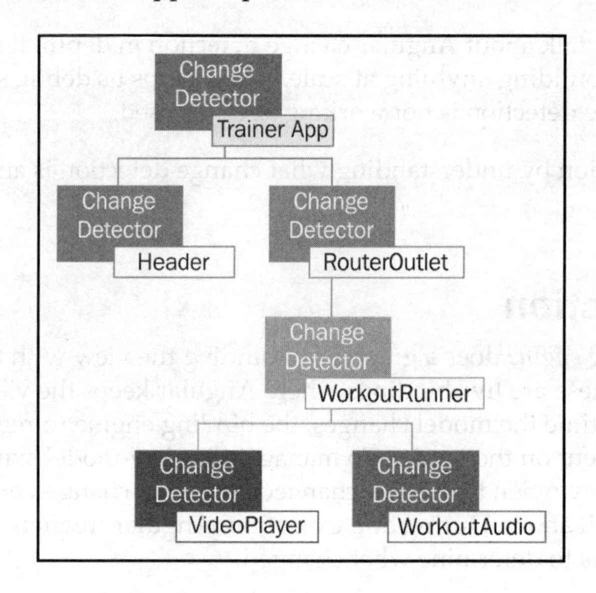

A *change detector* attached to a component has the responsibility of detecting changes in the component. It does that by parsing the binding on the component's template HTML and sets up the necessary change detection watches.

Remember, the detector only sets up watches on model properties used in the template, not on all component properties.

Another important point worth highlighting here is that *change detection is set up one way, from model to view*. Angular does not have the concept of two-way data binding and hence the preceding figure is a directed tree without cycles. This also makes change detection more predictable. Interleaving model and view updates is disallowed.

## When does change detection kick in?

Does Angular constantly check for changes in the model data? Considering the fact that the component properties we bind the view to do not inherit from any special class, Angular has no way of knowing which property changed. The only way out for Angular is to constantly query each data-bound property to know its current value and compare it against its old value for changes. Highly inefficient to say the least!

Angular does better than that, as change detection runs get executed only at specific times during app execution. Think carefully about any web application; what causes a view to update?

View can get updated due to:

- **User input/ browser events**: We click on a button, enter some text, and scroll the content. Each of these actions can update the view (and the underlying model).
- **Remote XHR requests**: This is another common reason for view updates. Getting data from a remote server to show on the grid and getting user data to render a view are examples of this.
- **setTimeout and setInterval timers**: As it turns out, we can use `setTimeout` and `setInterval` to execute some code asynchronously and at specific intervals. Such code can also update the model. For example, a `setInterval` timer may check for stock quotes at regular intervals and update the stock price on the UI.

For obvious reasons, Angular change detection too kicks in only when any of these conditions occur.

The interesting part here is not when Angular's change detection kicks in but how Angular is able to intercept all *browser events, XHR requests,* and `setTimeout` and `setInterval` functions.

This feat in Angular is performed by a library called **zone.js**. As the documentation describes:

> *A Zone is an execution context that persists across async tasks.*

One of the basic abilities of this library is that it can hook into a piece of code and trigger callbacks when code execution starts and when it ends. The code being monitored could be a sequence of calls that are both synchronous and asynchronous in nature. Consider this example, which highlights the usage:

```
let zone = new NgZone({ enableLongStackTrace: false });
let doWork = function () {
  console.log('Working');
};

zone.onMicrotaskEmpty.subscribe((data:any) => {
  console.log("Done!");
});

zone.run(() => {
  doWork();
    setTimeout(() => {
        console.log('Hard');
        doWork();
    }, 200);
    doWork();
});
```

We wrap a piece of code inside a call to the `zone.run` call. This code calls the `doWork` function synchronously twice, interleaved with a `setTimeout` call that invokes the same function after a lapse of 200 milliseconds.

By wrapping this sequence inside `zone.run`, we can know when the call execution is complete. In zone terminology, these are **turns**. The code before `zone.run` sets up a subscriber that gets called when execution is complete, using the `zone.onMicrotaskEmpty` function:

If we execute the preceding code, the logs look like:

```
Working  // sync call
Working  // sync call
Done!   // main execution complete
Hard     // timeout callback
Working  // async call
Done!   // async execution complete
```

The `onMicrotaskEmpty` subscription is executed twice, once after the sequential execution completes (defined inside `run` callback) and one after the asynchronous `setTimeout` execution is complete.

Angular 2 change detection uses the same technique to execute our code within zones. This code could be an *event handler*, which internally makes more synchronous and asynchronous calls before completing. Or it could be a `setTimeout/setInterval` operation that may again require a UI update.

The Angular change detection framework subscribes to the `onMicrotaskEmpty` observable for the executing zone, and kicks in change detection whenever a turn is complete. The following diagram highlights what happens when code similar to the one just described is run on a button click:

```
function onClick() {
   this.doWork();
   setTimeout(() => {
       console.log('Hard');
       this.doWork();
   },
   200);
   this.doWork();
}
```

During the execution of the code block, if the zone library determines that the call is asynchronous in nature, it spawns a new micro task that has its own life cycle. It is the completion of these micro tasks that also triggers `onMicrotaskEmpty`.

If you want to know how the change detection trigger looks inside Angular, here is an excerpt from the Angular source code (simplified further):

```
class ApplicationRef_ {

  private _changeDetectorRefs:ChangeDetectorRef[] = [];

  constructor(private zone: NgZone) {
    this.zone.onMicrotaskEmpty
      .subscribe(() => this.zone.run(() => this.tick());
  }

  tick() {

    this._changeDetectorRefs.forEach((detector) => {
      detector.detectChanges();
    });
  }
}
```

The `ApplicationRef` class tracks all the change detectors attached throughout the app and triggers a change detection cycle when the application-level zone object fires the `onMicrotaskEmpty` event. We will shortly touch upon what happens during this change detection.

*Zonejs* gets the ability to track execution context across any asynchronous call because it overrides the default browser API. The override, also termed **monkey patching**, overrides the *event subscription, XHR requests*, and `setTimeout`/`setInterval` API. In the example highlighted previously, the `setTimeout` we invoke is a monkey-patched version of the original browser API.

Now that we know how change detectors are set up and when this activity kicks in, we can look at how it works.

# How does change detection work?

Once the change detectors are set up and the browser API is monkey-patched to trigger change detection, the real change detection kicks in. This is quite a simple process.

As soon as any of the asynchronous event callbacks is triggered (execution of an event handler is also an async activity), Angular first executes the application code we have attached to the callback. This code execution may result in some model updates. Post the execution of the callback, Angular needs to respond to the changes by triggering a *change detection run*.

In a change detection run, starting from the top of the component tree, every change detector evaluates its respective component's template bindings to see if the value of the binding expression has changed.

There are some things that we need to highlight regarding this execution:

- Angular does a strict equality check (using ===) to detect changes. Since it's not a deep comparison, for a binding that refers to an object Angular will only update the view when the object reference changes.
- The change detection flow is unidirectional (starting from root), from parent to child in a top-down fashion. The detectors on the parent component run before the child detectors.

By default, the change detection algorithm navigates the complete tree, irrespective of where the change was triggered in the tree. This implies all binding is evaluated on every change detection run.

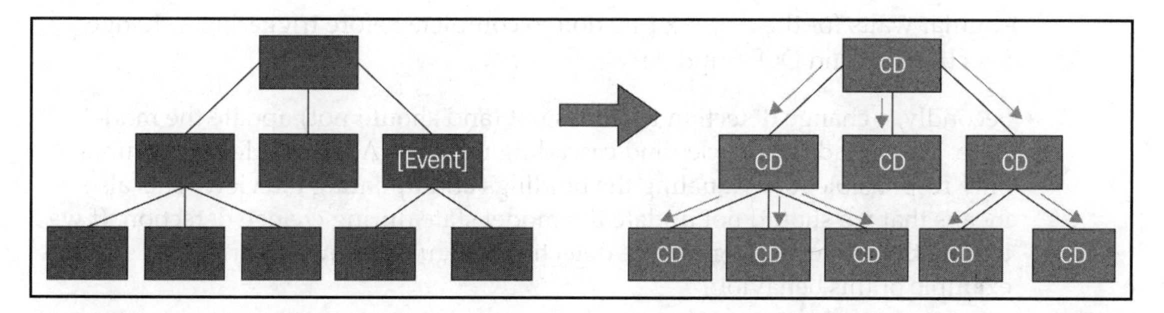

Binding evaluation on every run may seem inefficient but it is not. Angular employs some advance optimizations to make this check superfast. Still if we want to tweak this behavior, we do have some switches that can reduce the number of checks performed. We will touch upon this topic soon.

- Change detectors only track properties that are part of template bindings, not the complete object/component properties.
- To detect changes in the bound value, the change detectors need to track the previous value of the expression evaluated during the last change detection run. Clearly some amount of book keeping is required for every template binding we use.

The obvious next question would be: What happens when a change is detected (by a change detector)?

Since all the hard work of setting up change detection and identifying changes has already been done, this step just involves updating the component state and synchronizing the component DOM.

There are a few more observations worth highlighting here:

- First and foremost, Angular separates the model update step from the DOM update.

  Consider this code snippet, which is invoked when someone clicks on a button:

  ```
  doWork() {
     this.firstName="David";
     this.lastName="Ruiz";
  }
  ```

  Assuming that both `firstName` and `lastName` are bound to the component view, a change to `firstName` does not update the DOM binding immediately. Instead, Angular waits for the `doWork` function to complete before triggering a change detection run and DOM update.

- Secondly, a change detection run does not (and should not) update the model state. This avoids any cycles and cascading updates. A change detection run is only responsible for evaluating the bindings and updating the view. This also means that we should not update the model state during change detection. If we update the model during change detection, Angular throws an error. Let's see an example of this behavior:

  1. Open `start.html` and update the last div to:

     ```
     <div class="col-sm-3">
     Change detection done {{changeDetectionDone()}}
     </div>
     ```

  2. And add a `changeDetectionDone` function to the component implementation (`start.component.ts`), which looks like:

     ```
     times: number = 0;
     changeDetectionDone(): number {
         this.times++;
         return this.times;
     }
     ```

3. Run the app, load the start page, and then look at the browser console. Angular has logged a number of errors that look like:

```
EXCEPTION: Expression has changed after it was checked.
Previous value: 'Change
detection done 1'. Current value: 'Change detection done 2' ...
```

We are changing the state of the component when calling the `changeDetectionDone` function (inside an interpolation), and Angular throws an error because it does not expect the component state to update.

This change detection behavior is enabled only when **production mode** in Angular has not been enabled. Angular's *production mode* can be enabled by calling the `enableProdMode()` function before bootstrapping the application (in `bootstrap.ts`).
When enabled, Angular behaves a bit differently. It turns off assertions and other checks within the framework. Production mode also affects the change detection behavior.
In non-production mode, Angular traverses the component tree twice to detect changes. If on the second pass any binding expression has changed, it throws an error.
In contrast, when in production mode, change detection tree traversal is done only once. The change detection error that we saw in the console will not show up if we enable production mode. This can lead to an inconsistency between the model and view state. Something we should be aware of!
The bottom-line is that we cannot alter the state of a component when change detection is in progress.
A direct corollary: if we are using a function inside the binding expression, function executions should be stateless, without any side effects.

- And lastly, this change detection traversal from root to leaf node executes only once during the change detection run.

A pleasant surprise for folks with an Angular 1 background! *The digest cycle count is Angular 2 is 1*. Angular 2 developers will never face "the digest iterations exceeded exception!" A far more performant change detection system!

# Change detection performance

Let's talk about change detection performance. If you think checking the complete component tree every time for change is inefficient, you would be surprised to know how fast it is. Due to some optimization done in terms of how expressions are evaluated and compared, Angular can perform thousands of checks in a couple of milliseconds.

Under the hood, for every expression involved in the view binding, Angular generates a change detection function that specifically targets a particular binding. While it may seem counterintuitive at first, Angular does not have a common function for determining whether an expression has changed. Instead, it's like writing our own change detection function for every property that we bind to. This allows the JavaScript VM to optimize the code, resulting in improved performance.

Want to learn more about it? Check out this video by Victor Savkin: `https://youtu.be/jvKGQSFQf10`.

In spite of all this optimization, there may still be cases where traversing the complete component tree may not be performant enough. This is especially true when we have to render a large dataset on the view, keeping the bindings intact. The good news is that the Angular change detection mechanism can be tweaked.

The reason Angular needs to do the complete tree walk is that model changes at one place may trigger model changes at other places. In other words, a model change may have a cascading effect, where interconnected model objects are also updated. Since Angular has no way to know what exactly changed, it checks the complete component tree and associated model.

If we can help Angular determine what parts of the application state are updated, Angular can be pretty smart about what part of the component tree it traverses to detect changes. We do this by storing the app data in some special data structures that help Angular decide what components need to be checked for changes.

There are three ways in which we can make Angular change detection smarter.

# Using immutable data structures

**Immutable objects/collections** are objects that cannot be changed once created. Any property change results in a new object being created. This is what **immutable.js**, a popular library for immutable data structures, has to say:

> *Immutable data cannot be changed once created, leading to much simpler application development, no defensive copying, and enabling advanced memoization and change detection techniques with simple logic.*

Let's try to understand how immutable data structures help in the Angular context with an example.

Imagine we are building a set of components to collect employee information for a **Human Resource** (**HR**) software. The employee component view looks something like this:

```
<Employee>
<summary [model]="employee"></employee>
<personal [model]="employee.personal"></personal>
<professional
[model]="employee.professional"></professional>
<address [model]="employee.home"></address>
   <address [model]="employee.work"></address>
</Employee>
```

It has sections for taking personal, professional, and address information. The `summary` components provide a read-only view for employee data being entered. Each of the components has a property called `model`, highlighting what part of employee data these components manipulate. Each of these components' summary, professional, personal, and address internally may have other child components. This is how the component tree looks:

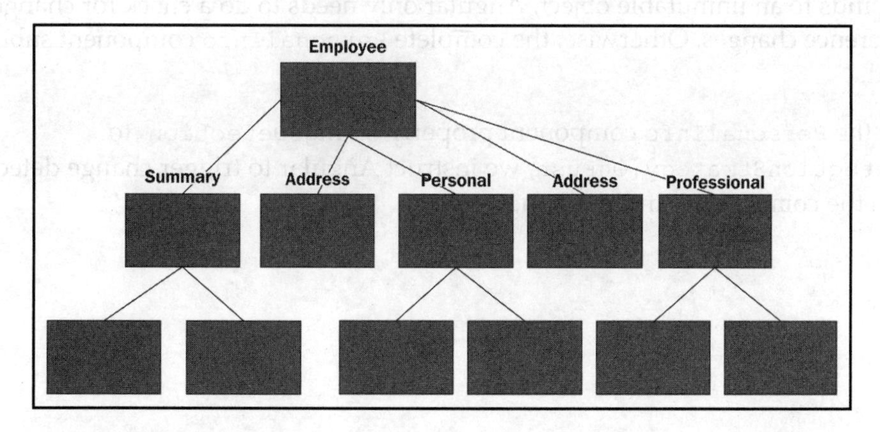

What happens when we update an employee's personal information? With standard objects (mutable), Angular cannot make any assumption about the shape of data and what all has changed; hence, it does the complete tree walk.

How does immutability help here? When using an immutable data structure, any change to an object's properties results in a new object being created. For example, if we create an immutable object using a popular library, *immutablejs*:

```
personalInfo = Immutable.Map({ name: 'David', 'age': '40' });
```

Changes to either the `name` or `age` property of `personalInfo` create a new object:

```
newPersonalInfo = personalInfo.set('name', 'Dan');
```

This immutability comes in handy if each of the employee model properties (`personal`, `professional`, `home` and `work`) is immutable.

Take for instance the `PersonalInfo` component definition that binds to personal info data:

```
@Component({
  selector:'personal',
  template: `
    <h2>{{model.name}}</h2>
    <span>{{model.age}}</span>`,
  changeDetection: ChangeDetectionStrategy.OnPush
})
class PersonalInfo {
  @Input() model;
}
```

Since the only thing `PersonalInfo` depends upon is the `model` property, and the `model` property binds to an immutable object, Angular only needs to do a check for changes if the `model` reference changes. Otherwise, the complete `PersonalInfo` component subtree can be skipped.

By setting the `PersonalInfo` component property `changeDetection` to `ChangeDetectionStrategy.OnPush`, we instruct Angular to trigger change detection only when the component's inputs change.

If we change the change detection strategy to `OnPush` for each of the Employee component children and update the employee's personal info, only the `PersonalInfo` component subtree is checked for changes:

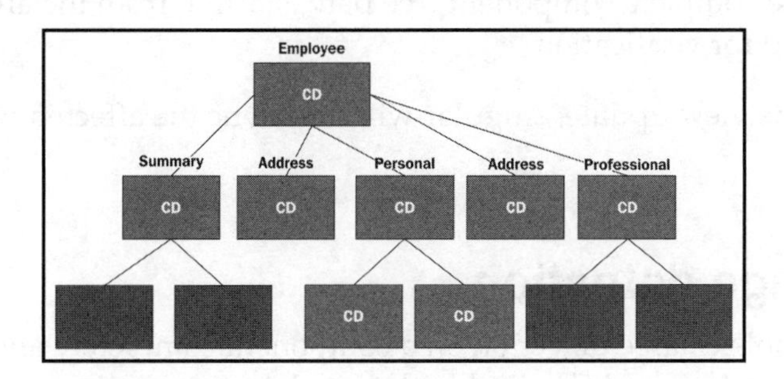


For a large component tree, such an optimization will improve the app/view performance manifold.

When set to `OnPush`, Angular triggers change detection only when the component's input property changes or there is an event raise inside the component or its children.

Developing applications using immutable data structures departs from the standard development paradigm where the applicate state is totally mutable. What we have highlighted in this section is how Angular takes advantage of immutable data structures to optimize the change detection process.

*Observables* are another kind of data structure that can help us optimize Angular change detection.

## Using Observables

**Observables** are data structures that trigger events when their internal state changes. The Angular *eventing infrastructure* extensively uses *observables* to communicate the components' internal state to the outside world.

While we have used observable *output properties* (the `EventEmitter` class) to raise events, *input properties* too can take observables. Such observable inputs can help Angular optimize change detection.

When using observables, the change detection switch still remains `ChangeDetectionStrategy.OnPush`. But this time, only if a component input triggers an event (as they are observables) will Angular perform the dirty check. When the input triggers an event, the complete component tree path, starting from the affected component to the root, is marked for verification.

When performing the view update, Angular will only sync the affected path and ignore the rest of the tree.

## Manual change detection

We can actually disable change detection on a component completely and trigger manual change detection when required. To disable change detection, we just need to inject the component-specific change detector (the `ChangeDetectorRef` class instance) into the component and call the `detach` function:

```
constructor(private ref: ChangeDetectorRef) {
    ref.detach();
}
```

Now the onus is on us to inform Angular when the component should be checked for changes.

We can reattach the component to the change detection tree by using the `reattach` function on `ChangeDetectorRef`.

The `ChangeDetectorRef` class has two functions that can be utilized for manual change detection:

- `markForCheck`: This marks the path from the detector to the root for dirty checking. Remember, the actual dirty check is only performed once all of the app code has executed, and not as soon as we call `markForCheck`. This snippet shows the use of this function:

```
this._userService.getUserDetails()
.subscribe((user)=>
{ this.user = user; ref.markForCheck();}
```

- `detectChanges`: This function actually performs change detection on the component on which it is called (and its children). Using `detectChanges` is like isolating the tree from the rest of the application and performing local change detection.

We seldom need to disable the standard change detector setup, unless there are situations where standard change detection becomes an expensive affair.

Take, for example, a public chatroom app, which is receiving messages from thousands of people connected to it. If we constantly keep pulling the messages and refreshing the DOM, the app may become unresponsive. In such a scenario, we can disable change detection on parts of the chat app component tree and manually trigger change detection to update the UI at specific intervals.

While we have seen three ways to tweak change detection behavior, the good thing is that these are not exclusive. Parts of the component tree can use immutable data structures, parts can use observables, parts can employ manual change detection, and the rest can still use the default change detection. And Angular will happily oblige!

Enough on change detection for now. We may never need it unless we are building some large views with a chatty UI. Such scenarios require us to squeeze every bit of performance out of the change detection system, and the system is ready for it.

Next, we will have a look at another common requirement that most apps invariably have: authenticating their users.

# Handling authentication and authorization

Most, if not all, apps have a requirement to authenticate/authorize their users. We may argue that authentication and authorization are more of a server concern than a client one, and that is correct. Still, the client side needs to adapt and integrate with the authentication and authorization requirement imposed by the server.

In a typical app execution workflow, the app first loads the partial views, then makes calls to pull data from the server, and finally binds data to its view. Clearly, the views and the remote data API are two assets that need to be secured.

To guard these resources, you need to understand how a typical application is secured on the server. There are primarily two broad approaches to securing any web applications: *cookie-based authentication* and *token-based authentication*. Each of them requires different handling on the client part. The next two sections describe how we can integrate with either of these approaches.

# Cookie-based authentication

This authentication mechanism is the easiest to implement if the server stack supports it. It's non-intrusive and may require bare minimum changes to the Angular application. **Cookie-based authentication** involves setting the browser cookie to track the user authentication session. The following sequence diagram explains a typical cookie-based authentication workflow:

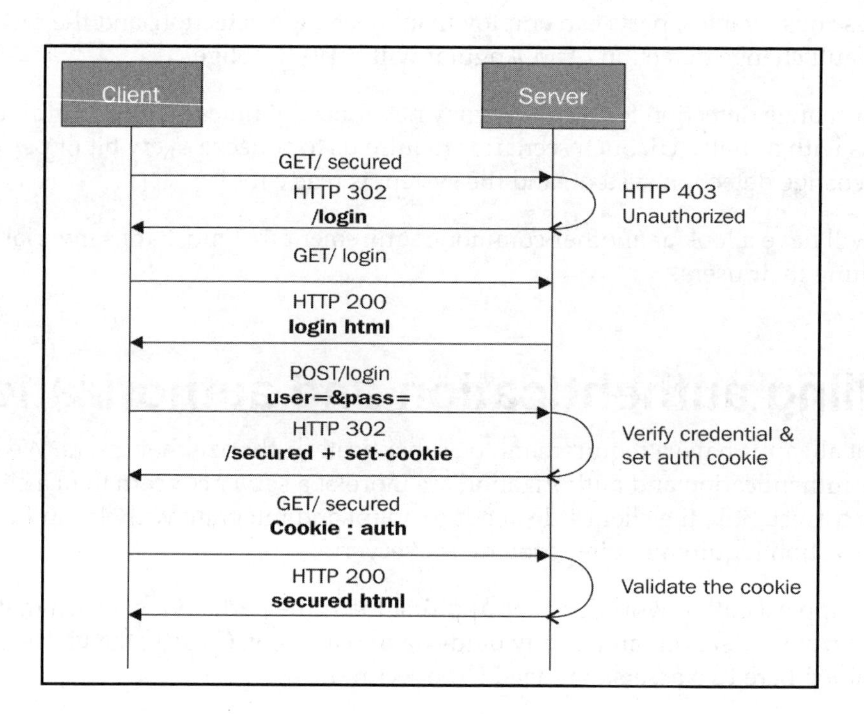

Here is how a typical authentication workflow works:

- When trying to access a secured resource from the browser, if the user is not authenticated, the server sends an HTTP 401 Unauthorized status code. As we will see later, a user request is an unauthorized request if there is no cookie attached to the request or the cookie is expired/invalid.
- This unauthorized response is intercepted by the server or, at times, by the client framework (Angular in our case) and it typically results in a 302 redirect (if intercepted by the server). The redirect location is the URL to the login page (the login page allows anonymous access).
- The user then enters the username and password on the login page and does a POST to the login endpoint.
- The server validates the credentials, sets a browser cookie, and redirects the user to the original requested resource.
- Henceforth, the authentication cookie is a part of every request (added by the browser automatically), and the server uses this cookie to confirm his identity and whether the user is authenticated.

This scenario assumes that the HTML and API exist under a single domain.

As we can see, with this approach, the Angular infrastructure is not involved; or the involvement is minimal. Even the login page can be a standard HTML page that just sends data to the login endpoint for authentication. If the user lands on the Angular app, it implicitly means that the user has been authenticated.

The cookie-based authentication flow may vary depending on the server framework, but the general pattern of setting a cookie and attaching a cookie with every subsequent request remains the same.

In a cookie-based application authentication, if the application wants to get the user context, a server endpoint (such as `/user/details` ) is exposed that returns the logged-in-user-specific data. The client application can then implement a service such as `UserService` that loads and caches the user profile data.

The scenario described here assumes that the API server (the server that returns data) and the site where the application is hosted are in a single domain. That may not be the case always. Even for *Personal Trainer*, the data resides on the *MongoLab* servers and the application resides on a different server (even if it is local). And we already know that this is a cross-domain access and it comes with its own set of challenges.

In such a setup, even if the API server is able to authenticate the request and send a cookie back to the client, the client application still does not send the authentication cookie on a subsequent request.

To fix this, we need to set a Boolean variable, `withCredentials`, to `true` on the XHR request. This can be enabled at the global level by overriding `BaseRequestOptions` (the `withCredentials` property). The framework uses the `BaseRequestOptions` class for the default HTTP request option. See the next section, *Token-based authentication*, to learn how to override `BaseRequestOptions`.

This can also be enabled on a per-request level by passing in the `withCredentials:true` flag in each HTTP request method as the last parameter:

```
this.httpService.get(url,{withCredentials:true});
```

The last parameter to every HTTP function, including `get`, `post`, and `put`, is a `RequestOptionsArgs` object. This allows us to override some properties of the request being made.

Once this flag is enabled, the client browser will start attaching the authentication cookie for the cross-domain requests.

The server too needs to have **cross-origin resource sharing** (**CORS**) enabled and needs to respond in a specific manner for the request to succeed. It should set the **access-control-allow-credentials** header to true and the **access-control-allow-origin** header to the host site making the request.

Check out the MDN documentation (`http://bit.ly/http-cors`) to learn about this scenario in detail.

Cookie-based authentication is definitely less work on the client side, but there are times when you have to revert to token-based access. This could be because:

- Cookies and cross-domain requests do not play nicely across browsers. Specifically, IE8 and IE9 do not support it.
- The server may not support generating cookies, or the server only exposes token-based authentication.
- Token-based solutions are easy to integrate with native mobile application and desktop clients.
- Tokens are not susceptible to cross-site request forgery (CSRF) attacks.

To know more about CSRF, look at the CRSF Prevention cheat sheet here at `http://bit.ly/csrf-cs`.

The next section talks about supporting token-based authentication.

# Token-based authentication

**Token-based access** is all about sending a token (typically in HTTP headers) with each request instead of a cookie. A simplified token-based workflow looks something like this:

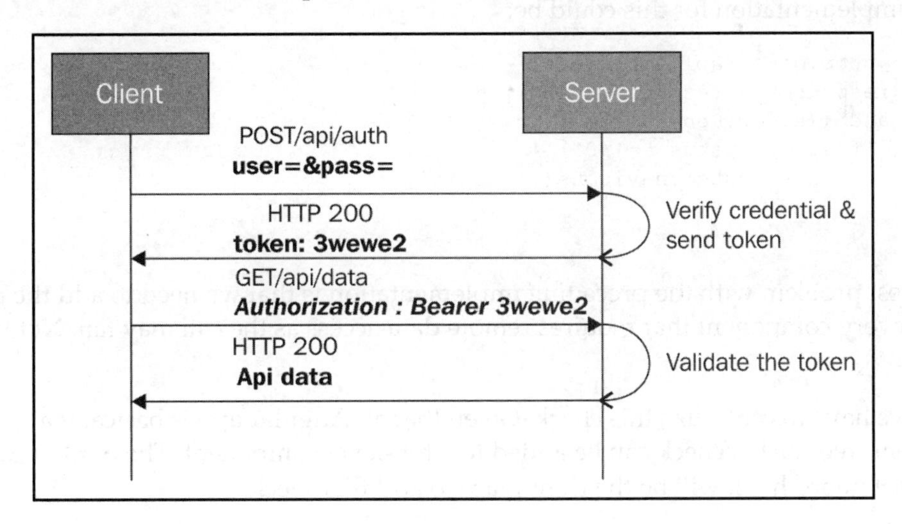

Many public APIs (such as *Facebook* and *Twitter*) use token-based authentication. The format of the token, where it goes, and how it is generated depends on the protocol used and the server implementation. Popular services that use token-based authentication implement the **OAuth 2.0** protocol for token generation and exchange.

In a typical token-based authentication setup, the views are available publically but the API is secured. If the application tries to pull data through API calls without attaching the appropriate token to the outgoing request, the server returns an *HTTP 401 Unauthorized* status code.

Integrating with a token-based authentication system requires a decent amount of setup on the client side too. Let's take the simplified example of a *Human Resource (HR)* system that supports token-based authentication so that you understand how the authentication workflow works with the Angular application as a client.

The HR system has a page showing a list of employees and a login page. It also has API endpoints to get a list of employees and generate access tokens. The API endpoint that returns the employee list is secured by token-based access.

The workflow starts with the user loading the employee list page. The view is loaded but the API call fails, with the server returning *HTTP 401 Unauthorized.*

On receiving a 401 HTTP error code, the app should respond by either routing the user to a login view (remember this is an SPA) or opening a login popup.

A naive implementation for this could be:

```
this._http.get('/api/employees')
.map(response => response.json())
    .catch((error:Response)=>{
        if(error.status == 401) {
            this.router.navigate(['/login']);
}
});
```

The biggest problem with the preceding implementation is that we need to add the catch block to every component that requires remote data access as the call may fail. Not very smart!

Instead we have to centralize this check. Given that an Angular app is basically a component tree, such a check can be added to a top-level component. The `canActivate` component guard hook will be the right place to add this check.

So given this route configuration on the app root:

```
@const routes: Routes = [
    { path: 'home', component: HomeComponent },
    { path: 'login', component: LoginComponent }
];
```

To block unauthorized access to `/home`, we can implement the `canActivate` guard class as:

```
export class AuthGuard implements CanActivate {
  canActivate() {
    // Check if there is auth token and return true.
    return true;
  }
}
```

And then extend the route definition for home to include this guard:

```
{ path: 'home', component: HomeComponent,
canActivate:[AuthGuard] },
```

To understand how the `canActivate` function can be implemented, we need to know what happens on the client during token authentication.

With token-based authentication, the login page looks similar to a cookie-based login page, but instead of cookies being set on the server as part of the login process, a token is returned. This token needs to be attached to all subsequent API requests that are secure. This token hence needs to be persisted in the *browser's local storage*, and all this is done the first time the token is received.

An *authentication service* can perform such chores. A sample implementation of the service looks like:

```
export class AuthService {
  authenticate(username: string, password: string) {
    return this.http.post('/login',
          JSON.stringify({ u: username, p: password }))
      .map((token: Response) => {
        localstorage.setItem('token',token);
        return true;
      });
  }
}
```

The service does an HTTP `post` to the login endpoint, and on receiving the authentication token, it stores the token in the browser's local storage.

Post authentication, the next task is to attach the token on subsequent requests that access the secured API resource. For most API implementations, this token needs to be added to the request header.

All `http` service functions: `get`, `post`, and others-take an extra parameter of type `RequestOptionsArgs`, used to pass some extra parameter to the request. We can use the `headers` property to set the *auth token* for HTTP requests that require authorization. The name of the token may vary based on the backend implementation, but this is how a typical request with a token looks:

```
this.http.get('secured/api/users',{
headers:{
'Accept': 'application/json',
'Authorization': 'Bearer ' + localStorage.getItem('token')
});
```

Although this approach works, it's quite verbose and cumbersome. Each call to the secured API now needs this extra parameter to be set. We need some better alternatives.

Angular has a `BaseRequestOptions` class that contains the default options for an HTTP request. We can replace `BaseRequestOptions` with our own options using Angular DI. This can be done by inheriting from the `BaseRequestOptions` class and overriding the DI provider for `BaseRequestOptions` during bootstrap:

```
class MyOptions extends BaseRequestOptions {
 header:Headers=new Header({
'Authorization': 'Bearer ' + localStorage.getItem('token')
 });
}

bootstrap(App, [HTTP_PROVIDERS, provide(RequestOptions, {useClass:
MyOptions})]);
```

Sadly, this doesn't work! Since the authorization token is not available till authentication is done, setting up `RequestOptions` during bootstrap will result in an empty `Authorization` header being set on all future HTTP requests.

Overriding `RequestOptions` during bootstrapping is only useful if we have the necessary content available during that phase.

What other options are we left with? Angular 2, unlike Angular 1, does not have**global interceptors** that can be used to inject the headers before the request is made. Therefore, the only viable alternative is to create a custom HTTP client service to communicate with the secure API. This service can attach the token transparently when available.

We have created a sample service implementation for this purpose and the implementation is available at `http://bit.ly/ng2-auth-svc`. We will be highlighting some relevant parts of this `AuthHttp` service.

This `AuthHttp` exposes the same interface as the HTTP service, with its function internally delegating the request to the original `http` service. The next snippets detail two such HTTP functions, `get` and `post`:

```
public get(url: string, options?: RequestOptionsArgs)
:Rx.Observable<Response> {
  return this._request(RequestMethod.Get, url, null,
  options);
    }

public post(url: string, body: string, options?:
RequestOptionsArgs) :Rx.Observable<Response> {
  return this._request(RequestMethod.Post, url, body,
  options);
}
```

Each of the HTTP function wrappers internally calls the `_request` private method. This function does most of the heavy lifting, which involves setting the HTTP authorization header, making a request, and receiving the response. The function implementation looks like this:

```
private _request(method: RequestMethod, url: string,
 body?: string, options?: RequestOptionsArgs):
 Rx.Observable<Response> {
let requestOptions = new RequestOptions(Object.assign({
            method: method,
            url: url,
            body: body
     }, options));
   if(!requestOptions.headers) {
      requestOptions.headers = new Headers();
   }
```

```
requestOptions.headers.set("Authorization"
,this._buildAuthHeader())

return Rx.Observable.create((observer) => {
this.process.next(Action.QueryStart);
this._http.request(new Request(requestOptions))
.map(res=> res.json())
.finally(() => {
this.process.next(Action.QueryStop);})
            .subscribe(
            (res) => {
            observer.next(res);
            observer.complete();},
            (err) => {
               switch (err.status) {
               case 401:
                  //intercept 401
                  this.authFailed.next(err);
                  observer.error(err);
                  break;
               default:
                  observer.error(err);
                  break;
}
          })
      })
   }
```

After merging the request options and setting the authorization header, the function creates a custom observable that makes a request using the `http` service `request` function. On receiving a response, the observer emits the response and is marked complete (no more events).

*401* error handling is a bit different in the preceding function. The function calls `this.authFailed.next(err)` and raises an event on an `EventEmitter` defined on the `AuthHttp` class:

```
authFailed: EventEmitter<any> = new EventEmitter<any>();
```

Then it triggers the standard error triggering mechanism with `observer.error(err);`

The usefulness of this event will be clear pretty soon.

The `_buildAuthHeader` implementation is simple as it pulls the authorization token from the browser's local storage:

```
private _buildAuthHeader(): string {
```

```
    return localStorage.getItem("authToken");
}
```

This service can now be injected and used to invoke any secure API, such as:

```
this.authHttp.get('/api/employees')
```

This call will add the authorization token if available to the API request.

What have we done thus far? The views have been secured using the `@CanActivate` decorator and the API endpoints are secured too, but there is still one scenario that needs to be handled. What happens when the token expires?

When the API token expires, any access to API endpoints again results in 401 errors. In such a scenario, the app should either redirect to the login page or show a login popup to continue.

To know when a request failed, we again use the `AuthHttp` service. The `authFailed` event on `AuthHttp` can be subscribed to. The best place to subscribe and react to this event would be in the *root component implementation*.

In the root component of the app, we just need to do:

```
ngOnInit() {
    this._authHttp.authFailed.subscribe((error)=>{
      this._router.navigate(['/login']);
      // or
      // this.showLoginDialog();
    });
}
```

And with that, we have now handled most of the scenarios related to token-based authentication.

Clearly, token-based authentication, even though flexible, requires a decent amount of setup and coordination among various components/services.

This walkthrough just outlines one mechanism to send a token to the server, but the process may vary based on the server stack too. Always refer to the backend/server documentation before implementing a token-based authentication in Angular.

While working on this book, there were not many libraries out there that implemented these standard chores required for token-based authentication. In future, before you start, do check whether there is a popular/mature community offering for the same.

We have taken care of authentication, but what about authorization? Once the user context is established, we still need to make sure that the user is only able to access parts that he/she is allowed to. *Authorization* is still missing.

# Handling authorization

Like authentication, authorization support too needs to be implemented on both the server and client side, more so on the server than the client. Remember, anyone can hack into the JavaScript code and circumvent the complete authentication/authorization setup. So, always tighten your server infrastructure irrespective of whether the client has the necessary checks in place or not.

This still does not mean that we do not do any authorization check, on the client. For standard users, this is the first line of defence against unwarranted access.

When working on an authorization requirement for any application, there are three essential elements that are part of the setup:

- The resources that need to be secured/authorized
- A list of roles and users that are part of these roles
- A mapping between the resources and the roles that defines who can access what

From an Angular app perspective, the resources are the pages and, sometimes, sections of pages that need to be restricted to specific roles. If the user is in a specific role, depending upon the role-resource mapping, they get access to some pages; else they are denied access.

While authorization in an Angular application can be implemented in a number of ways, we will outline a generic implementation that can be further customized to suite your needs in the future.

## Adding authorization support

To enable authorization, the first thing that we need to do is expose the logged-in user data, including his/her roles throughout the application.

## Sharing user authentication context

User context can be shared using an Angular service, which can then be injected into components that require the authorization context. Look at this service interface:

```
class SessionContext {
  currentUser():User { ...};
  isUserInRole(roles:Array<string>):boolean { ...};
  isAuthenticated:boolean;
}
```

The `SessionContext` service tracks the user login session and provides details such as:

- The logged-in user (`currentUser`)
- Whether the user is authenticated (`isAuthenticated`)
- The `isUserInRole` function, which returns `true` or `false` based on whether the user is part of any of the roles passed into the `roles` parameter

With such a service in place, we can add authorization for routes, thereby restricting access to some routes to specific roles only.

## Restricting routes

Like authentication, the `canActivate` guard check can also be used for authorization. Implement a class with the `CanActivate` interface and inject the `SessionContext` service into the constructor; then check whether the user belongs to a specific role in the `canActivate` function using the `SessionContext` service. Check out the following code snippet:

```
export class AuthGuard implements CanActivate {
  constructor(private session:SessionContext) { }
  canActivate() {
    return this.session.isAuthenticated &&
      session.isUserInRole(['Contributor', 'Admin']);
  }
}
```

Only users with roles of *Contributor* and *Admin* now have access to Home.

We register the preceding guard class with our routes in the same manner as we did earlier in the *Token-based Authentication* section.

And this is how we authorized access to our routes.

But what happens when a page has view elements that are rendered based on the user's role?

## Conditionally rendering content based on roles

Conditionally rendering content is easy to implement. We just need to show/hide HTML elements based on the user role. We can build a *structural directive* such as `ng-if` that can verify that the user belongs to a role before rendering the content. The directive usage looks like:

```
<div id='header'>
<div> Welcome, {{userName}}</div>
<div><a href='#/setting/my'>Settings</a></div>
<div *a2beRolesAllowed='["admin"])'>
<a href='#/setting/site'>Site Settings</a>
</div>
</div>
```

The preceding code checks whether the user is in an admin role before rendering a **Site Setting** hyperlink.

The directive implementation mimics how `ng-if` works, except that our show/hide logic depends upon the `SessionContext` service. Here is a sample implementation for the `a2beRolesAllowed` directive:

```
@Directive({ selector: '[a2beRolesAllowed]' })
export class RolesAllowedDirective {
  private _prevCondition: boolean = null;
  constructor(private _viewContainer: ViewContainerRef,
    private _templateRef: TemplateRef, private SessionContext _session) { }

  @Input() set a2beRolesAllowed(roles: Array<string>) {
    if (this._session.isUserInRole(roles)) {
      this._viewContainer
        .createEmbeddedView(this._templateRef);
    }
    else {
      this._viewContainer.clear();
    }
  }
}
```

This is a trivial implementation that uses `SessionContext` and the roles passed as input (`a2beRolesAllowed`) to show hide a fragment.

This brings us to the end of authentication and authorization implementation. The reference implementation walkthrough should help us build authentication and authorization into our apps. With this basic understanding in place, any setup can be tweaked to handle other custom authentication/authorization scenarios.

It's now time to address the elephant in the room: migrating from Angular 1 to Angular 2. If you are starting afresh on Angular 2, you can very well skip the next section.

# Migrating Angular 1 apps

If you have been working extensively on Angular 1, Angular 2 poses some pertinent questions:

- Should I migrate my Angular 1 apps to Angular 2?
- Is the framework ready for prime time?
- When should the migration happen?
- Is the migration one-shot or can it be done in an incremental fashion?
- What is the effort involved?
- Can I do something today that helps with the migration in the future?
- I am starting with a new Angular 1 app today. What should I do to make the migration seamless in the future when Angular 2 is released?

Every such query needs be addressed to make sure the transition is as smooth as possible. No one like surprises later in the game! In the coming sections, we will try to answer a number of such questions. As part of the learning, we will also walk you through migrating the Angular 1 version of the Trainer app (developed for the first version of this book) to Angular 2. This will help everyone make some informed decisions on when and how to migrate to Angular 2.

“Should I migrate or not?” is something that we will address first.

# Should I migrate?

Just because Angular 2 is here does not mean Angular 1 is gone. Angular 1 is still being actively developed in parallel with Angular 2. Google is committed to supporting Angular 1 for a good amount of time, and there have been a steady number of releases in Angular 1, with Angular 1.5 released in Feb 2016. Given that Angular 1 will not be going away any time soon, we can think from our app's perspective now.

First things first, what has Angular 2 to offer over its predecessor?

## Advantages of Angular 2

Angular 2 is designed for the future and overcomes a number of shortcomings of its predecessor. In this section, we emphasize on what makes Angular 2 a better framework than Angular 1.

Things you should be aware of while making a decision to move to Angular 2:

- **Better behavioral encapsulation**: Admittedly, while Angular 1 scopes seem to be god sent when we started learning Angular 1, we have now realized how difficult it is to manage the hierarchical nature of scopes. Component-based development in Angular 2 provides a better encapsulation in terms of the state of the application. A component manages its own state, takes input, and raises events: a clear demarcation of responsibilities that are easy to reason with!
- **Less of framework in app code**: You don't need special objects such as scope. DI works with annotation (in TypeScript). You don't set up watches. All in all, when reading a component code, you will not find framework-level constructs in it.
- **Smaller framework API to explore**: Angular 1 had a host of directives that one had to be aware of. With Angular 2 template syntax, directives related to browser events are gone. This reduces the number of directives that one needs to be aware of.
- **Performance**: Angular 2 is faster compared to its predecessor. A complete section of this book was dedicated to understanding what makes Angular 2 a high-performance framework.
- **Mobile-friendly**: Angular 2 tries to optimize the user's mobile experience by utilizing technologies such as server-side rendering and web workers. Angular 2 applications on mobile are more performant that those of its predecessor.
- **Cross-platform**: Angular 2 targets to run on most devices and across platforms. You can use Angular 2 to build applications for web and mobile. As we learned earlier, the separation of the rendering layer has open up a great number of possibilities in terms of where Angular 2 can be utilized.

In true sense, Angular 2 supersedes its predecessor, and in a perfect world, everyone should be working on a better framework/technology. But if you are the cautious type and want to try Angular 2 only once things become stable, we suggest you build your Angular 1 apps today in ways that allow easy migration to Angular 2.

The next section talks about the principles and practices to follow for Angular 1 today, allowing easy migration in future.

# Developing Angular 1 apps today for easy migration

Angular 2 is a paradigm shift and the way we develop components in Angular 2 is quite different from Angular 1. For easy migration, Angular 1 too should embrace component-based development. This can be achieved if we follow some guidelines/principles while building Angular 1 apps. The next few sections detail these guidelines.

The advices listed here are highly recommended even if you do not plan to migrate to Angular 2. These recommendation bits will make Angular 1 code more modular, organized, and testable.

## One component per file

This can be anything: an Angular 1 *controller, directive, filter,* or *service*. One component per file allows better organization of code and easy migration, allowing us to clearly identify how much progress has been made.

## Avoiding inline anonymous functions

Use named functions instead to declare controllers, directives, filters, and services. A declaration such as this:

```
angular.module('7minWorkout')
  .controller('WorkoutController',[...])

angular.module('app')
.directive('remoteValidator', [...])

angular.module('7minWorkout')
.filter('secondsToTime', function () { ... }

angular.module('7minWorkout')
.factory('workoutHistoryTracker', [...])
```

Should be converted to this:

```
function WorkoutController($scope, ...) { ... }
WorkoutController.$inject = ['$scope', ...];

function remoteValidator($parse) {...}
remoteValidator.$inject=[$parse];

function secondsToTime() {...}

function workoutHistoryTracker($rootScope, ...) { ...}
workoutHistoryTracker.$inject = ['$rootScope',...];
```

The advantages of using named functions are ease of debugging and ease of migration to TypeScript. Using named functions also requires that the dependencies be registered using the `$inject` function property.

`$inject`-based dependency declaration safeguards against minification and adds to the readability of the functions.

To avoid exposing global name functions with this approach, it is advisable to wrap the function in an **Immediately Invoked Function Expression** (**IIFE**):

```
(function() {
function WorkoutController($scope, ...) { ... }
WorkoutController.$inject = ['$scope', ...];

   angular
        .module('7minWorkout')
        .controller('WorkoutController', WorkoutController);

})();
```

## Avoiding $scope!

Yes, you read it right; avoid the $scope object or using scopes directly!

The biggest problem with Angular 1 scopes is their hierarchical nature. Accessing the parent scope from the child scope gives us tremendous flexibility but it comes at a cost. This can unknowingly create unwarranted dependencies that make the app really hard to debug and, of course, migrate. In contrast, in Angular 2 a view is bound to its component implementation and cannot access data outside its boundary implicitly. Therefore, if you plan to migrate to Angular 2, *avoid scopes at all costs*.

There are a number of techniques that can be used to remove the `$scope` object dependency. The next few subsections elaborate on some techniques that can help us avoid Angular 1 scopes.

## Using controller as (controller aliasing) syntax everywhere

Angular 1.3+ has the *controller as* syntax for *controller, directive,* and *routes. controller as* allows Angular 1 databinding expressions to bind to controller instance properties instead of the current scope object properties. With the controller as a paradigm in place, we never need to interact with the scope directly, and hence future migration becomes easy.

While controller aliasing gets rid of scope access, scopes are still there in Angular 1. The complete Angular 1 databinding infrastructure depends upon scopes. Controller aliasing just puts an indirection between our code and scope access.

Consider the following syntax for *controller as* in views:

```
<div ng-controller="WorkoutListController as workoutList">
   <a ng-repeat="workout in workoutList.workouts"
      href="#/workout/{{workout.name}}">
</div>
```

And the corresponding controller implementation:

```
function WorkoutListController($scope, ...) {
  this.workouts=[];
}
```

`WorkoutListController` as `workoutList` creates an alias `workoutList` for `WorkoutListController` on the current scope, hence allowing us to bind to the `workouts` property defined on the controller.

Route definition too allows controller aliasing using the `controllerAs` property in a *route definition object*:

```
$routeProvider.when('/builder/workouts', {
...
   controller: 'WorkoutListController',
   controllerAs: 'workoutList'
 });
```

Finally, directives too can use `controllerAs`, and together with the `bindToController` property on the *directive definition object*, we can get rid of any direct scope access.

Look at the angular documentation on controller, routes, and directive to get a basic understanding of the controller as syntax. Also look at the following posts for some more detailed samples on this topic:
`http://bit.ly/ng1-controller-as`
`http://bit.ly/ng1-bind-to`

## Avoiding ng-controller

If scopes can be avoided, so can controllers!

This may again seem counterintuitive, but the approach has real benefits. What we ideally want to do is emulate component behavior in Angular 1. Since the closest thing to components in Angular 1 is *element directives* (with `restrict='E'`), we should utilize *element directives* everywhere.

An Angular 1 element directive with its own template and isolated scope can very well behave like an Angular 2 component and only be dependent on its internal state for its view binding. We just don't need `ng-controller`.

Consider the usage of `ng-controller` for audio tracking from the Angular 1 version of the *Personal Trainer* app:

```
<div id="exercise-pane" class="col-sm-7">
...
  <span ng-controller="WorkoutAudioController">
    <audio media-player="ticksAudio" loop autoplay
     src="content/tick10s.mp3"></audio>
    <audio media-player="nextUpAudio"
     src="content/nextup.mp3"></audio>
    ...
  </span>
```

Instead of using `WorkoutAudioController`, an element directive can encapsulate the workout audio's view and behavior. Such a directive can then replace the complete `ng-controller` declaration and its view:

```
<div id="exercise-pane" class="col-sm-7">
...
<workout-audio-component></workout-audio-component>
```

When replacing `ng-controller` with an element directive, the scope variables that the controller depends upon should be passed to the directive using the `bindToController` property on the *directive definition object*. Something like this:

```
bindToController: {
   name: '=',
   title: '&'
}
```

This topic has been extensively covered in these two blog posts by Tero:

- `http://bit.ly/ng2-no-controllers`
- `http://bit.ly/ng2-refactor-to-component`

These are must-read posts with a wealth of information!

## Building using the Angular 1.5+ component API

Angular 1.5+ has a **component API** that allows us to create directives that can be easily migrated to Angular 2. The component API is preconfigured with sensible defaults, hence incorporating the best practices when it comes to building truly isolated + reusable directives.

Look at the component API (`http://bit.ly/ng1-dev-guide-components`) and this informative post by Tod Motto (`http://bit.ly/1MahwNs`) to learn about the component API.

To reiterate what has been emphasized earlier, these steps are not just targeted towards easy Angular 2 migration but also towards making Angular 1 code better. Component-based UI development is a better paradigm than what we are used to with Angular 1.

We highly recommend that you go through the Angular 1 style guide (`http://bit.ly/ng2-styleguide`). This guide contains a wealth of tips/patterns that allow us to build better Angular 1 apps, and is in sync with the guidelines provided previously for easy Angular 2 migration.

Finally, if you have decided to migrate, it's time to decide what to migrate.

## What to migrate?

For an app in maintenance mode, where most of the development activity revolves around bug fixes and some enhancements, it would be prudent to stick to Angular 1. Remember the old saying "*If it ain't broke, don't fix it.*"

If the app is being actively developed and has a clear long-term roadmap, migrating to Angular 2 is worth considering. As we dig deeper into the intricacies of migration, we will realize the time and effort involved in the process. While the Angular team has worked really hard to make this migration smooth, by no stretch of imagination is this a trivial job. It is going to take a good amount of time and effort to perform the actual migration.

The silver lining here is that we do not need to migrate everything at once. We can work slowly towards migrating parts of the Angular 1 code base to Angular 2. Both the frameworks can coexist, and can depend on each other too. This also allows us to develop new parts of applications in Angular 2. How cool is that!

But again, this flexibility comes at a cost-the cost of bytes. As both frameworks are downloaded, the page bytes do increase, something that we should be aware of.

Also, while the coexistence of both the frameworks allows us to migrate without much disruption, we cannot make it a perpetual activity. Eventually, Angular 1 has to go, and the sooner it does the better.

During migration, the best thing that can be done is to carve out new SPA's within the existing application. For example, we can build the Admin area of an app entirely using Angular 2, with a separate host page, but still share the common infrastructure of style sheets, images, and even Angular 1 services if we refactor the code a bit. As we will learn later, migrating services to Angular 2 is the easiest.

Breaking an application into multiple smaller ones introduces full-page refreshes, but this is a cleaner approach when it comes to migration.

Taking all of this into consideration, if you have decided to migrate and identified areas of migration, you need to do the prep work for migration.

# Preparing for Angular 2 migration

Welcome to the big brave world of Angular 2 migration! A successful migration strategy involves making sure that we do the groundwork beforehand, hence avoiding any late surprises.

As a prep work, the first step is to analyze the application from a third-party library dependency perspective.

## Identifying third-party dependencies

Any third-party library that an Angular 1 app uses needs a migration strategy too. These could be either jQuery-based libraries or Angular 1 libraries.

### jQuery libraries

jQuery libraries in Angular 1 were consumed by creating a directive wrapper over them. We will have to migrate such directives to Angular 2.

### Angular 1 libraries

Migrating Angular 1 libraries is bit of a tricky affair. Angular 1 has a massive ecosystem, whereas Angular 2 is a new kid on the block. It will take some time for the Angular 2 community offerings to be as rich as Angular 1. Hence, for every Angular 1 library we use, we need to find a substitute in Angular 2 or create one, or get rid of the library altogether.

Take for example the every-so-popular UI framework **ui-bootstrap** (`http://bit.ly/ng1-ui-bootstrap`). There are multiple efforts going on to rewrite this library for Angular 2, but none are complete. If we have a dependency on *ui-bootstap*:

- We can either use the ports, assuming that the components we want to use have been migrated. The notable ports here are **ng-bootstrap** (`https://goo.gl/3dHkaU`) and **ng2-bootstrap** (`https://goo.gl/u4hOJn`).
- Or wait for the port to be complete.
- Or take a more radical approach of building our own bootstrap library in Angular 2.

Each of these choices has trade-offs in terms of time and complexity.

Another choice that needs to be made is the development language. Should we use TypeSript, ES2015, or plain old JavaScript (ES5)?

## Choice of language

We would definitely recommend TypeScript. It's a super awesome language, and it integrates very well with Angular 2, vastly reducing the verbosity of Angular 2 declarations. Also, given that it can coexist with JavaScript, it makes our life easier. Even without Angular 2, TypeScript is one language that we should embrace for the web platform.

In the coming sections, we will migrate the Angular 1 **Personal Trainer** app to Angular 2. The app is currently available on *GitHub* at `http://bit.ly/a1begit`. This app was part of the first version of this book, *AngularJS by Example,* and was built using JavaScript.

We are again going to follow the checkpoint-based approach for this migration. The checkpoints that we highlight during the migration have been implemented as GitHub branches. Only this time, there is not companion code base to download.
Since we will be interacting with a *git* repository for v1 code and using N*ode.js* tools for the build, please set up *git* and *nodejs* on your dev box before proceeding further.

# Migrating Angular 1's Personal Trainer

Before we even begin the migration process we need to set up the v1 *Personal Trainer* locally.

The code for the migrated app can be downloaded from the GitHub site at `https://github.com/chandermani/angularjsbyexample`. Since we migrate in chunks, we have created multiple checkpoints that map to **GitHub branches** dedicated to migration. Branches such as `ng2-checkpoint8.1`, `ng2-checkpoint8.2`, and so on highlight this progression. During the narration, we will highlight the branch for reference. These branches will contain the work done on the app up to that point in time.

The *7 Minute Workout* code is available inside the repository folder named `trainer`.

So let's get started!

## Setting up Angular 1's Personal Trainer locally

Follow these steps and you will be up and running in no time:

1. From the command line, clone the v1 GitHub repository:

```
git clone https://github.com/chandermani/angularjsbyexample.git
```

2. Navigate to the new Git repo and check out the `ng2-base` branch to get started:

```
cd angularjsbyexample
git checkout ng2-base
```

3. Since the app loads its workout data from **MongoDB** hosted in **mLab** (`https://mlab.com/`), you need an mLab account to host workout-related data. Set up an mLab account by signing up with them. Once you have an mLab account, you need to retrieve your API key from mLab's management portal. Follow the instructions provided in the API documentation (`http://bit.ly/mlab-docs`) to get your API key.

4. Once you have the API key, update this line in `app/js/config.js` with your API key:

   ```
   ApiKeyAppenderInterceptorProvider
   .setApiKey("<yourapikey>");
   ```

5. And add some seed workout data into your mLab instance. The instructions to add the seed data into mLab are available in the source code file, `app/js/seed.js`.
6. Next, install the necessary *npm packages* required for v1 *Personal Trainer*:

   ```
   cd trainer/app
   npm install
   ```

7. Install `http-server`; it will act as a development server for our v1 app:

   ```
   npm i http-server -g
   ```

Verify that the setup is complete by starting the `http-server` from the `app` folder:

```
http-server -c-1
```

And open the browser location `http://localhost:8080`.

The v1 *Personal Trainer* start page should show up. Play around with the app to verify that the app is working fine. Now the migration can begin.

# Identifying dependencies

The first step before we begin migrating v1 *Personal Trainer* is to identify the external libraries that we are using in the Angular 1 version of Personal Trainer.

The external libraries that we are using in v1 are:

- `angular-media-player`
- `angular-local-storage`
- `angular-translate`
- `angular-ui-bootstrap`
- `owl. carousel`

Libraries such as `angular-media-player` and `angular-local-storage` are easy to migrate/replace. We have already done this in the earlier chapters of this book.

The `angular-translate` library can be replaced with `ng2-translate`, and as we will see in the coming sections, it is not a very challenging task.

We use `angular-ui-bootstrap` for **modal dialogs** in *Personal Trainer*. *ng2-bootstrap* (`http://bit.ly/ng2-bootstrap`) is a worthy successor and plans to have a 1-1 parity with the older bootstrap version, but while writing this book, the modal dialog implementation was a work in progress; hence we had look for alternatives.

There are two other libraries available specifically targeting modal dialogs: **angular2-modal** (`http://bit.ly/ng2-modal`) and **ng2-bs3-modal** (`http://bit.ly/ng2-bs3-modal`). We can pick one of these and get rid of the `angular-ui-bootstrap` library.

The *owl. Carousel* is a jQuery library, and we can write an Angular 2 component to wrap this plugin.

Now that we have sorted out the external dependencies, let's decide the language to use.

While the existing code base is JavaScript, we love TypeScript. Its type safety, its terse syntax, and how well it plays with Angular 2 makes it our language of choice. Hence, it's going to be TypeScript all the way.

Another thing that tilts the decision in favor of TypeScript is that we do not need to migrate the existing code base to TypeScript. Anything we migrate/build new, we build it in TypeScript. Legacy code still remains in JavaScript.

Let's start. As a first migration task, we need to set up a module loader for our v1 Personal Trainer.

# Setting up the module loader

Since we are going to create a number of new Angular 2 components spread across numerous small files, adding direct script reference is going to be tedious and error-prone. We need a module loader. A module loader can help us with:

- Creating isolated/reusable modules based on some common module formats
- Managing the script loading order based on dependencies
- Allowing bundling/packaging of a module and on-demand loading for dev/production deployments

We use the **SystemJS** module loader for this migration too.

Install SystemJS from the command line using:

```
npm i systemjs --save
```

All the commands need to be executed from the `trainer/app` folder.

We open `index.html` and remove all the script references of our app scripts. All script references with the source as `src='js/*.*'` should be removed, except `angular-media-player.js` and `angular-local-storage.js`, as they are external libraries.

Note: We are not removing script references for third-party libraries but only app files.

Add SystemJS configurations after all third-party script references:

```
<script src="js/vendor/angular-local-storage.js"></script>
<script src="node_modules/systemjs/dist/system.src.js">
</script>
<script>
  System.config({ packages: {'js': {defaultExtension: 'js'}}});
  System.import('js/app.js');
</script>
```

Remove the `ng-app` attribute on the body tag, keeping the `ng-controller` declaration intact:

```
<body ng-controller="RootController">
```

The `ng-app` way of bootstrapping has to go as we switch to the `angular.bootstrap` function. Manual bootstrapping helps when we bring Angular 2 into the mix.

The preceding `SystemJS.import` call loads the application by loading the first app module defined in `js/app.js`. We are going to define this module shortly.

Create a new file called `app.module.js` in the same folder as `app.js` and copy the complete contents of `app.js` into `app.module.js`.

Remember to get rid of the `use strict` statement. The TypeScript compiler does not like it.

All the app module definitions are not in `app.module.js`.

Next, clear `app.js` and add the following imports and bootstrap code:

```
import  './app.module.js';
import  './config.js';
import  './root.js';
import './shared/directives.js';
import './shared/model.js';
import './shared/services.js';
import './7MinWorkout/services.js';
import './7MinWorkout/directives.js';
import './7MinWorkout/filters.js';
import './7MinWorkout/workout.js';
import './7MinWorkout/workoutvideos.js';
import './WorkoutBuilder/services.js';
import './WorkoutBuilder/directives.js';
import './WorkoutBuilder/exercise.js';
import './WorkoutBuilder/workout.js';

angular.element(document).ready(function() {
  angular.bootstrap(document.body, ['app'],
{ strictDi: true });
});
```

We have added *ES6 import statements* to `app.js`. These are the same scripts that were earlier referenced in `index.html`. SystemJS now loads these script files when loading `app.js`.

Moving all of theAngular 1 module declaration into a new file `app.module.js` and importing it first in `app.js` makes sure that the Angular 1 modules are defined before any of the `import` statements are executed.

Do not confuse between *ES6 modules* and *Angular 2 modules* defined/accessed using `angular.module('name')`. These two are altogether different concepts.

The last few lines bootstrap the Angular 1 application using the `angular.bootstrap` function.

Module loading is enabled now; let's enable TypeScript too.

# Enabling TypeScript

To enable TypeScript, install the TypeScript compiler using *npm*:

```
npm i typescript -g
```

We can also install the TypeScript *type definition manager*, which allows us to use type definition for the libraries we use:

```
npm i typings --save-dev
```

**Type definitions** or **typings** are files that define the public interface for a TypeScript/JavaScript library. These type definitions help IDEs provide intellisense around the library functions. Typings are available for most of the popular JavaScript libraries and for frameworks/libraries written in TypeScript.

Next, open `package.json` and add these lines inside the script configuration:

```
"scripts": {
    "test": "echo "Error: no test specified" && exit 1"
    "tsc": "tsc -p . -w",
    "typings": "typings"
}
```

The two `scripts` properties that we just added are used to provide shortcuts for commonly executed scripts. See the following example, where we use the `typings` command to install `typings` (`npm run typings`).

Install the *type definitions* for Angular 1 and jQuery. From the command line, run:

```
npm run typings install dt~jquery dt~angular dt~angular-route dt~angular-
resource dt~angular-mocks dt~core-js dt~angular-ui-bootstrap -- --save --
global
```

If you are having trouble installing the *typings*, make sure that the installed `typings` package is the latest one, and then try again. We can check the latest version of the package using this command:

```
npm show typings version
```

Update `package.json` with the latest version and call `npm install` from the command line.

We now need to set up some configurations for the TypeScript compiler. Create a file called `tsconfig.json` (in the `trainer/app` folder) and copy the configuration from the `ng2-checkpoint8.1` repo branch (also available remotely at `http://bit.ly/ng2be-8-1-tsconfig`): Run the compiler using:

```
npm run tsc
```

This should start the TypeScript compiler and there should be no errors reported.

Keep this command running in a separate console window at all times during development. The compiler will continuously watch for changes to the TypeScript file and rebuild the code if changes are detected.

Change the extension for the `app.js` and `app.module.js` files to `app.ts` and `app.module.ts`. The TypeScript compiler detects these changes and compiles the TypeScript files. Post compilation, the compiler produces two files for each TypeScript file. One is the compiled JavaScript file (such as `app.js`) and the other is a map file (`app.js.map`) for debugging purposes.

We have not set up an elaborate build for this exercise as our primary focus is around migration.
For your own apps, the initial setup steps may vary depending upon how the build is already set up.

Before we test our new changes, config.js needs to be fixed because we have enabled *strict DI check* in Angular 1 through:

```
    angular.bootstrap(document.body, ['app'],
  { strictDi: true });
```

Replace the config.js content with updated content available in ng2-checkpoint8.1 or at http://bit.ly/ng2be-8-1-configjs (*and remember to set the API key again*). The update fixes the config function and makes it minification-friendly. Time to test the app!

Make sure the TypeScript compiler is running in one console; run http-server -c-1 on a new console window.

Navigate to http://localhost:8080; the app start page should load.

Commiting your local changes

If things work fine, you can even commit your local changes to your git repo. This will help you track what has changed over the time as we migrate the app piece by piece.
The implementation till this point is available in the ng2-checkpoint8.1 GitHub branch.
If you are facing problems, compare the ng2-base and ng2-checkpoint8.1 branches to understand the changes made. Since the code is hosted in GitHub, we can use the *GitHub compare* interface to compare commits in a single branch. See the documentation on how to do it here: http://bit.ly/github-compare.
The http://bit.ly/ng2be-compare-base-8-1 link shows a comparison between ng2-base and ng2-checkpoint8.1.
You can ignore the diff view for app.js and app.module.js, generated as part of the TypeScript compilation.

Time to introduce Angular 2!

# Adding Angular 2

We start by installing Angular 2 and dependent *npm modules* for our app. We will update the package.json file with the necessary packages first.

Copy all of the updated package file from `http://bit.ly/ng2be-8-2-package-json` into your local installation.

`package.json` now references some new packages related to Angular 2. Install the referenced packages by calling:

```
npm install
```

If you are having trouble with installing packages with `npm install`, delete the `node_modules` folder and run `npm install` again.

Then add a few library references that Angular 2 is dependent upon (and not loaded using SystemJS) in `index.html` before the `system.src.js` script reference (three in total):

```
<script src="/node_modules/core-js/client/shim.min.js"></script>
<script src="/node_modules/zone.js/dist/zone.js"></script>
<script src="/node_modules/reflect-metadata/Reflect.js"></script>
<script src="/node_modules/systemjs/dist/system.src.js"></script>
```

As it stands now, the SystemJS configuration is present in the `index.html` file itself. Since Angular 2 requires some decent amount of configuration, we are going to create a separate *SystemJS configuration file* instead, and reference that in `index.html`.

Add this script reference after the `system.src.js` reference:

```
<script src="systemjs.config.js"></script>
```

Now clear the script section containing the call to the `System.config` function and replace it with:

```
<script>System.import('app');</script>
```

Copy the `systemjs.config.js` from `http://bit.ly/ng2be-migrate-systemjs-config` and place it in the same folder as `package.json`.

This configuration file derives from the *Quick Start Guide* for Angular available at `http://bit.ly/ng2-qsguide`. You can learn more about this configuration from the online guide.

Also update `tsconfig.json` and add a new property called `moduleResolution` to `compilerOptions`:

```
"removeComments": false,
"moduleResolution": "node"
```

This instructs TypeScript to look for type definitions in the `node_modules` folder. Remember, Angular 2 typings are bundled as part of the Angular 2 library itself, and hence separate type definition import is not required.

Now that the Angular 2-specific references have been added, we need to modify the existing bootstrap process to also load Angular 2.

The Angular team has come up with an Angular 2 service, **UpdateAdapter**, that allows Angular 2 to bootstrap within an Angular 1 setup. The `UpgradeAdapter` service enables a number of common migration use cases. Using `UpgradeAdapter`, we can:

- Bootstrap an app with both the Angular 1 and Angular 2 frameworks loaded. This is the first thing we are going to do.
- Incorporate an Angular 2 component in an Angular 1 view.
- Incorporate an Angular 1 component in an Angular 2 view, albeit with some limitations.
- Register an Angular 1 service with Angular 2 and vice versa.

The sole purpose of the `UpgradeAdpater` service is to allow the gradual migration of artefacts from Angular 1 to Angular 2. As we make progress with our migration efforts, the role of `UpgradeAdpater` becomes clear.

Let's learn how to bootstrap the hybrid Angular 1 and Angular 2 app using `UpgradeAdpater`.

# Bootstrapping the hybrid app

Since we are starting with Angular 2, we need to define a root app module.

Create a new file, `app-ng1.module.js`, and copy the complete content of `app.module.ts` to the new file. Also remember to update the `import` statement in `app.ts`:

```
import  './app-ng1.module.js';
```

Let's now add the Angular 2 root module definition (`AppModule`) to `app.module.ts`.

Replace the content of `app.module.ts` with the Angular 2 module definition. Copy the new definition from `ng2-checkpoint8.2` (GitHub location: `http://bit.ly/ng2be-8-2-app-module-ts`).

Next, create a file called `upgrade-adapter.ts` in the same folder as `app.ts` and add a *global export* for the `UpdateAdapter` instance:

```
import {UpgradeAdapter} from '@angular/upgrade';
import {AppModule} from './app.module';
export const upgradeAdapter = new UpgradeAdapter(AppModule);
```

This instance of `UpgradeAdpater` (named `upgradeAdapter`) is now available globally, and can be used to bootstrap the hybrid app.

Exporting an instance of `UpgradeAdapter` allows us to reuse the same instance across the app. This is a mandatory requirement for interoperability between the frameworks.

Import the file we just created into `app.ts`. Add this import statement after other imports in `app.ts`:

```
import {upgradeAdapter} from './upgrade-adapter';
```

And replace the bootstrap code with:

```
angular.element(document).ready(function() {
    upgradeAdapter.bootstrap(document.body, ['app'], { strictDi: true });
});
```

Refresh your app and make sure it works as before. Do watch out for errors in TypeScript compiler console window.

Congratulations! We now have a hybrid app up and running. Both frameworks are now working in tandem.

Look at the `ng2-checkpoint8.2` branch if you are facing issues upgrading to Angular 2.
Again, you can also compare these git branches `ng2-checkpoint8.1` and `ng2- checkpoint8.2` to understand what has changed (`http://bit.ly/ng2be-compare-8-1-8-2`).

The migration process can start now. We can start by migrating a part of an Angular 1 view/directive to Angular 2.

# Injecting Angular 2 components into Angular 1 views

The most common migration pattern involves migrating lower-level Angular 1 directives/views to Angular 2 components. If we visualize the Angular 1 HTML view structure as a tree of directives, we start at the leaf. We migrate parts of a directive/view to an Angular 2 component and embed the component inside the Angular 1 view template. This Angular 2 component is injected into the Angular 1 view as an *element directive*.

The closest thing to Angular 2 components that Angular 1 has is element directives. During migration, we are either migrating element directives or controller (`ng-controller`)-view pairs.

This is a bottom-up approach to migrating view/directives to Angular 2 components. The following diagram highlights how the Angular 1 view hierarchy gradually transforms into an Angular 2 component tree:

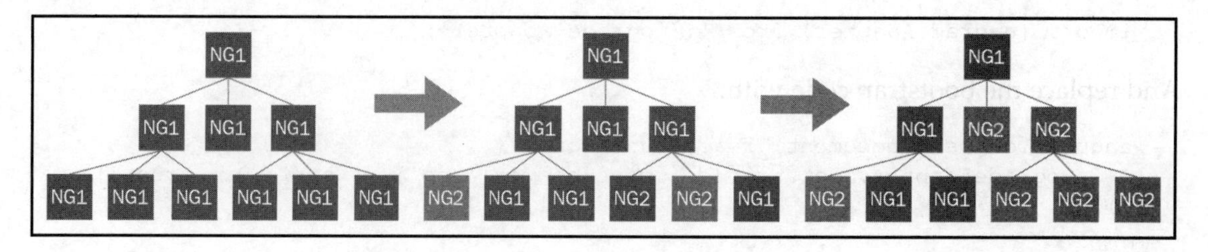


Let's migrate something small and get a feel of how things work. `ExerciseNavController` and its corresponding view fit the bill.

# Migrating our first view to Angular 2 component

`ExerciseNavController` is part of *workout builder* and is located inside `trainer/app/js/WorkoutBuilder/exercise.js`. The corresponding view is served from `trainer/app/partials/workoutbuilder/left-nav-exercises.html`.

The primary purpose of this this controller-view pair is to show the list of available exercises when building a workout (available user path `http://localhost:8080/#/builder/workouts/new`):

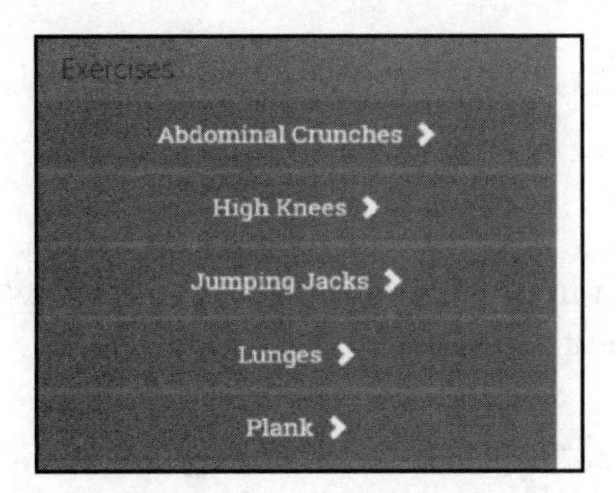


Clicking on any of these exercise names adds the exercise to the workout being constructed.

Let's start with creating a component for the above view.

Before starting on the new component, add a new workout builder module (`WorkoutBuilderModule`) to the application. Copy the module definition from `ng2-checkpoint8.3` in the `WorkoutBuilder` folder (GitHub location: `http://bit.ly/ng2be-8-3-workout-builder-module-ts`). Also import the newly created module in `app.module.ts`.

Instead of inlining the complete code here, we suggest copying the `exercise-nav-component.ts` file from GitHub branch `ng2-checkpoint8.3` (`http://bit.ly/ng2be-8-3-exercisenavts`) and adding it to the `WorkoutBuilder` folder locally. Since there is a decent amount of boilerplate, we will highlight only the relevant parts.

Contrasting template syntax

The Angular team has published an excellent reference (`http://bit.ly/ng2-a1-a2-quickref`) that details the common view syntaxes in Angular 1 and their equivalents in Angular 2. Highly recommended when migrating an Angular 1 app!

To start with, if you look at the `exercise-nav-component.ts` file, the component template is similar to `left-nav-exercises.html` used in Angular 1, except there is no `ng-controller` and the template bindings are Angular 2-based:

```
template: `<div id="left-nav-exercises">
           <h4>Exercises</h4>
           <div *ngFor="let exercise of exercises" class="row">
           ...
           </div>`
```

If we focus on the component implementation (`ExercisesNavComponent`), the first striking thing is the component's dependencies:

```
constructor(
@Inject('WorkoutService') private workoutService: any,
@Inject('WorkoutBuilderService') private workoutBuilderService: any)
```

`WorkoutService` and `WorkoutBuilderService` are Angular 1 services injected into Angular 2 components.

Nice! If that is your initial reaction, we can't blame you. It is pretty cool! But the story is still incomplete. There is no magic happening here. Angular 2 cannot access Angular 1 services unless it is told where to look. It's `UpgradeAdapter` that does most of the heavy lifting here.

## Injecting Angular 1 dependencies into Angular 2

`UpgradeAdapter` has an API method that allows us to register an *Angular 1 service* with Angular 2. Open `app.ts` and add these lines after the `upgrade-adapter` import:

```
upgradeAdapter.upgradeNg1Provider('WorkoutService');
upgradeAdapter.upgradeNg1Provider('WorkoutBuilderService');
```

The `updateNg1Provider`, as its name suggests, registers an Angular 1 dependency with the *Angular 2 root injector*. Once registered, the dependency is available throughout the Angular 2 application.

**Sharing functionality with services**
`UpgradeAdapter` will also make sure that the same instance of the service is shared across the two frameworks. This makes services a prime candidate for sharing behavior across the frameworks. And, as we will see soon, it works the other way round too.

Dependencies at times have other dependencies, and hence it's better if we bring in all the service dependencies from Angular 1 at one go. Copy the list of Angular 1 dependency registrations (done using `upgradeNg1Provider`) from `http://bit.ly/ng2be-8-3-appts` into your local `app.ts`. Remember to delete the two extraneous declarations that we have already imported above.

Back to component integration! As the `ExercisesNavComponent` is rendered inside an Angular 1 view, it needs to be registered as an *Angular 1 directive.*

## Registering Angular 2 components as directives

`ExercisesNavComponent` is an Angular 2 component, but it can be converted into an Angular 1 directive. Open `app.ts` and add the highlighted lines:

```
import {ExercisesNavComponent} from './WorkoutBuilder/exercise-nav-
component'
import {upgradeAdapter} from './upgrade-adapter';
angular.module('WorkoutBuilder').directive('exerciseNav',
upgradeAdapter.downgradeNg2Component(ExercisesNavComponent) as
angular.IDirectiveFactory);
```

This time the `UpgradeAdapter` function used is `downgradeNg2Component`. This function returns a *factory function* containing the *directive definition object*. We register the component as an Angular 1 directive, `exerciseNav`.

Every Angular 2 component is registered as an *element directive* when used in Angular 1.

The component implementation is complete. We now need to clean up the old code and inject the new directive in the view.

Open `app.ts` and add the import statement to import the newly created component:

```
import './WorkoutBuilder/exercise-nav-component';
```

Delete the definition of `ExercisesNavController` from `exercise.js`, and replace the content of `left-nav-exercises.html` (located in the `partials` folder) with:

```
<exercise-nav></exercise-nav>
```

And we are good to go.

Angular 1 still loads `left-nav-exercises.html` as part of the route transition, but the view inside is an Angular 2 component.

Go ahead and try out the new implementation. Create a new workout and try to add exercises from the left nav. The functionality should work as before.

Look at `ng2-checkpoint8.3` in case you are facing issues upgrading to Angular 2.
You can compare the git branches `ng2-checkpoint8.2` and `ng2-checkpoint8.3` to understand what has changed (`http://bit.ly/ng2be-compare-8-2-8-3`).

While we have only migrated a trivial component, this exercise highlights how easy it is to convert/downgrade an Angular 2 component to the Angular 1 directive and use it in Angular 1 view. The overall encapsulation of an Angular 2 component makes this chore easy.

This downgraded component can even take an input from the parent scope using the all-so-familiar Angular 2 property binding syntax:

```
<exercise-nav [exercises]='vm.exercises'></exercise-nav>
```

Add to that, the event raised by the component can be subscribed by the Angular 1 container scope too:

```
<exercise-nav (onExerciseClicked)='vm.add(exercise)'></exercise-nav>
```

We now have an Angular 2 component running inside Angular 1 using services initially designed for Angular 1. A promising start to our migration journey!

Before we move any further, it's time to highlight how this collaboration works and the rules of engagement.

# Rules of engagement

The migration story from Angular 1 to Angular 2 is only possible because these frameworks can coexist, and possibly share data. There are some touch points where the boundaries can be crossed. To have a better sense of how a hybrid application works and what is achievable in such a setup, we need to understand the areas of collaboration between the two frameworks.

There are three areas that need discussion:

- Template interleaving in DOM
- Dependency injection
- Change detection

Since Angular 2 components and Angular 1 directives can coexist in a DOM, the question we need to answer is: Who owns what parts of the DOM?

## Angular 1 directives and Angular 2 components

When it comes to ownership of a DOM element, the golden rule is:

*Every DOM element is owned/managed by exactly one of the Angular frameworks.*

Take our previous migration example. The view that is part of `ExercisesNavComponent` is managed by Angular 2, whereas the container view (`left-nav-exercises.html`) is managed by Angular 1.

Things get a bit tricky at the boundaries of these directives and components. Consider the declaration inside `left-nav-exercises.html`:

```
<exercise-nav></exercise-nav>
```

Who owns this? The short answer is Angular 1.

While this is an Angular 2 component, the host element is owned by Angular 1. This means all Angular 1 template syntax works:

```
<exercise-nav ng-if='showNav'></exercise-nav>
<exercise-nav ng-repeat='item in items'></exercise-nav>
```

As these components and directives coexist in the same view, they often need to communicate. There are two ways to manage this communication:

- Using templating capabilities of Angular 1 and Angular 2:
    - An Angular 2 component embedded inside an Angular 1 view can take inputs from the parent scope using event and property binding
    - In a similar fashion, if a directive is injected into an Angular 2 component view, it too can get inputs from the parent component and call the parent component function (through its isolated scope)
- Using shared services. We saw an example of this previously as we injected the `WorkoutService` and `WorkoutBuilderService` Angular 1 services into `ExercisesNavComponent`.

Injecting Angular 1 directives into Angular 2 is a bit tricky. To be able to inject an Angular 1 directive into an Angular 2 template, the directive needs to abide by some rules. We will talk about these rules in the coming sections.

Sharing functionality using services is far more flexible compared to sharing through view templates. Injecting services across framework boundaries requires us to register the service across both frameworks and let Angular take care of the rest. Let's learn how dependency injection works across boundaries.

# Resource sharing and dependency injection

How dependencies are registered in a hybrid app is driven by how DI works in these two frameworks. For Angular 1, there is only one global injector, whereas Angular 2 has a concept of hierarchical injectors. In a hybrid environment, the least common denominator is the global injector that both the frameworks support.

## Sharing an Angular 1 service

Dependencies defined in Angular 1 can be used in Angular 2 once they are registered with Angular 2's *app injector* (root injector). The `UpgradeAdapter` function `upgradeNg1Provider` handles this:

```
UpdateAdapter.upgradeNg1Provider(name:string,
  options?: {asToken: any;}))
```

Since dependency injection in Angular 1 is *string-token-based*, the first parameter is the name of the service (the string token). The second parameter allows us to register a custom Angular 2 token for the v1 service.

This is what we did in the exercise `nav` component migrated earlier:

```
upgradeAdapter.upgradeNg1Provider('WorkoutService');
```

When it comes to injecting the dependency in Angular 2, we require the `Inject` decorator (*with string token*) for injection:

```
constructor(
  @Inject('WorkoutService') private workoutService: any,
```

The `WorkoutService` is a generic provider registers with Angular 1. Had this been a TypeScript class, the registration could have been done using an extra class token:

```
upgradeAdapter.upgradeNg1Provider('WorkoutService',
  {asToken:WorkoutService});
```

And injected using the all-so-familiar type injection with no decorators required:

```
constructor(private workoutService: WorkoutService,
```

## Sharing an Angular 2 service

Services from Angular 2 too can be injected into Angular 1. Since Angular 1 only has a global injector, the dependency is registered at the global level. The `UpgradeAdapter` function that does this is:

```
UpgradeAdapter.downgradeNg2Provider(token:any):Function
```

The `downgradeNg2Provider` creates a factory function than can be consumed by the Angular 1 module's `factory` API:

```
angular.module('app').factory('MyService',
  UpgradeAdapter.downgradeNg2Provider(MyService))
```

`MyService` can now be injected across the Angular 1 app, like any other service.

`UpgradeAdapter` makes sure that only a single instance of the dependency is created and shared across the frameworks.

Look at the following diagram; it summarizes what we have discussed:

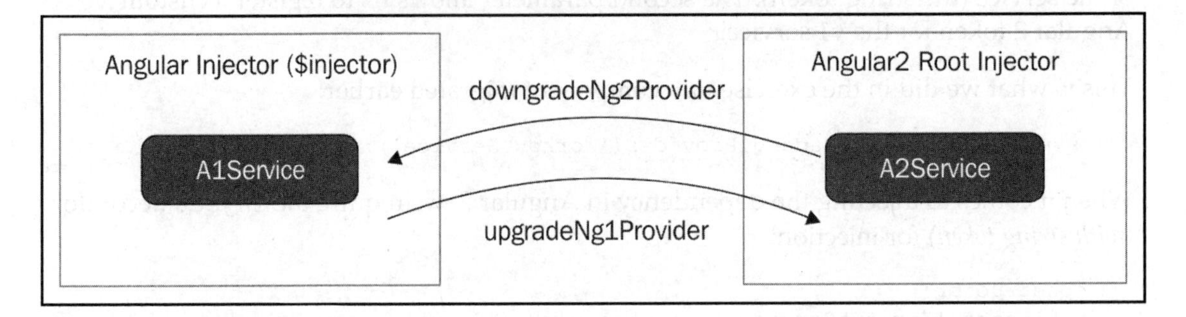


One last topic of this discussion is *change detection*.

## Change detection

In a hybrid application, change detection is managed by Angular 2. If you are used to calling `$scope.$apply()` in your code, you don't need to in a hybrid application.

We have already discussed how Angular 2 change detection works. The Angular 2 framework takes care of triggering Angular 1 change detection by internally calling `$scope.$apply()` on standard triggering points.

Now that we understand the rules of engagement, it is easier to comprehend how things work, what's feasible, and what's not.

Let's set some bigger/meatier targets and migrate the start and finish pages of the v1 app.

# Migrating the start and finish pages

The finish page is easy to migrate, and I suggest you do it yourself. Create a folder called `finish` inside the `js` folder, and create three files, `finish-component.ts`, `finish-component.tpl.html`, and `finish.module.ts`, for the component's code and view template and the module definition. Implement the component.

Import the finish module into `AppModule` (`app.module.ts`). And then fix the route for finish to load the finish component.

```
$routeProvider.when('/finish', { template: '<finish></finish>' });
```

Lastly, remember to delete the finish html template from the `partials/workout` folder.

If you are stuck in migrating the finish page, compare the `ng2-checkpoint8.3` and `ng2-checkpoint8.4` git branches to understand what has changed in the `8.4` branch (`http://bit.ly/ng2be-compare-8-3-8-4`).

The finish page was easy, the start page is not! While the start page seems to be an easy target, there are some challenges that require some head-scratching.

The biggest issue with the start page is that it uses a third-party library, *angular-translate,* to localize the content of the page. Since we are migrating the complete page/view to Angular 2, we need a mechanism to handle these Angular 1 library dependencies.

*angular-translate* comes with a *filter* (*pipe* in the Angular 2 world) and a directive, both named `translate`. Their job is to translate string tokens into localized string literals.

Now that the *start* page becomes an Angular 2 component, we need to convert the filter into an Angular 2 pipe and, in some way, make the `translate` directive work in Angular 2. The migration choices we have in this case are as follows:

- Create a new filter and upgrade the v1 `translate` directive using `UpgradeAdapter`.
- Find a suitable replacement for angular-translate in Angular 2 world.

Although the first choice seems to be the easiest, it has some serious limitations. Angular 2 `UpgradeApapter` comes with an `upgradeNg1Component` function, which can upgrade any Angular 1 directive. Not really! There are some stringent requirements around which a directive can be upgraded to Angular 2.

Upgrade of an Angular 1 component does not mean the component has been migrated. Angular 2 instead allows us to use an Angular 1 element directive as-is inside Angular 2 component views.

# Angular 1 directive upgrade

At times, the parts of an application may be migrated in a top-down fashion; a higher order view is converted into a component. In such a case, instead of migrating all the custom directives that are part of the Angular 1 view, we just upgrade them to Angular 2 components using the `UpgradeAdpater` function `upgradeNg1Component`. The following diagram illustrates this migration path:

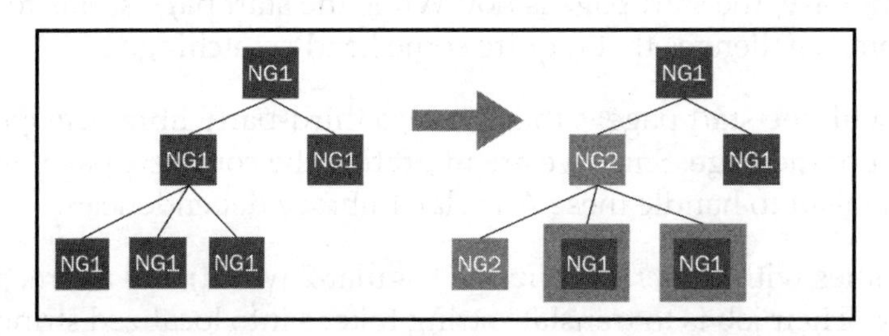


The Angular 2 framework puts some restrictions around what can be upgraded to an Angular 2 component. Here is an excerpt from the Angular 2 migration guide.

> *To be Angular 2 compatible, an Angular 1 component directive should configure these attributes:*
> `restrict: 'E'`. *Components are usually used as elements.*
> `scope: {}` *– an isolated scope. In Angular 2, components are always isolated from their surroundings, and we should do this in Angular 1 too.*
> `bindToController: {}`. *Component inputs and outputs should be bound to the controller instead of using the $scope.*
> `controller` *and* `controllerAs`. *Components have their own controllers.*
> *template or templateUrl. Components have their own templates.*
> *Component directives may also use the following attributes:*
> `transclude: true`, *if the component needs to transclude content from elsewhere.*
> `require`, *if the component needs to communicate with some parent component's controller.*
> *Component directives* **may not** *use the following attributes:*
> `compile`. *This will not be supported in Angular 2.*
> `replace`*: true. Angular 2 never replaces a component element with the component template. This attribute is also deprecated in Angular 1.*
> `priority` *and* `terminal`. *While Angular 1 components may use these, they are not used in Angular 2 and it is better not to write code that relies on them.*

The only Angular 1 directives that can be upgraded to Angular 2 are element directives, given that all other conditions are met.

With this sizeable laundry list, upgrading an Angular 1 directive to Angular 2 is difficult, when compared to an Angular 2 component downgrade. More often than not, we have to do an actual code migration of an Angular 1 directive if the parent view has been migrated to Angular 2.

Looking at the *angular-translate* source code, we realize it uses the `$compile` service; therefore, the upgrade option is ruled out. We need to find an alternative library.

We do have an internationalization library for Angular 2, `ng2-translate` (`http://bit.ly/ng2-translate`).

## Replacing angular-translate with ng2-translate

*ng2-translate* is an internationalization library that targets Angular 2. This library can replace v1 *angular-translate*.

Install the npm package for `ng2-translate`:

```
npm i ng2-translate --save
```

Update `systemjs.config.js` to include the *ng2-translate* library. Add entries to the `map` and `packages` properties:

```
var map = {...
  'ng2-translate': '/node_modules/ng2-translate/bundles'
var packages = { ...
  'ng2-translate': { defaultExtension: 'js' }
```

`ng2-translate` needs to be configured at the module level, so update `app.module.ts` with the highlighted code:

```
@NgModule({
 imports: [BrowserModule, ...,
   HttpModule, TranslateModule.forRoot(),],
 providers: [TranslateService,
   {
     provide: TranslateLoader,
     useFactory: (http: Http) => new TranslateStaticLoader(http, 'i18n',
'.json'),
     deps: [Http]
```

```
    }]
})
```

The preceding provider declaration sets up a loader that loads the translation files (`.json`) from the `i18n` folder. The `HttpModule` import is required for the translate library to load translations from server.

These statements require imports to keep the TypeScript compiler happy. Add these import statements to `app.module.ts`:

```
import {TranslateModule, TranslateService,                    TranslateLoader,
TranslateStaticLoader} from 'ng2-translate/ng2-translate';
```

The `ng2-translate` library is now ready to be used.

The first thing that we are going to do is set the default translation language as soon as the application bootstraps.

## Using a bootstrap-ready callback for initialization

With Angular 2, luckily the bootstrap function on `UpdateAdapter` has a `ready` callback function just for this purpose. It is invoked after both the frameworks bootstrap.

Update the bootstrap function in `app.ts` with the following code snippet:

```
upgradeAdapter.bootstrap(document.body, ['app'],
  { strictDi: true })
  .ready((updateApp: UpgradeAdapterRef) => {
    var translateService =
      updateApp.ng2Injector.get(TranslateService);

    var userLang = navigator.language.split('-')[0];
    userLang = /(fr|en)/gi.test(userLang) ? userLang : 'en';

    translateService.setDefaultLang('en');

    translateService.use(userLang);
  });
```

And add an import for `TranslateService`:

```
import {TranslateService} from 'ng2-translate/ng2-translate';
```

The code tries to determine the current browser language and sets the current language for translations accordingly. Make note of how we get hold of `TranslateService`. The `UpgradeAdapterRef` object holds the reference to the Angular 2 *root injector*, which in turn loads `ng2-translate`'s `TranslateService`.

Next copy the three files related to start page implementation, from the `ng2-checkpoint8.4` branch (`http://bit.ly/ng2be-8-4-start`) into a new folder `app/js/start`.

The start component now needs to be registered as an Angular 1 directive before use. Add this statement to `app.ts`:

```
import {StartComponent} from './start/start-component';
angular.module('start').directive('start',
upgradeAdapter.downgradeNg2Component(StartComponent) as
angular.IDirectiveFactory);
```

The start template file now uses the `translate` pipe (*the name of the pipe is the same as the Angular 1 filter* `translate`).

The page also has some pipes, which are used to search and sort the workout list shown on the page:

```
<a *ngFor="#workout of workouts|search:'name':searchContent|orderBy:'name'"
href="#/workout/{{workout.name}}">
```

We now need to add implementation for the pipes `orderBy` and `search`. Copy the complete code from `ng2-checkpoint8.4` (`http://bit.ly/ng2be-8-4-pipests`) and add it to a new file, `js/shared/pipes.ts` locally. We will not dwell into any of the pipe implementation here as we have done that already in earlier chapters.

Yet again we create a new Angular 2 module to share these pipes across the application. Copy the module definition from `ng2-checkpoint8.4` (`http://bit.ly/ng2be-shared-module-ts`) into the local `js/shared` folder, and import it into `app.module.ts`.

We have previously migrated the `secondsToTime` (available in `js/7MinWorkout/filters.js`) filter to Angular 2, and the implementation is available in the `pipes.ts` file.

Start and finish component implementation is complete. Let's integrate them into the app.

## Integrating the start and finish pages

Start/finish views are loaded as part of route change; hence we need to fix the route definition.

Open `app.ts` and add imports for start and finish:

```
import './start/start-component';
import './finish/finish-component';
```

The route definitions are in `config.js`. Update the start and finish route definitions to:

```
$routeProvider.when('/start',
{ template: '<start></start>' });
$routeProvider.when('/finish',
{ template: '<finish></finish>' });
```

The route template html is a part of the Angular 1 view. Since we have registered both `StartComponent` and `FinishComponent` as Angular 1 directives, the route loads the correct components.

If you have already migrated the finish page, you do not need to redo the import and route setup for finish as described.

A few more fixes are pending before we can test the implementation.

Update `app-ng1.module.ts` with modules for `start` and `finish`:

```
angular.module('app', ['ngRoute', ... ,
'start', 'finish']);
...
angular.module('start', []);angular.module('finish', []);
```

Finally copy the translation files `de.json` and `en.json` from the `ng2-checkpoint8.4` folder `i18n` (`http://bit.ly/ng2-8-4-i18n`). Now we are ready to be test what we have developed.

If not started, start the TypeScript compiler and *HTTP-server*, and then launch the browser. The start and finish pages should load just fine. But the translations do not work! Clicking on the language translation links on the top nav has no affect. Content always renders in English.

If you are stuck, compare the git branches `ng2-checkpoint8.3` and `ng2-checkpoint8.4` to understand what changed (`http://bit.ly/ng2be-compare-8-3-8-4`).

Translations still do not work because the top nav code (`root.js`) that enables translation is still using the older library. We need to get rid of angular-translate (the v1 library) altogether. Having two libraries doing the same work is not something we want, but removing it is also not that simple.

## Getting rid of angular-translate

To get rid of angular-translate (v1) library we need to:

- Remove the angular-translate's directive/filter references from all Angular 1 views.
- Get rid of any code that uses this library.

Getting rid of the v1 directive/filter altogether is a difficult task. We can neither add the v2 ng2-translate pipe in the Angular 1 view nor can we migrate every view using the v1 directive/filter to Angular 2 at one shot.

Why not write a new Angular 1 filter that uses the ng2-translate's translation service (`TranslateService`) for translations and then use the new filter everywhere? Problems solved!

Let's call this filter `ng2Translate`. We replace all references to the `translate` filter in v1 view with `ng2Translate`. All v1 `translate` directive references should also be replaced with `ng2Translate` filter.

Here is how the filter implementation looks like:

```
import {TranslateService} from 'ng2-translate';

export function ng2Translate(ng2TranslateService: TranslateService) {
   function translate(input) {
    if (input && ng2TranslateService.currentLang) {
      return ng2TranslateService.instant(input);
    }
  }
  translate['$stateful'] = true;
  return translate;
}
```

```
ng2Translate.$inject = ['ng2TranslateService'];
angular.module('app').filter("ng2Translate", ng2Translate);
```

Create a file called `filters.ts` in the `shared` folder and add the preceding implementation. The filter uses `TranslateService` (registered in Angular 1 as `ng2TranslateService`) to map string tokens to localized content. To test this implementation, there are a few more steps needed:

- Replace all references to `translate` (directive and filter) across the Angular 1 view with `ng2Translate`. There are references in these files: `description-panel.html`, `video-panel.html`, `workout.html` (in the folder `partials/workout`), and `index.html`. Replacing the filter in the interpolation is a simple exercise, and for the `translate` directive, replace it with interpolation. For example in `partials/workout/description-panel.html`, the line of code is as follows:

  ```
  <h3 class="panel-title" translate>RUNNER.STEPS</h3>
  ```

  It then becomes the following:

  ```
  <h3 class="panel-title">{{'RUNNER.STEPS'|ng2Translate}}</h3>
  ```

  Remember to quote the string token (`'RUNNER.STEPS'`) inside the interpolation.

- Import the filter into `app.ts`:

  ```
  import './shared/filters'
  ```

- The `ng2Translate` filter depends upon the `TranslateService`; hence needs to be registered with the Angular 1 injector using (in `app.ts` again):

  ```
  angular.module('app').factory('ng2TranslateService',
  upgradeAdapter.downgradeNg2Provider(TranslateService));
  ```

Angular 2's `TranslateService` is registered as `ng2TranslateService` in Angular 1.

- Finally copy the updated `root.js` from `http://bit.ly/ng2-migrate-root-no-trasnlate`. We have replace all references to `$translate` service with `ng2TranslateService`, and refactored the code to use the new service. `root.js` contains the implementation for v1 `RootController`.

We are good to go now. Try out the new implementation, the app should load translation with using ng2-translate library.

We can now delete all references to angular-translate. There are references in `index.html`, `app.module.ts` and `config.js`.

The migration of the start and finish pages is complete and it's time to look at some other targets.

Compare the branches `ng2-checkpoint8.4` and `ng2-checkpoint8.5` to understand the new changes in `ng2-checkpoint8.5` (`http://bit.ly/ng2be-compare-8-4-8-5`).

The next few migrations that we detail in the coming sections will not be as descriptive as earlier migrations. We strongly recommend you look at the relevant commits on the `angular2-migrate-ts` branch to understand how things move forward. We will only highlight relevant details for the coming migrations.

# Replacing the ui-bootstrap library

One thing that we have learned while migrating the start and finish pages to Angular 2 is how cumbersome it is to migrate third-party dependencies. Migrating an Angular 1 view that uses external libraries without migrating those libraries themselves is a challenge. We have also learned that it is much easier to embed an Angular 2 component inside an Angular 1 view than the other way round.

Given these observations, it becomes imperative that we migrate/replace third-party libraries first while migrating to Angular 2.

One such library that we want to get rid of is the **ui-bootstrap** (`http://bit.ly/ng1-ui-bootstrap`) library. While we only use the modal dialog service from ui-bootstrap, moving away from it is going to be a challenge.

Calling this modal dialog a service (`$uibModal`) would be a misnomer. While it is injected like a service in Angular 1, it actually manipulates the DOM and therefore cannot be upgraded using the `upgradeNg1Provider` function.

We again need an alternate implementation for modal dialogs in Angular 2. The library we choose for this exercise is angular2-modal (`http://bit.ly/ng2-modal`).

Personal Trainer uses angular2-modal in two places, in the top nav to show the workout history and during workout execution to show the exercise-related video.

As part of migrating to angular2-modal, we also migrate **top nav** (declared inside `index.html`) and the **video panel** (`partials/workout/video-panel.html`).

Look at the `ng2-checkpoint8.6` GitHub branch to understand what artefacts were changed during this migration. We will only *highlight* things that made the migration challenging.
You can also compare this branch with the previous one (`ng2-checkpoint8.5`) on GitHub, at `http://bit.ly/ng2be-compare-8-5-8-6`, to know what the changes are.
*This section will detail the code in the context of the* `ng2-checkpoint8.6` *GitHub branch.*

The biggest challenge with migrating to our new modal dialog library, *angular2-modal*, was that it required access to the root component to properly render the dialog in the middle of the screen. While this is not a problem in a standard Angular 2 app, for a hybrid app, the library failed to locate the root component as there was none.

Look at the *Using angular2-modal dialog library* section in `Chapter 3`, *More Angular 2 – SPA, Routing, and Data Flows in Depth*, to understand how to install and configure the library. You can also compare the branches `ng2-checkpoint8.5` and `ng2-checkpoint8.6` to determine changes specific to angular2-modal.

To get around these limitations, we first had to restructure the Angular 1 app, such that we have an Angular 2 root component. Such a component then encompasses the complete Angular 1 view. The new rendered html structure now looks like this:

```
ng1 ▼<body class="ng-scope">
  ng2▼ <ng2-root #ng2root id="NG2_UPGRADE_0_ng2Root_c0">
    ng1▼ <ng1-root class="ng-isolate-scope">
        ▶<div ng-controller="RootController" class="ng-scope">…</div>
        </ng1-root>
        <!--ng1 insertion point-->
      </ng2-root>
    </body>
```

Open `index.html`, the `ng2-root` (`Ng2RootComponent`) tag is an Angular 2 component that wraps the complete Angular 1 view. The existing Angular 1 view html itself is now wrapped inside a directive (`ng1Root`) using the `component` API. Look at files `ng1-root-component.ts`, `ng1-root-component.tpl.html` and `ng2-root-component.ts` to understand how these components are structured now and to provide angular2-modal `ng2-root` container reference in the `Ng2RootComponent` 's constructor.

The restructuring of root elements also employs another migration pattern. The Angular 2 component (`Ng2RootComponent`) transcludes the Angular 1 element directive (`ng1Root`). Check the view template of `Ng2RootComponent`:

```
@Component({
  selector: 'ng2-root',
  template: `<ng-content></ng-content>`
})
```

And it's use in `index.html`:

```
<ng2-root>
      <ng1-root></ng1-root>
</ng2-root>
```

In such a setup, while the `ng1RootComponent` is embedded inside the Angular 2 `Ng2RootComponent`, it derives its context from the parent Angular 1 view and hence can access the parent scope.

There were numerous other small changes made to the app as part of this migration, and comparing this branch against `ng2-checkpoint8.5` with help you understand what has changed.

We will stop here and direct you to the other GitHub branches pertaining to migration. All branches starting with `ng2-checkpoint*` are the migration branches. Try to migrate the pending views and compare them with the GitHub branch changes. Remember, a working version of the app has already been developed in Angular 2, and hence there is a good reference point. Look at the `README.md` file for each branch to know what part of the application was migrated to Angular 2.

Meanwhile, let's summarize our learnings from the migration that we did.

# Learnings

We hope this migration exercise has provided enough insight into the process. You can now gauge the complexity, the time, and the effort required to migrate elements from Angular 1 to Angular 2. Let's highlight our learning as part of this process:

- **Migration is time-consuming**: Migration by no stretch of imagination is a trivial exercise. Each page/view presents its own challenges that we need to overcome. Some elements are easy to migrate and some are not. The best thing you can do today if you are developing in Angular 1 would be to follow the advices from the *Developing Angular 1 apps today for easy migration* section.

- **Migrate third-party libraries first**: Migrating third-party libraries can be quite challenging. The reasons are manifold:
  - Such libraries are used across pages
  - They may not be upgradable to Angular 2 (using `UpgradeAdapter`)
  - Migrating each view that uses such a library may not be feasible when the library is extensively used

It's better to identify all third-party dependencies in your app and find a suitable alternative for them in the Angular 2 world. If possible, develop some proof of concept (POC) with the new library to understand how different the new library is from the existing implementation.

- **Libraries with overlap may exist**: While migrating, there could be scenarios where both Angular 1 and Angular 2 versions of a library coexist. Minimize this time period and migrate to the newer version as soon as possible.
- **It is easier to integrate Angular 2 components into Angular 1 than the other way round**: While migrating, migrate the complete view to Angular 2. Due to the restriction imposed by Angular 2, it becomes very difficult to have a parent Angular 2 component with embedded Angular 1 element directives.

With such limitations, a bottom-up approach to migrating works better than a top-down approach.

- **Anything non-UI-related is easy to migrate**: For *Personal Trainer,* we migrate the services last as they can be easily migrated.
- **Feature parity better Angular 1 and Angular 2**: Angular 2 may not have every feature that Angular 1 supports. In such a case, we need workarounds to achieve the desired behaviour.

That completes our migration story. With this, it's time to conclude the chapter and summarize our learnings from it.

# Summary

In this chapter, we gained some useful insight into a number of practical issues surrounding Angular development. These tips/guidelines can be extremely handy when building real-life applications using the framework.

We started the chapter by exploring the concept of *seed projects* and how these projects can get us up and running in no time. We looked at some popular seed projects that can serve as a base for any new Angular app development. We also touched upon *Yeoman* and *angular-cli,* a suite of tools that helps us kick-start new projects.

In spite of being a server-side concern, authentication and authorization do affect the client implementation. The section on authentication/authorization covered how to handle authentication in a cookie and token-based setup.

We looked at the ever-so-important topic of performance, where you learned ways to optimize an Angular app's performance.

Finally, we migrated the v1 *Personal Trainer* to Angular 2. The gradual migration process taught us the intricacies of migration, the challenges faced, and the workaround done.

The book is coming to a close, but for everyone reading it, the journey has just begun. It's time to put theories into practice, hone our newly acquired skills, build something useful with Angular, and share it with the world. The more you invest in Angular, the more rewarding the framework is. Let's get started!

# Index

# B



# C

# D

# E



# F



# G

## H



## I



## J



## K



## L



## M

## N



## O



## P

# Q



# R



# S

# T

# Y



# Z